Mathematical Essays
Dedicated to A. J. Macintyre

Mathematical Essays
Dedicated to
A. J. Macintyre

Edited by
H. Shankar

OHIO UNIVERSITY PRESS · ATHENS · OHIO

iv

Contents

ACKNOWLEDGMENTS   ix

LIFE AND LIST OF PUBLICATIONS OF A. J. MACINTYRE
    N. A. Bowen   xi

THE PADÉ TABLES AND CONTINUED FRACTIONS GENERATED BY
    TOTALLY POSITIVE SEQUENCES
    Robert J. Arms and Albert Edrei   1

SOME INTEGRABILITY THEOREMS FOR POWER SERIES WITH POSITIVE
    COEFFICIENTS
    R. Askey and R. P. Boas, Jr.   23

COMPLETELY INVARIANT DOMAINS OF ENTIRE FUNCTIONS
    I. N. Baker   33

ON THE ASYMPTOTIC PATHS OF ENTIRE FUNCTIONS REPRESENTED BY
    DIRICHLET SERIES
    F. Sunyer Balaguer   37

ON THE GROWTH OF ENTIRE FUNCTIONS
    P. D. Barry   43

INTEGERS—NO THREE IN ARITHMETIC PROGRESSION
    P. S. Chiang and H. Shankar   61

THE COMPOSITION OF ENTIRE AND MEROMORPHIC FUNCTIONS
    J. Clunie   75

INTERPOLATION BY ENTIRE FUNCTIONS OF EXPONENTIAL TYPE
    Richard F. DeMar   93

COMPLEMENTS TO SOME THEOREMS OF BOWEN AND MACINTYRE ON THE
    RADIAL GROWTH OF ENTIRE FUNCTIONS WITH NEGATIVE
    ZEROS
    David Drasin and Daniel F. Shea   101

Some Extremal Problems in Combinatorial Number Theory
   Paul Erdős   123

A Theorem Concerning the Real Part of a Power Series
   T. M. Flett   135

Quasiconformal Mappings which Hold the Real Axis Pointwise
Fixed
   F. W. Gehring   145

Some Examples related to the cos $\pi\rho$ Theorem
   W. K. Hayman   149

On a Theorem of Study Concerning Conformal Maps with
Convex Images
   Maurice Heins   171

On Groups of Automorphisms of Certain Sequence Spaces
   V. Ganapathy Iyer   177

On the Phragmén-Lindelöf Theorem, the Denjoy Conjecture,
and Related Results
   James A. Jenkins   183

A Strengthened Form of the $\frac{1}{4}$ Theorem for Starlike Univalent
Functions
   F. R. Keogh   201

An Alternative to Boundedness
   Bo Kjellberg   213

On a Result of A. J. Macintyre
   Thomas Kövari   217

Functions of Bounded Indices in One and Several Complex
Variables
   J. Gopala Krishna and S. M. Shah   223

The Minimum Modulus of Functions of Slow Growth in the
Unit Disk
   C. N. Linden   237

On the "Pits Effect" for Standard Integral Functions of
Finite Nonzero Order
   J. E. Littlewood   247

Contents

A DISTORTION THEOREM FOR A CLASS OF CONFORMAL MAPPINGS
A. J. Lohwater and Frank Ryan   257

SOME EXAMPLES IN THE THEORY OF GROUPS
I. D. Macdonald   263

EXCEPTIONAL VALUES OF $f^{(n)}(z)$, ASYMPTOTIC VALUES OF $f(z)$, AND LINEARLY ACCESSIBLE ASYMPTOTIC VALUES
Gerald R. MacLane   271

THE SPECTRA OF STEP-FUNCTION POTENTIALS
J. B. McLeod   289

ON CONVOLUTIONS AND GROWTH OF TYPICALLY REAL FUNCTIONS
E. P. Merkes   299

THE DERIVATIVE OF A FUNCTION OF BOUNDED CHARACTERISTIC
George Piranian   305

QUASI-SUBORDINATE FUNCTIONS
M. S. Robertson   311

CONVEX MEROMORPHIC FUNCTIONS
W. C. Royster   331

NONMAXIMAL PRIME IDEALS IN THE RING OF HOLOMORPHIC FUNCTIONS
L. A. Rubel   341

MEROMORPHIC FUNCTIONS WITH ONE VALIRON DEFICIENT VALUE
S. M. Shah and H. Shankar   343

THE BASIS OF VITALI'S THEOREM
J. M. Whittaker   353

COEFFICIENT DENSITY AND THE DISTRIBUTION OF VIRTUALLY ISOLATED SINGULARITIES ON THE CIRCLE OF CONVERGENCE
R. Wilson   359

ON HADAMARD COMPOSITION AND THE VIRTUALLY ISOLATED SINGULARITY
R. Wilson   371

Acknowledgments

The publication of this volume was originally planned to celebrate the sixtieth birthday of Professor A. J. Macintyre. His sudden death on August 4, 1967 came as a great shock to all those who knew and admired him. It is a fitting tribute to this great mathematician and inspiring teacher that this volume should now appear as a memorial. It contains research papers dedicated to him by his friends, collaborators, colleagues and students.

I wish to acknowledge the monetary contributions made to this volume by the following individuals: Rita Britt, P. S. Chiang, Cecil Craig, R. F. DeMar, Mary Jane Britt, H. C. Liu, Allister and Susan Macintyre, H. Shankar, G. M. Merriman, Lee Suyemoto, and H. S. Tung. I owe special thanks to Dr. H. D. Lipsich, Vice Provost, University of Cincinnati, for arranging a generous donation from the Charles Phelps Taft Memorial Fund, and to the Ohio University Press.

Many people assisted and encouraged me in planning and preparing this volume. I welcome this opportunity to express my gratitude to all of them, particularly to the referees who devoted their precious time to reading the contributed papers. I would also like to thank Professor A. J. Lohwater and Professor S. M. Shah for their expert advice and suggestions so generously and promptly given whenever needed. The editor is also grateful to Professor N. A. Bowen who prepared the biographical sketch of Professor A. J. Macintyre.

The editor would also like to thank the Ohio University Press staff and the printers for their cooperation, courtesy, and technical assistance and especially his wife for her patience, persistent encouragement, and confidence.

I should like to commend the contributing authors for the quality of their papers and to thank them sincerely for the spirit with which they cooperated in the making of this book.

January 1970
Athens, Ohio

H. Shankar,
Editor

Professor A. J. Macintyre, M.A., Ph.D., F.R.S.E.

N. A. Bowen

Department of Mathematics
The Royal University of Malta
MSIDA Malta

The death of A. J. Macintyre, Research Professor of Mathematics at the University of Cincinnati, Ohio, on August 4, 1967, came as a severe shock to all who knew the fifty-nine year old professor through personal contacts or by professional repute.

During my thirteen year association with him at King's College, Aberdeen, Scotland, I naturally came to know him well and to admire his many fine attributes. His popularity and the high esteem accorded to him by both staff and students were a response to his good nature and kindness, his sense of humor, his patience in explaining mathematical problems to those less quick than he to understand those problems, his ability to imbue his students with the feeling that mathematics was a living and growing subject rather than a defunct language of invariable symbols, and his encyclopedic knowledge of mathematical research, especially in the varied branches of mathematical analysis.

Another aspect of his character that impressed itself early on his mathematical friends, especially on his research collaborators, was his high professional standard—he would not countenance publication unless the proposed article was of superior content and exposition. However, what impressed me most about Professor A. J. Macintyre was his humility. Although he told me once that he did not regard himself as an original thinker, but merely one who could sometimes push further ahead with the ideas and methods of others, all who have worked in the field of analysis will surely agree that such an opinion of his work does him less than justice. I have no doubt that many of us would be extremely pleased and gratified to produce work of his quality.

Professor A. J. Macintyre was born in Sheffield, July 3, 1908, the only son of William E. G. A. and Mary Macintyre, both in the scholastic profession. He was educated at the Central Secondary School, Sheffield (later known as the High Storrs Grammar School), after which he entered Magdalene College, Cambridge, as a scholar in October, 1926, where he took his Mathematical Tripos, being awarded a First Class in the Mathematical Tripos (Part I) in June, 1927, receiving a First Class in the Mathematics Preliminary Examination in June, 1928 (for which he was awarded the Davidson prize for Mathematics), and becoming a Wrangler with Distinction in Schedule B in the Mathematical Tripos (Part II) in June, 1929. His mathematical contemporaries at Cambridge included Professor H. Davenport (Trinity), Professor S. Verblunsky, Dr. J. Cossar and Dr. D. W. Babbage (all of Magdalene), as well as his tutor and director at Magdalene A. S. Ramsey, the author of a well-known mathematical text book and father of the brilliant mathematical logician Frank Ramsey.

In 1929 A. J. Macintyre began research on integral and meromorphic functions under Sir Edward Collingwood at Cambridge, and in 1930 he taught applied mathematics and theoretical physics at Swansea University College as an Assistant Lecturer attached to the Mathematics Department, whose Head was then the late Professor A. R. Richardson. The staff at Swansea also included R. Wilson, now Professor Emeritus of Mathematics at the University of Wales. A. J. Macintyre began a firm friendship with R. Wilson which was to lead to a significant output of research—some a result of their collaboration—on a variety of problems arising in classical analysis.

In 1931 A. J. Macintyre was appointed as an Assistant Lecturer and in 1935 promoted to Lecturer in the Mathematics Department at the University of Sheffield. Naturally he had continued with his research work since leaving Cambridge and in 1933 was awarded a Ph.D. by that school for his thesis entitled "Some Problems of Integral and Meromorphic Functions."

In 1936 A. J. Macintyre accepted the post of Lecturer in Mathematics at King's College, Aberdeen, in a department headed by Professor E. M. Wright, co-author with G. H. Hardy of a well-known work on the theory of numbers. In 1946 A. J. Macintyre was elected a Fellow of the Royal Society of Edinburgh and in 1959, when Senior Lectureships were introduced into the University of Aberdeen, he was at once promoted even though at that time he was on leave of absence from Aberdeen as a Visiting Research Professor in Mathematics at the University of Cincinnati. In the fall of 1959 he resigned his Aberdeen post in order to return to a permanent post as Research Professor in Mathematics at Cincinnati, where in September 1963 he became the Charles Phelps Taft Professor of Mathematics and earned the distinction of being elected a Fellow of the Graduate School—a post he held until his death. During the academic year 1965–66 he held a Visiting Professorship at the University of California at Davis.

Macintyre's mathematical interest was varied and his output of research is shown in his papers listed in this volume. In his chosen area of analysis he made significant contributions in complex function theory. Most of his work dealt with the study of asymptotic paths, flat regions, Laplace transformation and integral functions, interpolation series, gap power series, extremum problems, Tauberian theorems and the Whittaker and Bloch constants. In collaboration with R. Wilson he focused his attention on the study of the coefficient theory of the Taylor series and produced a large amount of remarkably basic work. At his death he was working on the manuscript of a book on resonance in nonlinear oscillations. Unfortunately, the research tract on the coefficient theory of the Taylor series which was being prepared in collaboration with R. Wilson was also left unfinished. This volume illustrates the great debt the study of mathematics owes, and will owe, to the fertility of Macintyre's mind.

Mathematics itself was not the only cause to which A. J. Macintyre devoted his talents. A. J. Macintyre's mathematical insight, coupled with his fertile imagination, led to an interest in practical applications of theory. He was deeply engaged in mechanical problems of aircraft controls, of heat engines, and in designs for wind turbines, sail boats, and desalinization plants.

A. J. Macintyre's family life was a happy one. In 1940 he married Sheila Scott of Edinburgh, whose father, also a schoolmaster, later accompanied them to America where he settled until his death in 1963. A Girtonian and a Wrangler in Mathematics, Mrs. Macintyre had taught at St. Leonards School (St. Andrews) and at Stowe School. Like her husband, she was a good mathematician and after her marriage not only lectured in Mathematics at the Universities of Aberdeen and of Cincinnati, but also published a number of research papers on mathematical analysis dealing with the convergence of Abel's and other series and with the Whittaker constant. Like her husband, she was always popular with both staff and students in her profession and was loved by her family and circle of friends.*

The Macintyres had three children: Allister William born in 1944; Douglas Scott, born in 1946, who died of enteritis in March 1949; and Susan Elizabeth, born in 1950 and since 1967 Mrs. J. Gaines of Cincinnati.

Sincere thanks are due to those who helped to prepare this biographical sketch,† particularly to H. Shankar, J. M. Whittaker, and R. Wilson.

---

* For Mrs. Macintyre's obituary see J. London Math. Soc. 36 (1961), 254–256.

† Surveys of A. J. Macintyre's mathematical work have appeared in the Bull. London Math. Soc. 1, Pt. 3 (1969) and in the Proc. Edinburgh Math. Soc. (Dec. 1969).

Publications of A. J. Macintyre

1. Un théorème sur l'ultraconvergence, C.R. Acad. Sci. Paris 199 (1934), 598–599.
2. On the order of interpolated integral functions and of meromorphic functions with given poles (with R. Wilson), Quart. J. Math. Oxford Ser. 5 (1934), 211–220.
3. On the asymptotic paths of integral functions of finite order, J. London Math. Soc. 10 (1934), 34–39.
4. Elementary proof of theorems of Cauchy and Mayer, Proc. Edinburgh Math. Soc. (2) 4 (1934–6), 112–117.
5. A theorem concerning meromorphic functions of finite order, Proc. London Math. Soc. (2) 39 (1935), 282–294.
6. Two theorems on Schlicht functions, J. London Math. Soc. 11 (1936), 7–11.
7. On Bloch's Theorem, Math. Zeit. 44 (1938), 536–540.
8. Wiman's method and the flat regions of integral functions, Quart. J. Math. Oxford Ser. 9 (1938), 81–88.
9. On the minimum modulus of integral functions of finite order, Quart. J. Math. Oxford Ser. 9 (1938), 182–184.
10. Laplace's transformation and integral functions, Proc. London Math. Soc. (2) 45 (1939), 1–20.
11. On a theorem concerning function regular in an annulus, Recueil Math. (Mat. Sb.) N.S. 5 (47) (1939), 307–308.
12. Inequalities for the logarithmic derivatives of a polynomial (with W. H. J. Fuchs), J. London Math. Soc. 15 (1940), 162–168.
13. Coefficient density and the distribution of singular points on the circle of convergence (with R. Wilson), Proc. London Math. Soc. (1), 47 (1940), 60–80.
14. Some converses of Fabry's Theorem (with R. Wilson), J. London Math. Soc. 16 (1941), 220–229.
15. Logarithmic derivatives and flat regions of analytic functions (with R. Wilson), Proc. London Math. Soc. (2) 47 (1942), 404–435.
16. Some elementary inequalities in function theory (with W. W. Rogosinski), Edinburgh Math. Soc. Notes 35 (1945), 1–3.

17. Associated integral functions and singular points of power series (with R. Wilson), J. London Math. Soc. 22 (1947), 289–304.
18. Note on the preceding paper,* J. London Math. Soc. 23 (1948), 209–211.
19. Euler's limit for e^x and the exponential series, Edinburgh Math. Soc. Notes 37 (1949), 26–28.
20. Extremum problems in the theory of analytic functions (with W. W. Rogosinski), Acta Math. 82 (1950), 275–325.
21. An oscillation theorem of Tauberian Type (with N. A. Bowen), Quart. J. Math. Oxford Ser. (2) 1 (1950), 243–247.
22. Some theorems on integral functions with negative zeros (with N. A. Bowen), Trans. Amer. Math. Soc. 70 (1951), 114–126.
23. Asymptotic paths of integral functions with gap power series, Proc. London Math. Soc. (3) 2 (1952), 286–296.
24. Theorems on the convergence and asymptotic validity of Abel's series (with S. S. Macintyre), Proc. Royal Soc. Edinburgh Sect. A, 63 (1952), 222–231.
25. Operational methods and the coefficients of certain power series (with R. Wilson), Math. Ann. 127 (1954), 243–250.
26. Integral functions with gap power series (with P. Erdös), Proc. Edinburgh Math. Soc. (2) 10 (1954), 62–70.
27. Interpolatory methods for theorems of Vitali and Montel type (with N. A. Bowen), Proc. Royal Soc. Edinburgh Sect. A, 64 (1954), 71–79.
28. Interpolation series for integral functions of exponential type, Trans. Amer. Math. Soc. 76 (1954), 1–13.
29. The history and the theory of the aeroplane and the sailplane, Aberdeen University Rev. 36 (1956), 241–250.
30. A direct proof of Morera's theorem, Arch. Math. 8 (1957), 374–375.
31. A overconvergence theorem of G. Bourion and its application to the coefficients of certain power series, Ann. Acad. Sci. Fenn. Ser. A, I. Mathematica No. 250/23 (1958), 3–11.
32. Size of gaps and region of overconvergence, Collect. Math. XI 3 (1959), 165–174.
33. On an extension of a theorem of Bernstein to meromorphic functions (with S. M. Shah), J. Math. Anal. Appl. 3 (1961), 351–354.
34. Cartwright's theorem on functions bounded at the integers (with H. C. Liu), Proc. Amer. Math. Soc. 12 (1961), 460–462.
35. Studies on overconvergence (with H. G. Mushenheim), Collect. Math. XV (1963), 3–36.
36. Upper bounds for a Bloch Constant (with P. S. Chiang), Proc. Amer. Math. Soc. 17 (1966), 26–31.

* A functional inequality (by Sheila Scott Macintyre), J. London Math. Soc. 23 (1948), 202–209.

37. Convergence of $i^{i^{i^{\cdots}}}$, Proc. Amer. Math. Soc. 17 (1966), 67.
38. e and e^2 as continued fractions, (with C. I. Lubin), Amer. Math. Monthly, (2) 73 (1966), 1124, prob. 5445.
39. Inequalities for functions regular and bounded in a circle (with Cecil Craig), Pacific J. Math. 20 (1967), 449–454.
40. A proof of the power series expansion without differentiation theory (with Willy John Wilbur), Proc. Amer. Math. Soc. 18 (1967), 419–424.
41. Some theorems of Bloch type (with P. S. Chiang), Proc. Amer. Math. Soc. 18 (1967), 953–954.
42. Integers, no three in arithmetic progression (with P. S. Chiang), Math. Mag. 41 (1968), 128–130.
43. Entire functions related to the Dirichlet divisor problem (with N. A. Bowen), in *Entire Functions and Related Parts of Analysis*, Proceedings of Symposia in Pure Mathematics, vol. 11, American Mathematical Society, Providence, R.I., 1968, pp. 66–78.

Mathematical Essays
Dedicated to A. J. Macintyre

The Padé Tables and Continued Fractions Generated by Totally Positive Sequences

Robert J. Arms

Computer Services Division
National Bureau of Standards
Washington, D. C. 20234
U.S.A.

Albert Edrei*

Department of Mathematics
Syracuse University
Syracuse, New York 13210
U.S.A.

Introduction

Let

(1) $$a_0 + a_1 z + a_2 z^2 + \cdots = f(z) \qquad (a_0 \neq 0),$$

be a formal power series (its radius of convergence may be zero).

In 1892, H. Padé associated with (1) an infinite array of rational functions

(2)
$$
\begin{array}{cccc}
R_{00} & R_{10} & R_{20} & \cdots \\
R_{01} & R_{11} & R_{21} & \cdots \\
R_{02} & R_{12} & R_{22} & \cdots \\
\cdots & \cdots & \cdots & \cdots
\end{array}
$$

defined by the following extremal property:

* The research of the second author was supported by a grant from the National Science Foundation GP-7507.

1

Let the ordered pair of nonnegative integers (m, n) be given. Consider all the rational functions, regular at the origin, of the form

$$(3) \qquad R(z) = \frac{P(z)}{Q(z)} = c_0 + c_1 z + c_2 z^2 + \cdots,$$

where

(i) $P(z)$ is a polynomial of degree not greater than m;

(ii) $Q(z)$ is a polynomial of degree not greater than n and does not vanish identically;

(iii) subject to these restrictions select $P(z)$ and $Q(z)$ so as to maximize the integer N such that

$$(4) \qquad a_j = c_j \qquad (j = 0, 1, 2, \ldots, N);$$

(iv) denote by $P_{mn}(z)$ and $Q_{mn}(z)$ the polynomials $P(z)$ and $Q(z)$ thus determined. Put

$$(5) \qquad R_{mn}(z) = \frac{P_{mn}(z)}{Q_{mn}(z)}$$

and place $R_{mn}(z)$ at the intersection of the nth row $(n \geq 0)$ and mth column $(m \geq 0)$ of the configuration (2).

The array of rational functions thus defined is the Padé table of the series (1).

The existence and uniqueness of the Padé table is well known [6, p. 237, Theorem 5.1] and need not be considered here.

Put

$$(6) \qquad a_{-k} = 0 \qquad (k = 1, 2, 3, \ldots);$$

the Hankel determinants

$$(7) \qquad A_m^{(n)} = \begin{vmatrix} a_m & a_{m-1} & \cdots & a_{m-n+1} \\ a_{m+1} & a_m & \cdots & a_{m-n+2} \\ \cdots & \cdots & \cdots & \cdots \\ a_{m+n-1} & a_{m+n-2} & \cdots & a_m \end{vmatrix} \quad (m \geq 0, n \geq 1), A_m^{(0)} = 1 (m \geq 0),$$

which play an important role in many questions of complex analysis,* are closely connected with the Padé table.

In particular, if for all $m \geq 0$ and all $n \geq 0$,

$$(8) \qquad A_m^{(n)} \neq 0,$$

---

* For instance in the moment problem and in the study of singularities of analytic functions [3].

Padé's table is said to be *normal*; in this case polynomials Q_{mn} and P_{mn} which determine the rational *approximant* R_{mn} are given by the formulas

$$(9) \quad Q_{mn}(z) = \frac{1}{A_m^{(n)}} \begin{vmatrix} 1 & z & z^2 & \cdots & z^n \\ a_{m+1} & a_m & a_{m-1} & \cdots & a_{m-n+1} \\ a_{m+2} & a_{m+1} & a_m & \cdots & a_{m-n+2} \\ \cdots & \cdots & \cdots & \cdots & \cdots \\ a_{m+n} & a_{m+n-1} & a_{m+n-2} & \cdots & a_m \end{vmatrix}$$

$$(m \geq 0, n \geq 1), \quad Q_{m0}(z) \equiv 1,$$

and

$$(10) \quad \begin{cases} P_{mn}(z) = \dfrac{1}{A_m^{(n)}} \displaystyle\sum_{j=0}^{m} \begin{vmatrix} a_j & a_{j-1} & a_{j-2} & \cdots & a_{j-n} \\ a_{m+1} & a_m & a_{m-1} & \cdots & a_{m-n+1} \\ a_{m+2} & a_{m+1} & a_m & \cdots & a_{m-n+2} \\ \cdots & \cdots & \cdots & \cdots & \cdots \\ a_{m+n} & a_{m+n-1} & a_{m+n-2} & \cdots & a_m \end{vmatrix} z^j \\[2em] \hspace{5em} (m \geq 0, n \geq 1), \\[1em] P_{m0}(z) = \displaystyle\sum_{j=0}^{m} a_j z^j. \end{cases}$$

In a *normal* Padé table all the polynomials P_{mn} and Q_{mn} are uniquely determined by their extremal character and the normalization

$$(11) \qquad\qquad Q_{mn}(0) = 1 \qquad (m, n = 0, 1, 2, \ldots).$$

We always impose this normalization so that whenever we consider a normal table, the *Padé polynomials* P_{mn}, Q_{mn} will be defined without ambiguity and necessarily given by (10) and (9). We shall say that P_{mn} is a *Padé numerator* and Q_{mn} a *Padé denominator*.

This note is concerned with the "convergence problem" of the Padé table. In the three editions of his classical treatise on continued fractions (1914, 1929, 1957), Perron remarks that little is known about the question.

The convergence properties of the tables of a few isolated functions have been investigated successfully, in particular the tables of

$$(12) \qquad e^z, \qquad a_0 + \sum_{n=1}^{\infty} \frac{z^n}{\Gamma(a + n + 1)}, \qquad \prod_{n=1}^{\infty}(1 - q^n z),$$

$$A + \sum_{n=1}^{\infty} \frac{q^n}{1 + q^n} z^n, \qquad \sum_{n=0}^{\infty} q^{n^2} z^n,$$

where a_0, a, A, and q $(0 < |q| < 1)$ are constants [6, pp. 246–248; 1, pp. 249–257].

In this paper we consider the Padé tables of the members of the following class $(\mathscr{S})$ of meromorphic functions.

Definition of the class ($\mathscr{S}$). *An analytic function $f(z)$ belongs to the class* ($\mathscr{S}$) *if it is representable in the form*

(13)
$$f(z) = a_0\, e^{\gamma z}\, \frac{\displaystyle\sum_{j=1}^{\infty}(1 + \alpha_j z)}{\displaystyle\sum_{j=1}^{\infty}(1 - \beta_j z)},$$

where

(14)
$$\left\{\begin{array}{l} a_0 > 0, \quad \gamma \geqq 0, \\ \alpha_j \geqq 0, \quad \beta_j \geqq 0 \quad \text{(for all } j \geqq 1), \quad \sum_j (\alpha_j + \beta_j) < +\infty. \end{array}\right.$$

We shall always assume that

$$\alpha_1 \geqq \alpha_2 \geqq \alpha_3 \geqq \cdots, \qquad \beta_1 \geqq \beta_2 \geqq \beta_3 \geqq \cdots,$$

and do not exclude the possibility that some or all of the quantities α and β be zero.

Our main result is

Theorem 1. *Let* (1) *be the expansion of a nonrational member of the class* ($\mathscr{S}$). *Then*

I. *The Padé table of the series in* (1) *is always normal; moreover*

(15)
$$A_m^{(n)} > 0,$$

for all $m \geqq 0$ and all $n \geqq 0$.

II. *Let $\{m_\lambda\}_{\lambda=1}^{\infty}$ and $\{n_\lambda\}_{\lambda=1}^{\infty}$ be two sequences of positive integers such that*

(16)
$$m_\lambda \to \infty, \qquad n_\lambda \to \infty,$$

and such that

(17)
$$\lim_{\lambda \to \infty} \frac{m_\lambda}{n_\lambda} = \omega \qquad (0 \leqq \omega \leqq +\infty).$$

Then, as $\lambda \to \infty$,

(18)
$$P_{m_\lambda, n_\lambda}(z) \to \quad a_0\, e^{[\omega/(1+\omega)]\gamma z} \prod_{j=1}^{\infty}(1 + \alpha_j z),$$

(19)
$$Q_{m_\lambda, n_\lambda}(z) \to \quad e^{-[1/(1+\omega)]\gamma z} \prod_{j=1}^{\infty}(1 - \beta_j z)$$

uniformly in any bounded region of the complex plane.

In the special case $\gamma = 0$, the assumption (17) *may be omitted.*

In the analytic theory of continued fractions there exists an extensive literature concerning the representation of the series (1) (with the normalization $a_0 = 1$), by infinite fractions of the form*

(20) $\quad 1 + \dfrac{k_1 z \;\big|}{\big|\; 1 + l_1 z} + \dfrac{k_2 z^2 \;\big|}{\big|\; 1 + l_2 z} + \dfrac{k_3 z^2 \;\big|}{\big|\; 1 + l_3 z} + \dfrac{k_4 z^2 \;\big|}{\big|\; 1 + l_4 z} + \cdots$

$$(k_j \neq 0, j = 1, 2, 3, \ldots),$$

or of the closely related form

(21) $\quad 1 + \dfrac{\kappa_1 z \;\big|}{\big|\; 1} + \dfrac{\kappa_2 z \;\big|}{\big|\; 1} + \dfrac{\kappa_3 z \;\big|}{\big|\; 1} + \cdots \qquad (\kappa_j \neq 0, j = 1, 2, 3, \ldots).$

Such representations are not always possible. For instance (1) is representable in the form (20) if and only if $A_n^{(n)} \neq 0$ $(n = 1, 2, 3, \ldots)$. For the representation (21), the necessary and sufficient condition is

$$A_n^{(n-1)} A_n^{(n)} \neq 0 \qquad (n = 1, 2, 3, \ldots; A_1^{(0)} = 1).$$

Hence, assertion I of Theorem 1 leads immediately to

Corollary 1.1. *All nonrational members of* $(\mathscr{S})$ *(normalized by the condition* $a_0 = 1$*) are represented by continued fractions of the forms* (20) *and* (21).

Since the *approximants* (also known as *convergents*) of the fractions (20) and (21) are those entries of the Padé table which we have denoted by

$$R_{nn}, \qquad R_{n, \, n-1},$$

we see that Theorem 1 not only proves the possibility of the expansions (20) and (21), but also yields complete information concerning their convergence. In this manner we obtain

Theorem 2. *Let* $f(z)$ *be a meromorphic function of the class* $(\mathscr{S})$, *normalized by the condition* $f(0) = 1$. *Let* (1) *be its Taylor expansion and* (20) *its associated continued fraction.*
Denote by

$$P_n(z) = 1 + p_{n1} z + p_{n2} z^2 + \cdots + p_{nn} z^n,$$

and

$$Q_n(z) = 1 - q_{n1} z + q_{n2} z^2 - \cdots + (-1)^n q_{nn} z^n,$$

the numerator and denominator of the nth approximant of (20).

---

* Following Perron [6], we use Pringsheim's notation for continued fractions.

I. *Then, all the coefficients p and q are strictly positive and*

(22)
$$\begin{cases} P_n(z) \to & e^{(\gamma/2)z} \prod_{j=1}^{\infty} (1 + \alpha_j z), \\ Q_n(z) \to & e^{-(\gamma/2)z} \prod_{j=1}^{\infty} (1 - \beta_j z), \end{cases}$$

uniformly in any bounded region of the complex plane.

The elements k and l of (20) satisfy the following relations:

(23) $$k_j > 0 \qquad (j = 1, 2, 3, \ldots),$$

(24) $$\lim_{j \to \infty} k_j = 0,$$

(25) $$-a_1 < \sum_{j=1}^{n} l_j < 0 \qquad (n = 1, 2, 3, \ldots),$$

(26) $$\sum_{j=1}^{\infty} l_j = -\frac{\gamma}{2} - \sum_{j=1}^{\infty} \beta_j.$$

II. *The order ρ of the meromorphic function $f(z)$ is given by*

(27) $$\limsup_{n \to \infty} \{k_1 k_2 \cdots k_n\}^{1/2n \log n} = e^{-1/\rho}.$$

III. *The quantity*

$$\tau = \limsup_{n \to \infty} \left(n^2 \sqrt[n]{k_1 k_2 \cdots k_n} \right)$$

is always finite and the genus of $f(z)$ is zero (that is $\gamma = 0$) if and only if $\tau = 0$.

We conclude this introduction by pointing out that the class $(\mathscr{S})$ was introduced and studied by Schoenberg in connection with smoothing operations [7, pp. 361–369].

Schoenberg associates with the real sequence

(28) $$a_0, a_1, a_2, \ldots \qquad (a_0 \neq 0),$$

the infinite matrix

(29)
$$\begin{matrix} a_0 & 0 & 0 & 0 & 0 & \cdots \\ a_1 & a_0 & 0 & 0 & 0 & \cdots \\ a_2 & a_1 & a_0 & 0 & 0 & \cdots \\ a_3 & a_2 & a_1 & a_0 & 0 & \cdots \\ \cdots & \cdots & \cdots & \cdots & \cdots & \cdots \end{matrix}$$

and says that the sequence (28) is *totally positive* if (29) has only nonnegative minors (of all finite orders, with any choice of rows and columns).

Schoenberg proved the remarkable

Theorem A. *Let $f(z)$ be a member of the class $(\mathscr{S})$ and let* (1) *be its Taylor expansion.*

Then the sequence (28) *is totally positive.*

Schoenberg conjectured the converse of Theorem A. This was later proved by one of us [4, 5], so that there is complete equivalence between the total positivity of (28) and the fact that the generating function $f(z)$ belongs to the class $(\mathscr{S})$.

Theorem A is the key to our proof of Theorem 1; the more difficult converse will not be needed.

1. Notations and basic formulas

It will be necessary in our proofs to consider simultaneously several functions f, g, ... as well as the Hankel determinants and Padé polynomials associated with these functions. We shall write, with their obvious meaning,

$$A_m^{(n)}[f], \; A_m^{(n)}[g], \ldots,$$

or

$$\begin{cases} P_{mn}(z\,|\,f), & P_{mn}(z\,|\,g), \ldots, \\ Q_{mn}(z\,|\,f), & Q_{mn}(z\,|\,g), \ldots \end{cases}$$

This is to be understood as follows: if f has the expansion (1) and if

$$g(z) = \sum_{l=0}^{+\infty} d_l z^l,$$

then

$$A_m^{(n)}[f] = A_m^{(n)}, \quad A_m^{(n)}[g] = \begin{vmatrix} d_m & d_{m-1} & \cdots & d_{m-n+1} \\ d_{m+1} & d_m & \cdots & d_{m-n+2} \\ \cdots & \cdots & \cdots & \cdots \\ d_{m+n-1} & d_{m+n-2} & \cdots & d_m \end{vmatrix} = D_m^{(n)}.$$

Similarly $Q_{mn}(z\,|\,g)$ and $P_{mn}(z\,|\,g)$ are obtained by replacing a and A in the right-hand sides of (9) and (10) by d and D.

Following Pólya and Szegő, we write

$$G(z) \ll H(z)$$

to indicate that the power series $G(z)$ and $H(z)$ (with a common center) have real nonnegative coefficients and that the coefficients of $G(z)$ are not greater than the corresponding coefficients of $H(z)$.

We now turn our attention to some elementary identities which play an important role in this note.

From (9), (10), and a straightforward computation, we deduce first

$$(1.1) \qquad f(z)Q_{mn}(z) = \frac{1}{A_m^{(n)}} \sum_{j=0}^{\infty} \begin{vmatrix} a_j & a_{j-1} & \cdots & a_{j-n} \\ a_{m+1} & a_m & \cdots & a_{m-n+1} \\ a_{m+2} & a_{m+1} & \cdots & a_{m-n+2} \\ a_{m+n} & a_{m+n-1} & \cdots & a_m \end{vmatrix} z^j$$

$$= P_{mn}(z) + z^{m+n+1} S_{mn}(z),$$

where

$$(1.2) \qquad S_{mn}(z) = \frac{1}{A_m^{(n)}} \sum_{j=m+n+1}^{\infty} z^{j-m-n-1} \begin{vmatrix} a_j & a_{j-1} & \cdots & a_{j-n} \\ a_{m+1} & a_m & \cdots & a_{m-n+1} \\ a_{m+2} & a_{m+1} & \cdots & a_{m-n+2} \\ \cdots & \cdots & \cdots & \cdots \\ a_{m+n} & a_{m+n-1} & \cdots & a_m \end{vmatrix}$$

Now consider simultaneously (1) and

$$(1.3) \qquad b_0 + b_1 z + b_2 z^2 + \cdots = \frac{1}{f(-z)},$$

and assume that both the tables of $f(z)$ and $1/f(-z)$ are normal; then

$$(1.4) \qquad a_0 = \frac{1}{b_0} \neq 0,\ A_m^{(n)}[f(z)] \neq 0,\ A_n^{(m)}[1/f(-z)] = B_n^{(m)} \neq 0.$$

Relations (1.1) and (1.3) imply

$$\frac{P_{mn}(-z)}{f(-z)} - Q_{mn}(-z) = (-1)^{m+n} z^{m+n+1} S_{mn}(-z) \sum_{j=0}^{\infty} b_j z^j.$$

The smallest power of z which appears in the right-hand side of this formula is z^{m+n+1} so that, in view of the uniqueness of the normalized Padé polynomials of normal tables, we obtain the fundamental identity

$$(1.5) \qquad \frac{1}{a_0} P_{mn}(-z) \equiv \frac{1}{B_n^{(m)}} \begin{vmatrix} 1 & z & z^2 & \cdots & z^m \\ b_{n+1} & b_n & b_{n-1} & \cdots & b_{n-m+1} \\ b_{n+2} & b_{n+1} & b_n & \cdots & b_{n-m+2} \\ \cdots & \cdots & \cdots & \cdots & \cdots \\ b_{n+m} & b_{n+m-1} & b_{n+m-2} & \cdots & b_n \end{vmatrix}$$

$$= Q_{nm}(z \mid 1/f(-z)).$$

Comparing the coefficients of z^m, we are led to

$$(1.6) \qquad \frac{1}{a_0} \frac{A_m^{(n+1)}}{A_m^{(n)}} = \frac{B_{n+1}^{(m)}}{B_n^{(m)}},$$

and hence, using (1.4), we find

$$(1.7) \qquad \frac{1}{a_0^n} A_m^{(n)} = \frac{1}{b_0^m} A_n^{(m)} \qquad (m \geq 0, n \geq 0).$$

Our proof of relations (1.5), (1.6), and (1.7) is only valid if $f(z)$ and $1/f(-z)$ have normal tables. However, this restriction is unnecessary. For instance, (1.7) is an identity (in some of the variables a_j) which holds provided $a_0 \neq 0$, and the b's are defined by (1.3). This is almost obvious and could be deduced from the fact that the quantities

$$a_0^{j+1} b_j \qquad (j = 0, 1, 2, \ldots)$$

are polynomials in the variables a. After multiplication by a suitable power of a_0, (1.7) becomes a relation which is readily seen to be a polynomial identity.

2. Connection between the Padé tables of $f(z)$, $(1 + \alpha z)f(z)$ and $f(z)/(1 - \alpha z)$

We consider the four formal power series

$$f_0(z) = a_0 + a_1 z + a_2 z^2 + \cdots,$$

$$f_1(z) = f_0(z)(1 + \alpha z) = g_0 + g_1 z + g_2 z^2 + \cdots,$$

$$f_2(z) = \frac{f_0(z)}{1 - \alpha z} = h_0 + h_1 z + h_2 z^2 + \cdots,$$

$$f_3(z) = \frac{1}{f_0(-z)} = b_0 + b_1 z + b_2 z^2 + \cdots.$$

If

$$(2.1) \qquad A_m^{(n)}[f_0(z)] A_m^{(n)}[f_1(z)] A_m^{(n)}[f_2(z)] A_m^{(n)}[f_3(z)] \neq 0$$

we may use formulas (9) and (10) to define

$$P_{mn}^{(l)}(z) = P_{mn}(z \mid f_l), \quad Q_{mn}^{(l)}(z) = Q_{mn}(z \mid f_l) \qquad (l = 0, 1, 2, 3).$$

Starting from the obvious relation

$$(2.2) \quad
\begin{vmatrix}
\alpha^n & -\alpha^{n-1} & \alpha^{n-2} & \cdots & (-1)^n \\
a_{m+1} & a_m & a_{m-1} & \cdots & a_{m-n+1} \\
a_{m+2} & a_{m+1} & a_m & \cdots & a_{m-n+2} \\
\cdots & \cdots & \cdots & \cdots & \cdots \\
\cdots & \cdots & \cdots & \cdots & \cdots \\
a_{m+n} & a_{m+n-1} & a_{m+n-2} & \cdots & a_m
\end{vmatrix}
$$

$$
=
\begin{vmatrix}
0 & 0 & 0 & \cdots & 0 & (-1)^n \\
g_{m+1} & g_m & g_{m-1} & \cdots & g_{m-n+2} & a_{m-n+1} \\
g_{m+2} & g_{m+1} & g_m & \cdots & g_{m-n+3} & a_{m-n+2} \\
\cdots & \cdots & \cdots & \cdots & \cdots & \cdots \\
\cdots & \cdots & \cdots & \cdots & \cdots & \cdots \\
g_{m+n} & g_{m+n-1} & g_{m+n-2} & \cdots & g_{m+n} & a_m
\end{vmatrix}
= A_{m+1}^{(n)}[f_1],
$$

we obtain, in view of (9) and (2.1),

(2.3) $$A^{(n)}_{m+1}[(1 + \alpha z)f_0] = \alpha^n A^{(n)}_m Q^{(0)}_{mn}(-1/\alpha).$$

Consider now the determinant of order $n + 2$

$$\Lambda = \begin{vmatrix} \alpha^{n+1} & -\alpha^n & \alpha^{n-1} & \cdots & (-1)^{n+1} \\ a_j & a_{j-1} & a_{j-2} & \cdots & a_{j-n-1} \\ a_{m+1} & a_m & a_{m-1} & \cdots & a_{m-n} \\ a_{m+2} & a_{m+1} & a_m & \cdots & a_{m-n+1} \\ \cdots & \cdots & \cdots & \cdots & \cdots \\ a_{m+n} & a_{m+n-1} & a_{m+n-2} & \cdots & a_{m-1} \end{vmatrix},$$

and let Λ_{lk} denote the cofactor, in Λ, of the element in the lth row and kth column.

By a well-known result of the theory of determinants [2, p. 33, Corollary 3]

(2.4) $$\begin{vmatrix} \Lambda_{11} & \Lambda_{1,n+2} \\ \Lambda_{21} & \Lambda_{2,n+2} \end{vmatrix} = (-1)^n \Lambda A^{(n)}_m,$$

and also, as in (2.2),

(2.5) $$\Lambda = \begin{vmatrix} g_j & g_{j-1} & \cdots & g_{j-n} \\ g_{m+1} & g_m & \cdots & g_{m-n+1} \\ g_{m+2} & g_{m+1} & \cdots & g_{m-n+2} \\ \cdots & \cdots & \cdots & \cdots \\ g_{m+n} & g_{m+n-1} & \cdots & g_m \end{vmatrix}.$$

Beside (2.1) assume

(2.6) $$A^{(n)}_{m-1}[f_0] \neq 0,$$

and notice that (2.2), (2.3), and (2.4) then yield

(2.7) $$\frac{\Lambda}{A^{(n)}_m[f_1]} = \frac{\alpha Q^{(0)}_{mn}(-1/\alpha)}{A^{(n)}_{m-1}Q^{(0)}_{m-1,n}(-1/\alpha)} \begin{vmatrix} a_{j-1} & a_{j-2} & \cdots & a_{j-n-1} \\ a_m & a_{m-1} & \cdots & a_{m-n} \\ a_{m+1} & a_m & \cdots & a_{m-n+1} \\ \cdots & \cdots & \cdots & \cdots \\ a_{m+n-1} & a_{m+n-2} & \cdots & a_{m-1} \end{vmatrix}$$

$$+ \frac{1}{A^{(n)}_m} \begin{vmatrix} a_j & a_{j-1} & \cdots & a_{j-n} \\ a_{m+1} & a_m & \cdots & a_{m-n+1} \\ a_{m+2} & a_{m+1} & \cdots & a_{m-n+2} \\ \cdots & \cdots & \cdots & \cdots \\ a_{m+n} & a_{m+n-1} & \cdots & a_m \end{vmatrix}.$$

Multiply (2.7) by z^j, let $j = 0, 1, 2, \ldots, m$, and sum the resulting identities; in view of (10) and (2.5) we obtain

$$(2.8) \qquad P_{mn}^{(1)}(z) = P_{mn}^{(0)}(z) + \frac{\alpha Q_{mn}^{(0)}(-1/\alpha)}{Q_{m-1,n}^{(0)}(-1/\alpha)} z P_{m-1,n}^{(0)}(z).$$

Let

$$(2.9) \qquad A_{m-1}^{(n)}[f_3] \neq 0;$$

this relation, analogous to (2.6), enables us to replace $f_0(z)$ by $f_3(z)$ in (2.8) and leads to

$$(2.10) \qquad P_{mn}(z \,|\, (1 + \alpha z)f_3(z)) = P_{mn}^{(3)}(z) + \frac{\alpha Q_{mn}^{(3)}(-1/\alpha)}{Q_{m-1,n}^{(3)}(-1/\alpha)} z P_{m-1,n}^{(3)}(z).$$

Since

$$\frac{1}{(1 - \alpha z)f_3(-z)} = \frac{f_0(z)}{1 - \alpha z} = f_2(z),$$

we deduce from (2.10) and (1.5)

$$(2.11) \qquad Q_{nm}^{(2)}(z) = Q_{nm}^{(0)}(z) - \frac{\alpha P_{nm}^{(0)}(1/\alpha)}{P_{n,m-1}^{(0)}(1/\alpha)} z Q_{n,m-1}^{(0)}(z).$$

3. Lower bounds for $A_m^{(n)}$

Let $f_0(z) = \sum a_j z^j$ be a function of the class $(\mathscr{S})$. Then by Theorem A all the minors of the matrix (29) are nonnegative. Hence if $\alpha \geq 0$ and if we expand, with respect to its first row, the first determinant in (2.2), we obtain a sum of nonnegative terms. One of these terms is $A_{m+1}^{(n)}$ and therefore

$$A_{m+1}^{(n)}[(1 + \alpha z)f_0] \geq A_{m+1}^{(n)}[f_0].$$

A repeated application of the argument and the replacement of $m + 1$ by m lead to

$$(3.1) \qquad A_m^{(n)}\left[\prod_{j=1}^{\infty}(1 + \alpha_j z)f_0\right] \geq A_m^{(n)}[f_0] \qquad (m \geq 0, n \geq 0, \alpha_j \geq 0),$$

and also to

$$(3.2) \qquad A_m^{(n)}[(1 + \gamma z/s)^s f_0] \geq A_m^{(n)}[f_0] \qquad (m \geq 0, n \geq 0, \gamma \geq 0),$$

where s is a positive integer. Letting $s \to +\infty$, we deduce from (3.2)

$$(3.3) \qquad A_m^{(n)}[e^{\gamma z}f_0] \geq A_m^{(n)}[f_0] \qquad (m \geq 0, n \geq 0).$$

Finally, since $1/f_0(-z)$ is also of the class $(\mathscr{S})$,

$$A_n^{(m)}[(1 + \beta z)/f_0(-z)] \geq A_n^{(m)}[1/f_0(-z)] \qquad (n \geq 0, m \geq 0, \beta \geq 0),$$

which, in view of (1.7), yields

(3.4) $A_m^{(n)}[f_0(z)/(1 - \beta z)] \geq A_m^{(n)}[f_0(z)]$ $(m \geq 0, n \geq 0, \beta \geq 0)$,

and hence

(3.5) $A_m^{(n)}\left[f_0(z) \middle/ \prod_{j=1}^{\infty} (1 - \beta_j z) \right] \geq A_m^{(n)}[f_0]$ $(m \geq 0, n \geq 0, \beta_j \geq 0)$.

Combining (3.1), (3.3), and (3.5), we obtain

Lemma 1. *If $f(z)$ and $f_0(z)$ belong to the class $(\mathscr{S})$ and if $f(0) = 1$, then*
$$A_m^{(n)}[f(z)f_0(z)] \geq A_m^{(n)}[f_0(z)]   (m \geq 0, n \geq 0).$$

Notice now that

(3.6) $A_m^{(n)}[e^z] = \dfrac{(n-1)!}{m!} A_{m+1}^{(n-1)}[e^z]$ $(m \geq n \geq 1)$.

This may be verified as follows: subtract from each column of $A_m^{(n)}[e^z]$ the previous column multiplied by a suitable factor; the factor is $m - k$ for the column whose entry, in the first row is $1/(m - k)!$.

Repeated applications of the reduction formulas (3.6) lead to

(3.7) $A_m^{(n)}[e^z] = \prod_{k=1}^{n} \dfrac{1}{k(k+1) \cdots (k + m - 1)}$ $(m \geq n \geq 1)$,

which, in view of (1.7), remains valid if the restriction $m \geq n$ is omitted.

Lemma 2. *Let $f(z)$ be of the form* (13). *Then*

(3.8) $A_m^{(n)}[f(z)] \geq a_0^n \max\{\gamma^{mn} A_m^{(n)}[e^z], (\alpha_1\alpha_2 \cdots \alpha_m)^n, (\beta_1\beta_2 \cdots \beta_n)^m\}$
$$(m \geq 0, n \geq 0).$$

(The quantities α, β, γ are real, nonnegative but some or all of them may be zero.)

Proof. Let

(3.9) $G(z) = \prod_{j=1}^{\infty} (1 + \alpha_j z)$ $(\alpha_j \geq 0)$,

(3.10) $H(z) = \prod_{j=1}^{\infty} (1 - \beta_j z)$ $(\beta_j \geq 0)$.

By assumption,

$$f(z) = a_0 e^{\gamma z} \frac{G(z)}{H(z)}   (\gamma \geq 0, a_0 > 0).$$

Obviously

$$A_m^{(n)}[f(z)] = a_0^n A_m^{(n)}[e^{\gamma z} G(z)/H(z)], \tag{3.11}$$

and also

$$A_m^{(n)}[e^{\gamma z}] = \gamma^{mn} A_m^{(n)}[e^z].$$

Hence, in view of Lemma 1,

$$A_m^{(n)}[f(z)] \geq a_0^n \gamma^{mn} A_m^{(n)}[e^z]. \tag{3.12}$$

Lemma 1 also shows that (3.9) and the obvious relation

$$A_m^{(n)}[(1 + \alpha_1 z)(1 + \alpha_2 z) \cdots (1 + \alpha_m z)]$$
$$= (\alpha_1 \alpha_2 \cdots \alpha_m)^n \qquad (n = 0, 1, 2, 3, \ldots),$$

imply

$$A_m^{(n)}[G(z)] \geq (\alpha_1 \alpha_2 \cdots \alpha_m)^n. \tag{3.13}$$

By (1.7) and (3.10)

$$A_m^{(n)}[1/H(z)] = A_n^{(m)}[H(-z)],$$

and as in (3.13)

$$A_n^{(m)}[H(-z)] \geq (\beta_1 \beta_2 \cdots \beta_n)^m.$$

Hence

$$A_m^{(n)}[1/H(z)] \geq (\beta_1 \beta_2 \cdots \beta_n)^m. \tag{3.14}$$

Lemma 2 is now an obvious consequence of (3.11), (3.12), (3.13), (3.14), and Lemma 1.

If $f(z)$ is transcendental, and of the class $(\mathscr{S})$, (15) clearly follows from (3.8). We have thus proved assertion I of Theorem 1.

4. Upper bounds for the Padé polynomials

Let $f(z)$ be a transcendental function of the form (13). We shall impose the normalization $a_0 = 1$; an inspection of the definitions (9) and (10) of the Padé polynomials shows that this is no restriction. By a known result [4, p. 89] we then have, uniformly in any bounded region of the complex plane,

$$Q_{\infty n}(z) = \lim_{m \to \infty} Q_{mn}(z|f(z)) = (1 - \beta_1 z)(1 - \beta_2 z) \cdots (1 - \beta_n z) \quad (\beta_{j+1} \leq \beta_j), \tag{4.1}$$

and hence

$$P_{m\infty} = \lim_{n \to \infty} P_{mn}(z \mid f(z)) = \lim_{n \to \infty} Q_{nm}(-z \mid 1/f(-z)) \tag{4.2}$$

$$= (1 + \alpha_1 z)(1 + \alpha_2 z) \cdots (1 + \alpha_m z) \qquad (\alpha_{j+1} \leq \alpha_j).$$

Putting

$$r_s(z) = a_s z^s + a_{s+1} z^{s+1} + a_{s+2} z^{s+2} + \cdots$$

we may write (1.1) as

$$f(z)Q_{mn}(z) - P_{mn}(z)$$

$$= \frac{1}{A_m^{(n)}} \begin{vmatrix} r_{m+n+1}(z) & zr_{m+n}(z) & z^2 r_{m+n-1}(z) & \cdots & z^n r_{m+1}(z) \\ a_{m+1} & a_m & a_{m-1} & \cdots & a_{m-n+1} \\ a_{m+2} & a_{m+1} & a_m & \cdots & a_{m-n+2} \\ \cdots & \cdots & \cdots & \cdots & \cdots \\ \cdots & \cdots & \cdots & \cdots & \cdots \\ a_{m+n} & a_{m+n-1} & a_{m+n-2} & \cdots & a_m \end{vmatrix},$$

which implies

(4.3) $|f(z)Q_{mn}(z) - P_{mn}(z)| \leqq \max\{|r_{m+n+1}(z)|, |r_{m+n}(z)|, \ldots,$

$$|r_{m+1}(z)|\} Q_{mn}(-|z|).$$

The radius of convergence of (1) being $1/\beta_1$, inequality (4.3) is valid for $|z| < 1/\beta_1$. Hence, for n fixed and $m \to \infty$,

(4.4) $$P_{mn}(z) \to f(z)Q_{\infty n}(z),$$

uniformly for $|z| \leqq \eta < 1/\beta_1$.

Similarly, for m fixed and $n \to \infty$,

(4.5) $$Q_{mn}(z) \to \frac{P_{m\infty}(z)}{f(z)},$$

uniformly for $|z| \leqq \zeta < 1/\alpha_1$.

Now formula (6) of [4] and the analogous formula

(4.6) $$Q_{mn}(z) - Q_{m,n-1}(z) = -z Q_{m-1,n-1}(z) \frac{A_{m+1}^{(n)}/A_m^{(n)}}{A_m^{(n-1)}/A_{m-1}^{(n-1)}},$$

yield, respectively,

(4.7) $$Q_{jn}(-z) \ll Q_{mn}(-z) \qquad (m \leqq j),$$

and

(4.8) $$Q_{mn}(-z) \ll Q_{mj}(-z) \qquad (n \leqq j).$$

Considering $1/f(-z)$, instead of $f(z)$, we see that (1.5) enables us to transform (4.7) and (4.8) into

(4.9) $$P_{nj}(z) \ll P_{nm}(z) \qquad (m \leqq j),$$

and

(4.10) $P_{nm}(z) \ll P_{jm}(z)$ $(n \leqq j)$.

Combining relations (4.1), (4.2), (4.4), (4.5), (4.7), (4.8), (4.9), and (4.10), we obtain

(4.11) $(1 + \beta_1 z)(1 + \beta_2 z) \cdots (1 + \beta_n z)$

$$\ll Q_{mn}(-z) \ll e^{\gamma z} \frac{(1 + \beta_1 z)(1 + \beta_2 z)(1 + \beta_3 z) \cdots}{(1 - \alpha_{m+1} z)(1 - \alpha_{m+2} z)(1 - \alpha_{m+3} z) \cdots},$$

and

(4.12) $(1 + \alpha_1 z)(1 + \alpha_2 z) \cdots (1 + \alpha_n z)$

$$\ll P_{nm}(z) \ll e^{\gamma z} \frac{(1 + \alpha_1 z)(1 + \alpha_2 z)(1 + \alpha_3 z) \cdots}{(1 - \beta_{m+1} z)(1 - \beta_{m+2} z)(1 - \beta_{m+3} z) \cdots}.$$

We now prove

Lemma 3. *Let $\{m_\lambda\}_{\lambda=1}^\infty$, and $\{n_\lambda\}_{\lambda=1}^\infty$ be any two sequences of positive integers such that $m_\lambda \to \infty$ and $n_\lambda \to \infty$.*

Assume that one of the two sequences

(4.13) $\{P_{m_\lambda, n_\lambda}(z)\}_{\lambda=1}^\infty$, $\{Q_{m_\lambda, n_\lambda}(z)\}_{\lambda=1}^\infty$,

converges at all points of some closed interval $[0, \varepsilon]$ of the positive axis. Then each of the two sequences (4.13) converges uniformly in any bounded region of the complex plane. The respective limit functions $\phi(z)$ and $\psi(z)$ are entire functions connected by the relation

(4.14) $\phi(z) = f(z)\psi(z).$

Proof. Assume, for instance, that if z is real and

$$0 \leqq z \leqq \varepsilon   (\varepsilon > 0),$$

$Q_{m_\lambda, n_\lambda}(z)$ converges as $\lambda \to \infty$. Using (4.11), the condition $m_\lambda \to \infty$ and Vitali's theorem, we see that this convergence is uniform in any bounded region of the complex plane. Now (4.3) shows that the following relation holds in the neighborhood of the origin:

$$\lim_{\lambda \to \infty} P_{m_\lambda, n_\lambda}(z) = f(z) \lim_{\lambda \to \infty} Q_{m_\lambda, n_\lambda}(z).$$

Hence, using (4.12), the condition $n_\lambda \to \infty$, and Vitali's theorem we easily verify all the assertions of our lemma.

5. Proof of Assertion II of Theorem 1

Let $f(z)$ be defined as in the previous section, and let $a_0 = 1$. In the special case $\gamma = 0$, relations (4.11) and (4.12) obviously imply

$$\lim_{\lambda \to \infty} P_{m_\lambda, n_\lambda}(z) = \prod_{j=1}^{\infty} (1 + \alpha_j z) \qquad (z \geq 0),$$

$$\lim_{\lambda \to \infty} Q_{m_\lambda, n_\lambda}(z) = \prod_{j=1}^{\infty} (1 - \beta_j z) \qquad (z \leq 0),$$

provided $m_\lambda \to \infty$ and $n_\lambda \to \infty$. Hence, for $\gamma = 0$, Theorem 1 follows from Vitali's theorem.

We now assume $\gamma > 0$ and first prove assertion II of our theorem for functions of the form

$$(5.1) \qquad \phi_N(z) = e^{\gamma z} \prod_{j=1}^{N} (1 + \alpha_j z) \qquad (\alpha_j > 0).$$

The Padé table of the exponential function has the properties expressed by (17), (18), and (19) [6, p. 248] so that, in view of (2.8) and (17),

$$(5.2) \qquad \lim_{\lambda \to \infty} P_{m_\lambda, n_\lambda}(z \,|\, (1 + \alpha z)e^{\gamma z}) = (1 + \alpha z)e^{[\omega/(1 + \omega)]\gamma z}.$$

Hence Lemma 3 yields

$$(5.3) \qquad \lim_{\lambda \to \infty} Q_{m_\lambda, n_\lambda}(z \,|\, (1 + \alpha z)e^{\gamma z}) = e^{-[1/(1 + \omega)]\gamma z},$$

as well as the uniformity of the convergence in both (5.2) and (5.3). Repeating our arguments a finite number of times, we prove Theorem 1 for functions of the form (5.1).

Now (2.8), (4.7), and (4.10) obviously imply

$$P_{mn}(z \,|\, f(z)) \ll P_{mn}(z \,|\, f(z)(1 + \alpha z)) \ll P_{mn}(z \,|\, f(z))(1 + \alpha z),$$

and hence

$$(5.4) \qquad P_{mn}(z \,|\, f(z)) \ll P_{mn}\!\left(z \,|\, f(z) \prod_{j=N+1}^{\infty} (1 + \alpha_j z) \right)$$

$$\ll P_{mn}(z \,|\, f(z)) \prod_{j=N+1}^{\infty} (1 + \alpha_j z).$$

Since relations (18) and (19) are proved for special functions of the form (5.1), we see that if z denotes a fixed, real, nonnegative number, (5.4) implies

$$e^{[\omega/(1+\omega)]\gamma z} \prod_{j=1}^{N} (1 + \alpha_j z) \leqq \liminf_{\lambda \to \infty} P_{m_\lambda, n_\lambda}\left(z \,|\, e^{\gamma z} \prod_{j=1}^{\infty} (1 + \alpha_j z)\right)$$

$$\leqq \limsup_{\lambda \to \infty} P_{m_\lambda n_\lambda}\left(z \,|\, e^{\gamma z} \prod_{j=1}^{\infty} (1 + \alpha_j z)\right) \leqq e^{[\omega/(1+\omega)]\gamma z} \prod_{j=1}^{\infty} (1 + \alpha_j z).$$

Letting $N \to \infty$, we obtain

$$\lim_{\lambda \to \infty} P_{m_\lambda, n_\lambda}\left(z \,|\, e^{\gamma z} \prod_{j=1}^{\infty} (1 + \alpha_j z)\right) = e^{[\omega/(1+\omega)]\gamma z} \prod_{j=1}^{\infty} (1 + \alpha_j z).$$

Hence, in view of Lemma 3, relations (18) and (19) are proved for functions of the form

(5.5)
$$e^{\gamma z} \prod_{j=1}^{\infty} (1 + \alpha_j z),$$

and, by the transformation $f(z) | 1/f(-z)$, also for functions of the form

$$e^{\gamma z} \Big/ \prod_{j=1}^{\infty} (1 - \beta_j z).$$

Then, one of our arguments shows that assertion II of Theorem 1 is true for functions of the form

$$u(z) = e^{\gamma z} \frac{\displaystyle\prod_{j=1}^{N} (1 + \alpha_j z)}{\displaystyle\prod_{j=1}^{\infty} (1 - \beta_j z)}.$$

We may now replace the function $f(z)$ in (5.4) by the function $u(z)$, so that the arguments which proved (5.5) show that assertion II of our theorem is true for all nonrational functions of the class $(\mathscr{S})$.

This completes the proof of Theorem 1.

6. Proof of Theorem 2

The positivity of the coefficients p and q (and in fact the positivity of all the similar coefficients of the Padé polynomials) was proved in [4, pp. 88–89]. Since the approximants in the diagonal of the Padé table coincide with the approximants of the continued fraction, relations (22) are special cases of (18) and (19).

Consider now the well-known recurrence formulas [6, pp. 128–129]

(6.1)
$$\begin{cases} P_n(z) = (1 + l_n z)P_{n-1}(z) + k_n z^2 P_{n-2}(z) & (n \geq 2); \\ \qquad\qquad P_0(z) = 1, \; P_1(z) = 1 + (k_1 + l_1)z, \\ Q_n(z) = (1 + l_n z)Q_{n-1}(z) + k_n z^2 Q_{n-2}(z) & (n \geq 2); \\ \qquad\qquad Q_0(z) = 1, \; Q_1(z) = 1 + l_1 z, \end{cases}$$

which yield

$$(6.2) \qquad p_{n1} = p_{n-1,1} + l_n, \qquad -q_{n1} = -q_{n-1,1} + l_n,$$

$$(6.3) \quad p_{n2} = p_{n-1,2} + l_n p_{n-1,1} + k_n, \qquad q_{n2} = q_{n-1,2} - l_n q_{n-1,1} + k_n,$$

and

$$(6.4) \qquad P_n(z)Q_{n-1}(z) - Q_n(z)P_{n-1}(z) = (-1)^{n-1} k_1 k_2 \cdots k_n z^{2n-1}.$$

The connections between the coefficients k_n and the Hankel determinants are also well known [6, p. 130];

$$(6.5) \qquad k_1 = a_1, \quad k_n = A_n^{(n)} A_{n-2}^{(n-2)} / \{A_{n-1}^{(n-1)}\}^2 \qquad (n \geq 2; A_0^{(0)} = 1).$$

The inequalities (23) are obvious consequences of the latter relations. It is immediately seen that

$$l_1 = -\frac{a_2}{a_1} < 0,$$

but we have no information* concerning the signs of the other l's. Hence, from (6.3) and (6.5) we can only conclude

$$(6.6) \qquad 0 < k_n \leq \max\{(p_{n2} - p_{n-1,2}), (q_{n2} - q_{n-1,2})\}.$$

In view of (22)

$$\lim_{n \to \infty} (p_{n2} - p_{n-1,2}) = 0, \qquad \lim_{n \to \infty} (q_{n2} - q_{n-1,2}) = 0,$$

so that (24) follows from (6.6).

Relations (6.2) and (6.5) yield

$$(6.7) \qquad \sum_{j=1}^{n} l_j = -q_{n1} = p_{n1} - k_1 = p_{n1} - a_1,$$

and lead to (25) because the coefficients p and q are positive. By (22)

$$\lim_{n \to \infty} q_{n1} = \frac{\gamma}{2} + \sum_{j=1}^{\infty} \beta_j,$$

and hence (26) follows from (6.7).

In order to prove assertion II of Theorem 2, we first notice that (6.4) implies

$$k_1 k_2 \cdots k_n r^{2n-1} \leq P_n(r)Q_{n-1}(-r) + Q_n(-r)P_{n-1}(r) \qquad (r \geq 0),$$

---

* In the symmetrical case $\alpha_j = \beta_j$ $(j = 1, 2, 3, \ldots)$ it is easy to deduce $0 = l_2 = l_3 = l_4 = \cdots$, from known results [6, pp. 137 and 141].

and therefore, by (4.11) and (4.12),

$$(6.8) \qquad k_1 k_2 \cdots k_n r^{2n-1} \leq 2e^{2\gamma r} \frac{\prod\limits_{j=1}^{\infty} (1 + \alpha_j r)(1 + \beta_j r)}{\prod\limits_{j=n}^{\infty} (1 - \alpha_j r)(1 - \beta_j r)}$$

provided

$$(6.9) \qquad \alpha_j r < 1, \qquad \beta_j r < 1 \qquad (j \geq n).$$

By elements of the theory of entire functions we know that

$$\lim_{n \to \infty} n(\alpha_n^{\rho+\varepsilon} + \beta_n^{\rho+\varepsilon}) = 0 \qquad (\varepsilon > 0),$$

and hence the inequalities (6.9) are valid with

$$(6.10) \qquad r = n^{1/(\rho+\varepsilon)} \qquad (n > n_0).$$

Assume that $\rho < 1$ (then γ is necessarily zero); choose $\varepsilon > 0$ such that $\rho + \varepsilon < 1$, let r be defined by (6.10) and use this value in (6.8). Straightforward estimates, the details of which will be left to the reader, readily yield

$$(6.11) \qquad \limsup_{n \to \infty} \{k_1 k_2 \cdots k_n\}^{1/2n \log n} \leq e^{-1/\rho}.$$

If $\rho = 1$ it is possible (but not necessary) that $\gamma > 0$; this prevents us from selecting r as in (6.10). We shall take, in (6.8), $r = n$, and, by estimates similar to those which lead to (6.11), we find

$$(6.12) \qquad \limsup_{n \to \infty} \{(k_1 k_2 \cdots k_n)^{1/n} n^2\} \leq e^{2\gamma},$$

provided we use the familiar fact that the growth of a canonical product of genus zero cannot exceed the minimum type of order one. The inequality (6.12) shows that (6.11) remains true for $\rho = 1$.

Inequalities in the opposite direction may be derived from Lemma 2. By (3.8) we have

$$(6.13) \qquad A_n^{(n)} \geq \max\{\alpha_n^{n^2}, \beta_n^{n^2}\}$$

as well as

$$(6.14) \qquad A_n^{(n)} \geq \gamma^{n^2} A_n^{(n)}[e^z].$$

If ρ_1 is the exponent of convergence of the zeros of $f(z)$ and ρ_2 the exponent of convergence of its poles, (6.13) yields

(6.15) $$\limsup_{n \to \infty} \{A_n^{(n)}\}^{1/n^2 \log n} \geqq \max \{e^{-1/\rho_1}, e^{-1/\rho_2}\}.$$

If $\gamma > 0$, (6.14) and (3.7) imply

(6.16) $$\liminf_{n \to \infty} \{A_n^{(n)}\}^{1/n^2 \log n} \geqq e^{-1}.$$

Elementary properties of sequences show that

(6.17) $$\limsup_{n \to \infty} \{A_n^{(n)}\}^{1/n^2 \log n} \leqq \limsup_{n \to \infty} \left(\frac{A_n^{(n)}}{A_{n-1}^{(n-1)}} \right)^{1/2n \log n},$$

and by (6.5)

(6.18) $$\frac{A_n^{(n)}}{A_{n-1}^{(n-1)}} = k_1 k_2 \cdots k_n \qquad (n = 1, 2, 3, \ldots).$$

Hence we deduce from (6.15), (6.16), (6.17), and (6.18),

(6.19) $$\limsup_{n \to \infty} \{k_1 k_2 \cdots k_n\}^{1/2n \log n} \geqq \max \{e^{-1/\rho_1}, e^{-1/\rho_2}\},$$

(6.20) $$\limsup_{n \to \infty} \{k_1 k_2 \cdots k_n\}^{1/2n \log n} \geqq e^{-1} \qquad (\gamma > 0).$$

Assertion II of Theorem 2 is an obvious consequence of (6.11), (6.19), and (6.20).

By (6.12), the quantity τ in assertion III of Theorem 2 is always finite; to complete the proof of the theorem there only remains to examine the conditions under which $\tau = 0$.

If $\gamma = 0$ we must have $\tau > 0$ by (6.14), (6.18), and (3.7).

The assumption $\gamma = 0$ yields $\tau = 0$ by an argument similar to the one which leads to (6.12). We obtain this sharper form by taking, in (6.8), $r = (n/\varepsilon)$ $(0 < \varepsilon)$ and by using the fact that the growth of $\prod_{j=1}^n (1 + \alpha_j z)$ $\times (1 + \beta_j z)$ is at most of the minimum type of order 1.

REFERENCES

1. M. B. Balk, The Padé interpolation process for certain analytic functions, *Issledovaniya po sovremennym problemam teorii funkcii kompleksnogo peremennogo*, 234–257, Gosudarstv. Izdat. Fiz.-Mat. Lit., Moscow, 1960.
2. M. Bôcher, *Introduction to higher algebra*, Macmillan, New York, 1915.
3. A. Edrei, Sur les déterminants récurrents et les singularités d'une fonction donnée par son développement de Taylor, *Compositio Math.* 7 (1939), 20–88.

4. ———, Proof of a conjecture of Schoenberg on the generating function of a totally positive sequence, *Canad. J. Math.* **5** (1953), 86–94.

5. ———, On the generating function of a doubly infinite, totally positive sequence, *Trans. Amer. Math. Soc.* **74** (1953), 367–383.

6. O. Perron, *Die Lehre von den Kettenbrüchen*, 3rd ed., vol. 2, B. G. Teubner, Stuttgart, 1957.

7. I. J. Schoenberg, Some analytic aspects of the problem of smoothing, in *Studies and essays presented to R. Courant on his 60th birthday, Jan.* 8, 1948, Interscience, New York, 1948, pp. 351–370.

(*Received March 1, 1968*)

Some Integrability Theorems for Power Series with Positive Coefficients*

R. Askey†
Department of Mathematics
University of Wisconsin
Madison, Wisconsin, 53706
U.S.A.

R. P. Boas, Jr.‡
Department of Mathematics
Northwestern University
Evanston, Illinois, 60201
U.S.A.

Let $f(x) = \sum_{k=0}^{\infty} a_k x^k$, $a_k \geq 0$, $0 \leq x < 1$; let s_n be the partial sums of the series at $x = 1$, $s_n = \sum_{k=0}^{n} a_k$; and (if $\sum a_k$ converges) let r_n be the remainders, $r_n = \sum_{k=n}^{\infty} a_k$. The simplest of all tauberian theorems states that $f(x)$ approaches a limit as $x \to 1-$ if and only if $\sum a_k$ converges, in other words that $f \in L^{\infty}(0, 1)$ if and only if $\{s_n\} \in l^{\infty}$. The corresponding theorem with index 1 is that $f \in L(0, 1)$ if and only if $\{n^{-2} s_n\} \in l$; this is also trivial. Askey [1] has proved the nontrivial intermediate theorem that, with $1 < p < \infty$, we have $f \in L^p(0, 1)$ if and only if $\{n^{-2/p} s_n\} \in l^p$; that is, an (unweighted) L^p property of f is equivalent to a weighted l^p property of $\{s_n\}$ (see Appendix A).

* Sponsored by the Mathematics Research Center, United States Army, Madison, Wisconsin, under Contract No.: DA-31-124-ARO-D-462.

† Mathematics Research Center.

‡ Research supported by National Science Foundation Grant GP 9637.

One might ask for a parallel theorem connecting an unweighted l^p property of $\{s_n\}$ with an integrability property of f; of course $\{s_n\}$ cannot belong to l^p when $a_n \geq 0$ (unless $a_n \equiv 0$), but it might if $a_n \geq 0$ ultimately. It amounts to the same thing to ask for a theorem connecting $\{r_n\} \in l^p$ with an integrability property of f. We shall prove the following theorem.

Theorem 1. *If $a_k \geq 0$, $\sum a_k$ converges, and $1 < p \leq \infty$ then $\{r_n\} \in l^p$ if and only if*

(1)
$$(1 - x)^{-2/p}[f(1) - f(x)] \in L^p(0, 1).$$

For $p = 1$, the theorem holds if we replace (1) by

$$(1 - x)^{-2}[f(1) - 2f(x) + f(x^2)] \in L.$$

Theorem 1 is almost trivial for $p = \infty$ and (with the appropriate modification) for $p = 1$. We confine ourselves to the case $1 < p < \infty$.

Suppose first that $\sum a_k$ converges and $\sum r_n^p < \infty$. Then

(2)
$$\int_0^1 (1 - x)^{-2}[f(1) - f(x)]^p \, dx$$

$$= \int_0^1 (1 - x)^{-2}\left\{\sum_{k=1}^{\infty} a_k(1 - x^k)\right\}^p dx$$

$$= \sum_{n=1}^{\infty} \int_{1-1/n}^{1-1/(n+1)} (1 - x)^{-2}\left\{\sum_{k=1}^{\infty} a_k(1 - x^k)\right\}^p dx.$$

Since $1 - x^k \leq k(1 - x)$, the right-hand side does not exceed

$$\sum_{n=1}^{\infty} n^2 \cdot n^{-2}\left\{\sum_{k=1}^{n} ka_k(n^{-1}) + \sum_{k=n+1}^{\infty} a_k\right\}^p \leq 2^{p-1}\sum_{n=1}^{\infty}\left\{n^{-1}\sum_{k=1}^{n} ka_k\right\}^p + 2^{p-1}\sum_{n=1}^{\infty} r_n^p.$$

The first sum on the right is at most a constant multiple of $\sum r_n^p$ (see [2, p. 38, Lemma 6.18], with $c = 0$, $s = 1$). Hence the left side of (2) is finite.

Now suppose that (1) holds. We have

$$\int_0^1 (1 - x)^{-2}[f(1) - f(x)]^p \, dx$$

$$= \int_0^1 (1 - x)^{-2}\left\{\sum_{k=1}^{\infty} a_k(1 - x^k)\right\}^p dx$$

$$= \sum_{n=1}^{\infty} \int_{1-1/n}^{1-1/(n+1)} (1 - x)^{-2}\left\{\sum_{k=1}^{\infty} a_k(1 - x^k)\right\}^p dx$$

$$\geq \sum_{n=1}^{\infty}\left\{\sum_{n=1}^{\infty} a_k[1 - (1 - (n + 1)^{-1})^k]\right\}^p$$

$$\geq \sum_{n=1}^{\infty} \left\{ \sum_{k=n}^{\infty} a_k(1 - 2e^{-1}) \right\}^p$$

$$\geq K \sum_{n=1}^{\infty} \left\{ \sum_{k=n}^{\infty} a_k \right\}^p,$$

since $(1 - 1/n)^k \leq 2/e$ for $k \geq n$. That is, $\sum r_n{}^p$ converges. This completes the proof of Theorem 1.

We note that the integrability condition (1) in Theorem 1 can be replaced by other conditions; for example, when $1 < p < \infty$ condition (1) is equivalent to

$$(1 - x)^{1 - 2/p} f'(x) \in L^p(0, 1)$$

and to

$$(1 - x)^{2 - 2/p} f''(x) \in L^p(0, 1).$$

For a proof see Appendix B.

There are various other directions in which these theorems can be extended. First we show that the results can be extended to $0 < p < 1$. Finally we give one sample theorem of a dual nature, where now $a_n = \int_0^1 x^n \, d\mu(x)$ with $d\mu(x) \geq 0$. All of this work is an extension of Hardy and Littlewood [3].

Theorem 2. Let $f(x) = \sum_{n=0}^{\infty} a_n x^n$, $a_n \geq 0$, $0 \leq x < 1$. Then if $0 < p < 1$, a necessary and sufficient condition that $\int_0^1 [f(x)]^p \, dx < \infty$ is that

$$\sum_{n=1}^{\infty} \left(\sum_{k=0}^{n} a_k \right)^p n^{-2} < \infty.$$

Let $f \in L^p(0, 1)$. Then as in [1] we have

$$\int_0^1 [f(x)]^p \, dx \geq \sum_{n=1}^{\infty} \int_{1-1/n}^{1-1/(n+1)} \left[\sum_{k=0}^{n} a_k(1 - (n + 1)^{-1})^k \right]^p dx$$

$$\geq A \sum_{n=1}^{\infty} n^{-2} \left(\sum_{k=0}^{n} a_k \right)^p.$$

In the other direction we have

$$\int_0^1 [f(x)]^p \, dx \leq \sum_{n=2}^{\infty} n^{-2} \left[\sum_{k=0}^{\infty} a_k(1 - 1/n)^k \right]^p$$

$$\leq A \sum_{n=2}^{\infty} n^{-2} \left(\sum_{k=0}^{n} a_k \right)^p + A \sum_{n=2}^{\infty} n^{-2} \left[\sum_{k=n}^{\infty} a_k e^{-k/n} \right]^p = I + J.$$

$$J = A \sum_{n=2}^{\infty} n^{-2} \left[\sum_{j=1}^{\infty} \sum_{k=jn}^{(j+1)n-1} a_k e^{-k/n} \right]^p$$

$$\leqq A \sum_{n=2}^{\infty} n^{-2} \sum_{j=1}^{\infty} e^{-jp} \left[\sum_{k=jn}^{(j+1)n} a_k \right]^p$$

$$\leqq A \sum_{n=2}^{\infty} n^{-2} \sum_{j=1}^{\infty} e^{-jp} \left[\sum_{k=0}^{(j+1)n} a_k \right]^p$$

$$= A \sum_{j=1}^{\infty} e^{-jp} \sum_{n=2}^{\infty} n^{-2} \left[\sum_{k=0}^{(j+1)n} a_k \right]^p$$

$$= A \sum_{j=1}^{\infty} (j+1)^2 e^{-jp} \sum_{n=2}^{\infty} [(j+1)n]^{-2} \left[\sum_{k=0}^{(j+1)n} a_k \right]^p$$

$$\leqq A \sum_{j=1}^{\infty} (j+1)^2 e^{-jp} \sum_{r=2}^{\infty} r^{-2} \left[\sum_{k=0}^{r} a_k \right]^p \leqq A \sum_{n=2}^{\infty} n^{-2} \left(\sum_{k=0}^{n} a_k \right)^p.$$

Since Theorem 1 is not the same for $p = 1$ and $p > 1$ it is to be expected that its extension to $p < 1$ will be somewhat different. Define $\Delta^j f(x)$ by

$$\Delta^j f(x) = f(1) - jf(x) + \frac{j(j-1)}{2} f(x^2) + \cdots + (-1)^j f(x^j).$$

It is easy to see that

(3) $$\Delta^j f(x) = \sum_{k=1}^{\infty} a_k (1 - x^k)^j.$$

The extension of Theorem 1 to $p < 1$ is

Theorem 3. *Let* $0 < p < 1$, $f(x) = \sum_{k=0}^{\infty} a_k x^k$, $a_k \geqq 0$; *let* $\Delta^j f(x)$ *be defined by* (3); *and let* $jp > 1$. *Then a necessary and sufficient condition that* $\sum_{n=1}^{\infty} \left(\sum_{k=n}^{\infty} a_k \right)^p < \infty$ *is that* $\int_0^1 |\Delta^j f(x)|^p (1 - x)^{-2} \, dx < \infty$.

As usual one direction is very easy. We have

$$\int_0^1 [\Delta^j f(x)]^p (1 - x)^{-2} \, dx \geqq \sum_{n=1}^{\infty} \int_{1-1/n}^{1-1/(n+1)} \left[\sum_{k=n}^{\infty} a_k (1 - x^k)^j \right]^p (1 - x)^{-2} \, dx$$

$$\geqq A \sum_{n=1}^{\infty} \left\{ \sum_{k=n}^{\infty} a_k \right\}^p.$$

In the other direction we have

$$\int_0^1 [\Delta^j f(x)]^p (1 - x)^{-2} \, dx \leqq A \sum_{n=1}^{\infty} \int_{1-1/n}^{1-1/(n+1)} \left[\sum_{k=0}^{n} a_k (1 - x^k)^j \right]^p (1 - x)^{-2} \, dx$$

$$+ A \sum_{n=1}^{\infty} \int_{1-1/n}^{1-1/(n+1)} \left[\sum_{k=n+1}^{\infty} a_k \right]^p (1 - x)^{-2} \, dx = I + J.$$

Clearly $J < A \sum_{n=1}^{\infty} \left[\sum_{k=n}^{\infty} a_k \right]^p$. To bound I, we need the following Lemma. For $p > 1$ it is Lemma 6.18 in [2].

Lemma 1. *Let $s > 0, 0 < p < 1$, and $a_k \geq 0$. There are constants $K = K(p, r, s)$ > 0 such that*

(4a)
$$\sum_{n=1}^{\infty} n^r \left[n^{-s} \sum_{k=1}^{n} k^s a_k \right]^p \geq K \sum_{n=1}^{\infty} n^r \left(\sum_{k=n}^{\infty} a_k \right)^p, \qquad r > -1;$$

(4b)
$$\sum_{n=1}^{\infty} n^r \left(\sum_{k=n}^{\infty} a_k \right)^p \geq K \sum_{n=1}^{\infty} n^r \left[n^{-s} \sum_{k=1}^{n} k^s a_k \right]^p, \qquad r < ps - 1.$$

We shall give a proof of (4b), which is the inequality we use below; (4a) is proved in the same way. First we remark that the Cauchy condensation test holds for series whose terms are quasi-monotone (see [8]). Then we have

$$\sum_{n=1}^{\infty} n^r \left[n^{-s} \sum_{k=1}^{n} k^s a_k \right]^p \leq A \sum_{n=1}^{\infty} 2^{n(1+r-sp)} \left(\sum_{k=1}^{2^n} k^s a_k \right)^p$$

$$\leq A \sum_{n=1}^{\infty} 2^{n(1+r-sp)} \left[\sum_{l=1}^{n-1} 2^{ls} \sum_{k=2^l}^{2^{l+1}} a_k \right]^p$$

$$\leq A \sum_{n=1}^{\infty} 2^{n(1+r-sp)} \sum_{l=1}^{n} 2^{lps} \left(\sum_{k=2^l}^{\infty} a_k \right)^p$$

$$= A \sum_{l=1}^{\infty} 2^{lps} \left(\sum_{k=2^l}^{\infty} a_k \right)^p \sum_{n=l}^{\infty} 2^{n(1+r-sp)}$$

$$= 2A \sum_{l=1}^{\infty} 2^{l+lr} \left(\sum_{k=2^l}^{\infty} a_k \right)^p \leq A' \sum_{n=1}^{\infty} n^r \left(\sum_{k=n}^{\infty} a_k \right)^p.$$

We need $1 + r - sp < 0$ to ensure the convergence of $\sum 2^{n(1+r-sp)}$.

The basic idea of this proof is due to Leindler [6] and Sunouchi [7]. We would like to thank Professor B. Rosser for pointing out that an alternate method of proof which was given in an earlier version of this paper was incorrect.

To estimate I we use

$$1 - (1 - 1/n)^k \leq k/n \quad \text{for} \quad k = 0, 1, \dots, n.$$

Then estimating I we have

$$I \leq A \sum_{n=1}^{\infty} \left[n^{-j} \sum_{k=0}^{n} a_k k^j \right]^p \leq A \sum_{n=1}^{\infty} \left(\sum_{k=n}^{\infty} a_k \right)^p$$

by Lemma 1 since $jp > 1$.

We need the condition $jp > 1$ for if we don't have it then the integral in Theorem 3 will be infinite in general, no matter what series condition we have. Consider

$$f(x) = \sum_{n=0}^{\infty} (x/2)^n = 2/(2 - x).$$

Clearly the sum $\sum_{n=1}^{\infty} (\sum_{k=n}^{\infty} 2^{-k})^p < \infty$ for all $p > 0$ but

$$\int_0^1 [f(1) - f(x)]^p (1-x)^{-2}\,dx = \int_0^1 \frac{2^p(1-x)^p}{(2-x)^p}\,\frac{dx}{(1-x)^2} = \infty \quad \text{if} \quad p \leq 1.$$

A similar calculation shows that $jp > 1$ is necessary for this function for all integers j.

Finally we give one sample theorem in the dual case.

Theorem 4. *Let* $a_n = \int_0^1 x^n\,d\mu(x)$, *where* $d\mu(x) \geq 0$ *and* $\int_0^1 d\mu(x)$ *exists. Then for* $0 < p < \infty$, $\sum a_n^p < \infty$ *if and only if* $\int_0^1 (1-x)^{-2}[\int_x^1 d\mu]^p\,dx < \infty$.

If $\sum a_n^p < \infty$ then we have

$$\sum_{n=1}^{\infty} a_n^p = \sum_{n=1}^{\infty} \left[\sum_{k=1}^{\infty} \int_{1-1/k}^{1-1/(k+1)} x^n\,d\mu(x) \right]^p$$

$$\geq \sum_{n=1}^{\infty} \left[\sum_{k=n}^{\infty} \int_{1-1/k}^{1-1/(k+1)} x^n\,d\mu(x) \right]^p \geq A \sum_{n=1}^{\infty} \left[\sum_{k=n}^{\infty} \int_{1-1/k}^{1-1/(k+1)} d\mu(x) \right]^p$$

$$= A \sum_{n=1}^{\infty} \left[\int_{1-1/n}^{1} d\mu(x)\,dx \right]^p \geq A \int_0^1 (1-x)^{-2} \left[\int_x^1 d\mu(y) \right]^p\,dx.$$

In the other direction we have

$$\sum_{n=1}^{\infty} a_n^p = \sum_{n=1}^{\infty} \left[\sum_{k=1}^{\infty} \int_{1-1/k}^{1-1/(k+1)} x^n d\mu(x) \right]^p$$

$$\leq A \sum_{n=1}^{\infty} \left[\sum_{k=1}^{n} \left[\int_{1-1/k}^{1-1/(k+1)} d\mu(x) \right] (1 - 1/k)^n \right]^p$$

$$+ A_p \sum_{n=1}^{\infty} \left[\sum_{k=n}^{\infty} \int_{1-1/k}^{1-1/(k+1)} d\mu(x) \right]^p$$

$$\leq I + A_p \int_0^1 (1-x)^{-2} \left[\int_x^1 d\mu(y) \right]^p\,dx$$

as above.

To estimate I, observe that $(1 - 1/k)^n = e^{n\log(1-1/k)} \leq A e^{-n/k}$, $k = 1, 2, \ldots, n$. Also notice that $e^{-x} \leq A/x^r$ for $x \geq 1$ for any fixed r, with some $A = A(r)$. Using these we have

$$I \leq A \sum_{n=1}^{\infty} \left[\sum_{k=1}^{n} \left(\frac{k}{n}\right)^r \int_{1-1/k}^{1-1/(k+1)} d\mu(x) \right]^p$$

$$\leq A \sum_{n=1}^{\infty} \left[\sum_{k=n}^{\infty} \int_{1-1/k}^{1-1/(k+1)} d\mu(x) \right]^p \leq A \int_0^1 (1-x)^{-2} \left[\int_x^1 d\mu(y) \right]^p\,dx$$

for $rp > 1$, by Lemma 6.18 in [2] and Lemma 1.

It is possible to extend these results in many more ways by putting in weight factors $(1 - x)^c L(x)$ where $L(x)$ is slowly varying, by generalizing the L^p norm to a more general functional norm, and in a different direction to similar theorems for Laplace transforms. There are references to the first two generalizations in [2, Chapter 8], and one reference for Laplace transforms is [5], where other references are given. To do this last extension a continuous analogue of Lemma 6.18 in [2] is needed. It can be proved by the same methods. There are also generalizations to several variables.

We close with a statement of one generalization of Theorem 2 which probably will not occur to the reader.

Theorem 5. *Let* $0 < p_i < \infty$, $f_i(x) = \sum\limits_{n=0}^{\infty} a_{n,i} x^n$, $a_{n,i} \geq 0$. *Then*

$$\int_0^1 \prod_{i=1}^{j} [f_i(x)]^{p_i}\, dx < \infty \qquad \text{if and only if} \qquad \sum_{n=1}^{\infty} n^{-2} \prod_{i=1}^{j} \left(\sum_{k=0}^{n} a_{k,i} \right)^{p_i} < \infty.$$

The proof is the same as above.

It is also possible to extend the other theorems in the same fashion. We need a new version of Lemma 1 which can be proved in the same way as above. This proof also works for $p > 1$, while the proof given for Lemma 6.18 in [2] doesn't work in this more general context in the same simple way.

Appendix A

S. Karlin has supplied the following elegant argument for one half of Askey's theorem [1] quoted at the beginning of this paper. Let

$$f(x) = \sum_{n=0}^{\infty} a_n x^n, \qquad 0 \leq x < 1,$$

$$s_n = a_0 + a_1 + \cdots + a_n.$$

By partial summation we have

$$f(x) = \sum_{n=0}^{\infty} s_n (x^n - x^{n+1}) = \sum_{n=0}^{\infty} s_n \phi_n(x),$$

where $\phi_n(x) \geq 0$ and $\sum \phi_n(x) = 1$. By Jensen's inequality,

$$|f(x)|^p \leq \sum_{n=0}^{\infty} |s_n|^p \phi_n(x), \qquad p \geq 1.$$

Integrating, we obtain

$$\int_0^1 |f(x)|^p\, dx \leq \sum_{n=0}^{\infty} |s_n|^p [(n+1)(n+2)]^{-1},$$

and the (unit) constant is best possible since there is equality when $f(x) = 1$, for each x, $0 \le x < 1$.

This part of Askey's theorem is now seem to be an abelian theorem since no condition had to be imposed on a_n. We can evidently replace $|f(x)|^p$ and $|s_n|^p$ by $\psi(|f(x)|)$, $\psi(|s_n|)$ with any convex increasing ψ.

The same argument can be used to give shorter proofs of improved versions of many of the theorems of the present paper. However, Karlin's argument does not work for $0 < p < 1$, and in fact the theorem does not hold for $0 < p < 1$ without some additional hypothesis on a_n.

Some examples and further comments will appear in a Mathematics Center Technical Report by Askey and Karlin.

Appendix B

We shall prove the equivalence of the integrability conditions of Theorem 1 and the conditions stated on page 25. This discussion has little connection with the rest of the paper, but has some independent interest. The results are largely independent of the analytic character of the functions involved.

Theorem 6. *If $p > 1$ and $-p - 1 < ps < -1$, and if f, f', and f'' are non-negative on $(0, 1)$, then*

$$(5) \qquad\qquad [f(1) - f(x)](1 - x)^s \in L^p$$

if and only if

$$(6) \qquad\qquad f'(x)(1 - x)^{s+1} \in L^p;$$

also

$$(7) \qquad\qquad f(x)(1 - x)^{s+1} \in L^p$$

if and only if

$$(8) \qquad\qquad f'(x)(1 - x)^{s+2} \in L^p.$$

In the application to Theorem 1 we use the equivalence of (5) and (6), and the equivalence of (7) and (8) with f replaced by f'; and we take $s = -2/p$. We have stated only the first two of a scale of equivalences involving successive derivatives, which could be proved in the same way.

We shall need two inequalities which follow by a change of variable from a variant of Hardy's inequality ([4, Theorem 330]).

Lemma 2. *If g is nonnegative and integrable on $(0, 1)$ and $p > 1$ then*
    (a) *if $r > 1$,*

$$(9) \qquad \int_0^1 (1 - x)^{-r} \left\{ \int_x^1 g(t)\, dt \right\}^p dx \le K \int_0^1 (1 - x)^{-r} [(1 - x)g(x)]^p\, dx;$$

(b) *if r < 1*,

$$(10) \quad \int_0^1 (1-x)^{-r} \left\{ \int_0^x g(t)\, dt \right\}^p dx \leqq K \int_0^1 (1-x)^{-r} [(1-x)g(x)^p]\, dx;$$

in each case K depends only on r and p.

We now prove Theorem 6. Suppose first that (6) holds. Then $f(1-)$ exists, since

$$\int_x^1 f'(t)\, dt = \int_x^1 (1-t)^{s+1} f'(t)(1-t)^{-(s+1)}\, dt$$

$$\leqq \left\{ \int_x^1 (1-t)^{(s+1)p} f'(t)^p\, dt \right\}^{1/p} \left\{ \int_x^1 (1-t)^{-(s+1)p'}\, dt \right\}^{1/p'}$$

$$< \infty$$

because $s < -1/p$ and so $(s+1)p' < 1$. We now can write, by Lemma 2,

$$\int_0^1 [f(1-) - f(x)]^p (1-x)^{sp}\, dx = \int_0^1 (1-x)^{sp} \left\{ \int_x^1 f'(t)\, dt \right\}^p dx$$

$$\leqq K \int_0^1 (1-x)^{sp}(1-x)^p f'(x)^p\, dx$$

$$= K \int_0^1 (1-x)^{(s+1)p} f'(x)^p\, dx < \infty,$$

since (6) was assumed to hold.

Conversely, if (5) holds we have, by Taylor's formula with remainder,

$$f(1-) - f(x) = (1-x)f'(x) + \tfrac{1}{2}(1-x)^2 f''(c), \qquad 0 < x < c < 1;$$

hence

$$f(1) - f(x) \geqq (1-x)f'(x),$$

so (6) holds.

We have

$$f((1+x)/2) - f(x) = \int_x^{(1+x)/2} f'(t)\, dt \geqq (1-x)f'(x)/2.$$

Then (8) follows immediately from (7).

If (8) holds we use

$$f(x) - f(0) = \int_0^x f'(t)\, dt.$$

Then by Lemma 2 we have

$$\int_0^1 [f(x)]^p (1-x)^{(s+1)p} dx \leq A \int_0^1 [f(0)]^p (1-x)^{(s+1)p} \, dx$$

$$+ A \int_0^1 \left[\int_0^x f'(t) \, dt \right]^p (1-x)^{(s+1)p} \, dx$$

$$\leq A[f(0)]^p + A \int_0^1 [f'(x)]^p (1-x)^{(s+2)p} \, dx,$$

since $(s+1)p > -1$.

REFERENCES

1. R. Askey, L^p behavior of power series with positive coefficients, *Proc. Amer. Math. Soc.* **19** (1968), 303–305.
2. R. P. Boas, Jr., *Integrability theorems for trigonometric transforms*, Ergebnisse der Mathematik und ihrer Grenzgebiete, Band 38, Springer-Verlag, Berlin, 1967.
3. G. H. Hardy and J. E. Littlewood, Elementary theorems concerning power series with positive coefficients and moment constants of positive functions, *J. für Math.* **157** (1927), 141–158.
4. G. H. Hardy, J. E. Littlewood, and G. Pólya, *Inequalities*, Cambridge University Press, Cambridge, England, 1952.
5. P. Heywood, Integrability theorems of Tauberian character, *Proc. London Math. Soc.* (3) 15 (1965), 471–494.
6. L. Leindler, Über verschiedene Konvergenzarten trigonometrischer Reihen, *Acta Sci. Math. (Szeged)* **25** (1964), 233–249.
7. G. Sunouchi, On the convolution algebra of Beurling, *Tôhoku Math. J.* **19** (1967), 303–310.
8. O. Szász, Quasi-monotone series, *Amer. J. Math.* **70** (1948), 203–206.

(*Received February 16, 1968 and, in revised form, June, 14, 1968*)

Completely Invariant Domains of Entire Functions

I. N. Baker

Department of Mathematics
Imperial College
London S.W.7
England

Let $f(z)$ be a function meromorphic in the plane. We say a plane domain D is invariant (under f) if $f(D) \subset D$ and completely invariant if, in addition, $f(z) = w \in D$ implies $z \in D$. Such domains exist. If d is any integer greater than 1 the function z^d has the two disjoint completely invariant domains $D_1 : |z| < 1$ and $D_2 : |z| > 1$. It has been shown by Fatou [3] that $f = z + 1 + e^{-z}$ has a completely invariant domain, whose description is, however, rather complicated and which has indeed some nonaccessible boundary points. Fatou [2] also showed that if f is rational, it has at most two disjoint completely invariant domains. His proof consists essentially in showing that any such domain must contain at least $(d-1)$ singularities of the inverse of the rational function f of order d and there are altogether only $2(d-1)$ such singularities. This method is of course not available when f is transcendental.

We shall prove the

Theorem. *If $f(z)$ is entire and transcendental, then $f(z)$ has no two mutually disjoint completely invariant domains.*

This result is of particular interest in the global iteration theory of entire functions. If $f(z)$ is rational or entire the iterates $f_n(z)$ defined by

$$f_1(z) = f(z), f_n(z) = f_{n-1}(f(z)), n = 2, 3, \ldots,$$

are rational, entire, or transcendental according as $f(z)$ is. The global iteration theory developed by Fatou [2, 3] and Julia [5] deals with the perfect plane set $\mathscr{F}(f)$, where the sequence $\{f_n(z)\}$ is nonnormal, and the way in which $\mathscr{F}(f)$ divides the plane. The set $\mathscr{F}(f)$ is completely invariant in the above sense but is not a domain. Among the domains which are the components of the complement $\mathscr{C}(f)$ of $\mathscr{F}(f)$ there may occur completely invariant ones. In the case of z^d discussed above the set $\mathscr{F}$ is the unit circumference $|z| = 1$ which forms the common boundary of the two completely invariant domains D_1 and D_2. In the example $f = z + 1 + e^{-z}$ the whole of $\mathscr{C}(f)$ is a single completely invariant domain. H. Töpfer [6] states that a result of Ahlfors [1] shows that there can be at most two completely invariant components G_i of $\mathscr{C}(f)$ when f is transcendental. P. Bhattacharya has pointed out to me that this paper of Ahlfors is a rather vaguely stated account of work which appeared later as part of his theory of covering surfaces and that in fact it seems necessary to suppose the domains to have disjoint closures in Ahlfors' work. This would not be true for completely invariant domains of $\mathscr{C}(f)$, each of which can easily be shown to have the common boundary $\mathscr{F}(f)$. However, our theorem shows at once that *at most one component of $\mathscr{C}(f)$ can be completely invariant, if f is entire and transcendental.*

Proof of the theorem.

Suppose f is entire and transcendental and has the two mutually disjoint completely invariant domains G_1 and G_2. Clearly each domain is infinite. Take a value z_1 in G_1, which is not a Picard value of $f(z)$ nor completely branched and take two branches $p(z)$ and $q(z)$ of the inverse function of $f(z)$ which are regular at z_1 and satisfy $p(z_1) \neq q(z_1)$. By Gross' star theorem [4] one can continue $p(z)$ and $q(z)$ analytically (regularly) to infinity along almost all rays emanating from z_1. We can therefore pick such a ray L which meets G_2. Denote by γ the segment of L joining z_1 to the point z_2 in G_2 and directed from z_1 to z_2. Then as z moves along γ the functions $p(z)$ and $q(z)$ trace out curves $p(\gamma)$ and $q(\gamma)$ which are disjoint, since neither p nor q has an algebraic singularity on γ. Also $p(\gamma)$ joins $p_1 = p(z_1)$ in G_1 to $p_2 = p(z_2)$ in G_2 and similarly $q(\gamma)$ joins $q_1 = q(z_1)$ in G_1 to $q_2 = q(z_2)$ in G_2.

Since G_1 is a domain we can join p_1 to q_1 by an arc β_1 in G_1 and similarly p_2 may be joined by q_2 by an arc β_2 in G_2. If β_i is oriented from p_i to q_i, let $p_i{}'$ denote its last intersection with $p(\gamma)$ and $q_i{}'$ its first intersection with $q(\gamma)$. Let $\beta_i{}'$ denote the subarc of β_i whose end points are $p_i{}'$ and $q_i{}'$, oriented from $p_i{}'$ to $q_i{}'$ and let π and κ denote the arcs $p_1{}'p_2{}'$ and $q_1{}'q_2{}'$ of $p(\gamma)$ and $q(\gamma)$, respectively, oriented from $p_1{}'$ to $p_2{}'$ and from $q_1{}'$ to $q_2{}'$. Then

$$\pi\beta_2{}'\kappa^{-1}(\beta_1{}')^{-1}$$

is a simple closed curve (an index -1 indicates a reversal of orientation). Denote this curve by Γ, and the interior of Γ by D. Now $f(z)$ is entire and hence $f(D)$ is a bounded set. Moreover the frontier of the domain $f(D)$ is contained in $f(\Gamma)$ and hence in $\gamma \cup f(\beta_1) \cup f(\beta_2)$.

The curve $f(\beta_i)$ is a closed curve lying in G_i and passing through z_i. Thus $f(\beta_1)$ and $f(\beta_2)$ are mutually disjoint and exterior to one another. Consider the unbounded component H of the complement of $f(\beta_1) \cup f(\beta_2)$. H meets γ and in fact if r is the last point of intersection of γ with $f(\beta_1)$ and s the first point of intersection of γ with $f(\beta_2)$ the segment rs of γ is a cross cut of H whose end points belong to different components of the frontier of H. It follows that rs does not disconnect H. Now in fact a point w of rs ($\neq r$ or s) is the image $f(z)$ of an interior point z in the arc π of Γ. In the neighborhood of z and inside Γ the function $f(z)$ takes an open set of values near w, some of which lie off γ and in $H - (rs)$. Then since the frontier of $\mathscr{F}(D)$ is contained in $f(\beta_1) \cup f(\beta_2) \cup \gamma$ we see that $f(D)$ must contain the whole of $H - (rs)$. But this contradicts the boundedness of $f(D)$. Thus there cannot exist two mutually disjoint domains such as G_1 and G_2 and the theorem is proved.

One might ask if the theorem remains true for meromorphic $f(z)$. In this case our proof, which depends on the boundedness of $f(D)$ for bounded D no longer applies.

REFERENCES

1. L. V. Ahlfors, Quelques propriétés des surfaces de Riemann correspondant aux fonctions méromorphes, *Bull. Soc. Math. France* **60** (1932), 197–207.
2. P. Fatou, Sur les équations fonctionelles, *Bull. Soc. Math. France* **47** (1919), 161–271; and **48** (1920), 33–94, 208–314.
3. ———, Sur l'itération des fonctions transcendantes entières, *Acta Math.* **47** (1926), 337–370.
4. W. Gross, Über die Singularitäten analytischer Funktionen, *Monat. Math. Physik* **29** (1918), 3–47.
5. G. Julia, Mémoire sur l'itération des fonctions rationelles, *J. Math. Pure Appl.* **1**, No. 8 (1918), 47–245.
6. H. Töpfer, Über die Iteration der ganzen transzendenten Funktionen, insbesondere von sin z und cos z, *Math. Ann.* **117** (1939), 65–84.

(Received December 22, 1967)

On the Asymptotic Paths of Entire Functions Represented by Dirichlet Series

F. Sunyer Balaguer[*]

Angel Guimera
36 Pral 2°
Barcelona 17
Spain

1. Introduction

Let

$$(1) \qquad \sum_0^\infty a_n e^{\lambda_n s} \qquad (\lambda_n < \lambda_{n+1}, \lim \lambda_n = \infty)$$

be a Dirichlet series absolutely convergent at every point s, and let $F(s)$ be the entire function represented by (1). Suppose that

$$(2) \qquad s(u) = \sigma(u) + it(u)$$

is a continuous function of the real variable $u\,(0 \leq u < \infty)$ such that $\lim_{u \to \infty} s(u) = \infty$. According to the theory of the almost periodic functions, only if $|\sigma(u)| \to \infty$ as $u \to \infty$ can the path (2) be asymptotic. It is evident that if $\sigma(u) \to -\infty$ then $F(s) \to a_0$; but these asymptotic paths are not interesting.

On the contrary when $\sigma(u) \to +\infty$ and $F(s(u)) \to c$ (c = a finite constant) we have an interesting asymptotic path. In Section 2, I give a theorem on

[*] Deceased December 27, 1967.

the number of these paths which are contained in a given strip and are distinct; when I say that two asymptotic paths are distinct I shall suppose that between these paths $F(s)$ is not bounded. The proof is obtained with the interesting method used by Macintyre [1] in the proof of the Denjoy-Carleman-Ahlfors theorem.

In Section 3, I obtain a translation to a class of Dirichlet series of a classical theorem of Wiman for the Taylor series.

2. Main Theorems

I again consider a continuous curve (not necessarily an asymptotic path)

(a) $$s(u) = \sigma(u) + it(u)$$

such that $s(u) \to \infty$ as $u \to +\infty$, and now I suppose that if $u_1 < u_2$ then $\sigma(u_1) < \sigma(u_2)$ and that $\sigma(0) = 0$. Then I define the strip

$$S = \{s = \sigma(u) + it: 0 \le u < +\infty, \, t(u) \le t \le t(u) + A\}.$$

I also write

$$M(\sigma, F) = \sup_{-\infty < t < +\infty} |F(\sigma + it)|$$

and

$$M(\sigma, F, S) = \sup_{\substack{-\infty < t < +\infty \\ s \in S}} |F(\sigma + it)|.$$

With these definitions we can state the following theorem:

Theorem I. *If*

$$F(s) = \sum a_n e^{\lambda_n s}$$

where $\sum a_n e^{\lambda_n s}$ is absolutely convergent at every point s and if $F(s)$ has n distinct asymptotic paths in S, then

$$\liminf_{\sigma \to \infty} \frac{\log M(\sigma, F)}{e^{\pi(n-1)\sigma/A}} \ge \liminf_{\sigma \to \infty} \frac{\log M(\sigma, F, S)}{e^{\pi(n-1)\sigma/A}} > 0.$$

Proof. Evidently without loss of generality we can suppose that the n asymptotic paths $l_k (k = 1, 2, \ldots, n)$ do not intersect.

Now consider the part S_Φ of S for which $\sigma \le \Phi$; then by a method used by Macintyre [1] we can map S cut by the n curves l_k on the rectangle

(3) $$0 \le x \le \Phi', \quad |y| \le A/2 \quad \text{of the plane} \quad z = x + iy$$

cut along n parallels to the axis of the x; this mapping will be represented by $z = \psi(s)$ and it is conformal except on the cuts. Then again following Macintyre we can prove

Lemma 1. *If L is the lower bound of the length in the s plane of all curves belonging to S and joining a point of the $\sigma = 0$ to a point of $\sigma = \Phi$ but not intersecting any curve l_k, then Φ' verifies the inequality*

$$\Phi' \geqq L^2/\Phi.$$

It is evident that the rectangle (3) is formed by at most $n + 1$ rectangles of which $n - 1$ are bounded by the n parallels corresponding to the l_k ($k = 1, 2, \ldots, n$). These rectangles will be denoted by Δ_k ($k = 1, 2, \ldots, n - 1$). Therefore at least one of these rectangles has a width not greater than $A/(n - 1)$; I denote it by $\Delta_{\Phi'}$.

Under the hypothesis that we suppose it is possible to prove that there exists a value $\theta > 0$ such that if $\Phi' > \theta$ we can determine an x_0 such that for every k

$$\sup_{x = x_0, \, z \in \Delta_k} |F(\psi^{-1}(z))| > 1,$$

where ψ^{-1} is the inverse function of ψ.

Therefore according to a theorem of Lindelöf we have

$$\liminf_{\Phi' \to \infty} \frac{\log M(\Phi', F(\psi^{-1}), \Delta_{\Phi'})}{e^{\pi(n-1)(\Phi'-x_0)/A}} > 0.$$

Since $L \geqq \Phi$, and since $\Phi' \geqq \Phi$ because of Lemma 1, and as

$$M(\Phi, F, S) \geqq M(\Phi', F(\psi^{-1}), \Delta_{\Phi'}),$$

Theorem 1 follows.

3.

Let $F(s)$ be an entire function represented by a Dirichlet series $\sum a_n e^{\lambda_n s}$ where the sequence $\{\lambda_n\}$ has an upper density D and is such that $\inf(\lambda_{n+1} - \lambda_n) > h$. Moreover suppose that we have defined the function $\rho(\sigma)$ such that

$$\lim \rho(\sigma) = \rho \qquad \rho'(\sigma)\sigma \to 0,$$

$$\log M(\sigma, F) \leqq e^{\rho(\sigma)\sigma},$$

where ρ is Ritt's order of $F(s)$.

On the other hand, a result of Mandelbrojt [2] shows that there exists a sequence $\{\sigma_n\}$ such that for every

$$s = \sigma_n + it$$

there exists a point s' such that

$$|s' - s| < \pi D + o(1),$$

$$\log |F(s')| > e^{\rho(\sigma_n)\sigma_n - \rho d}(1 - o(1))$$

where $d = D(7 - 3\log(hD))$.

Now I need a result of Milloux, namely,

Lemma 2. *Let $f(s)$ be a holomorphic function in $|s| \leq R$ such that*

$$\log|f(S)| \leq M;$$

if on a path joining $s = 0$ with a point of $|s| = R$ the function is bounded by

$$\log|f(s)| \leq M, \qquad m < M,$$

then for $|s| \leq r < R$,

$$\log |f(s)| < M - (M - m)\frac{2}{\pi}\arcsin\frac{R - r}{R + r}.$$

If $F(s)$ has an asymptotic path in which $\sigma \to +\infty$ and if we denote by s_n a point such that s_n belongs to the asymptotic path and $s_n = \sigma_n + it_n$, using Lemma 2 and the properties of $\rho(\sigma)$ we can prove that for every $R > \pi D$ we have

$$1 - \frac{2}{\pi}\arcsin\frac{R - \pi D}{R + \pi D} = e^{-\rho d - \rho R}.$$

Therefore if ρ_0 is a function of D and d such that there exists a value of $R > \pi D$ which verifies

$$1 - \frac{2}{\pi}\arcsin\frac{R - \pi D}{R + \pi D} = e^{-\rho_0 d - \rho_0 R},$$

then for the same value of R and for $\rho < \rho_0$ we shall have

$$1 - \frac{2}{\pi}\arcsin\frac{R - \pi D}{R + \pi D} < e^{-\rho d - \rho R},$$

and hence we have proved the following:

Theorem 2. *If*

$$F(s) = \sum a_n e^{\lambda_n s}$$

is a Dirichlet series convergent at every point s and if ρ_0 represents the function of D and d above defined, where D is the upper density of $\{\lambda_n\}$ and

$$d = D(7 - 3 \log(hD)),$$

then if Ritt's order ρ of $F(s)$ verifies $\rho < \rho_0$, the function $F(s)$ has no asymptotic path such that $\sigma \to +\infty$.

This is the analogue of a classical theorem of Wiman for the Dirichlet series.

REFERENCES

1. A. J. Macintyre, On the asymptotic paths of integral functions of finite order, *J. London Math. Soc.* **10** (1934), 34–39.
2. S. Mandelbrojt, *Series adhérentes, Régularisation des Suites, Applications* (Col. Monographies sur la theorie des fonctions), Gautiers-Villars, Paris, 1952.

(*Received February 1, 1968*)

On the Growth of Entire Functions

P. D. Barry

Department of Mathematics
Imperial College
London, S.W.7
England
and
Department of Mathematics
University College
Cork, Ireland

1.

Let $f(z)$ be a nonconstant entire function of order ρ and lower order λ. Let $M(r)$ and $\mu(r)$ denote the maximum and minimum of $|f(z)|$ on $|z| = r$, respectively. The classical $\cos \pi\rho$ theorem (see e.g. [4, p. 40]) states that if $\rho \leqq 1$, then

$$(1.1) \qquad \limsup_{r \to \infty} \frac{\log \mu(r)}{\log M(r)} \geqq \cos \pi\rho;$$

this was sharpened by Kjellberg [8] who showed that if $\lambda < 1$, then

$$(1.2) \qquad \limsup_{r \to \infty} \frac{\log \mu(r)}{\log M(r)} \geqq \cos \pi\lambda.$$

Examining cases of slowest growth and generalizing a previous result of Heins, Kjellberg [9] showed that if $0 < \lambda < 1$, then

$$(1.3) \qquad \log \mu(r) > \log M(r) \cos \pi\lambda$$

on an unbounded sequence of r, unless

(1.4) $$\frac{\log M(r)}{r^{\lambda}} \to \alpha \qquad (r \to \infty),$$

where α is positive or $+\infty$.

Theorems of several authors, e.g. Valiron [10, p. 89], V. Bernstein, Cartwright, Pfluger, Chebotarëv, and Meiman (see [4, pp. 52–54]) and Macintyre (see [5, p. 92]) compare the minimum of $\log |f(z)|$ for $|z| \leq R/k$ ($k > 1$) with $\log M(R)$, and obtain inequalities of the form

$$\log |f(z)| > -H \log M(R)$$

outside certain exceptional sets, the coefficient of $\log M(R)$ here being negative. We compare $\log \mu(r)$ and $\log M(kr)$ and seek generalizations of the $\cos \pi \rho$ theorem and its refinements. We also ask what inequality better than

$$\log \mu(r) > (\cos \pi \rho - \varepsilon)\log M(r) \qquad (\varepsilon > 0),$$

which follows from (1.1), is satisfied in the exceptional case in (1.4) when α is finite.

Theorem 1. *Let $f(z)$ be a nonconstant entire function of order ρ and lower order λ.*

(a) *If $0 < \rho < \frac{1}{2}$, then for any $k > 0$,*

(1.5) $$\limsup_{r \to \infty} \frac{\log \mu(r)}{\log M(kr)} \geq k^{-\rho} \cos \pi \rho.$$

(b) *If $0 < \lambda < \frac{1}{2}$, then for any $k > 0$ such that $k^{-\lambda} \cos \pi \lambda \leq 1$,*

(1.6) $$\limsup_{r \to \infty} \frac{\log \mu(r)}{\log M(kr)} \geq k^{-\lambda} \cos \pi \lambda;$$

if for any such k,

(1.7) $$\log \mu(r) \leq (k^{-\lambda} \cos \pi \lambda)\log M(kr) \qquad (r \geq r_0),$$

then

$$\frac{\log M(r)}{r^{\lambda}} \to \alpha \qquad (r \to \infty)$$

where $\alpha > 0$ or $\alpha = +\infty$.

We can also show that if $\frac{1}{2} < \lambda < 1$ and all the zeros of f lie on the negative real axis, then for any $k > 0$ such that $1 + k^{1-\lambda} \cos \pi \lambda > 0$, (1.6) holds.

Theorem 2. *Let $h(r)$ be positive and continuous for $r \geq r_0$ and, for each $s > 0$,*

$$(1.8) \qquad \frac{h(sr)}{h(r)} \to 1 \qquad (r \to \infty).$$

Suppose that $h(r) \to 0$ $(r \to \infty)$ and $h'(r) > -O(r^{-1})$ $(r \to \infty)$.
If $f(z)$ is an entire function of order ρ $(0 < \rho < \frac{1}{2})$ and mean type, and

$$(1.9) \qquad \int^{\infty} h(t) \frac{dt}{t} = \infty,$$

then

$$\log \mu(r) > \cos \pi\rho \{1 - h(r)\} \log M(r)$$

on a sequence of $r \to \infty$.
If

$$\int^{\infty} h(t) \frac{dt}{t} < \infty,$$

there is an entire function of order ρ $(0 < \rho < \frac{1}{2})$ and mean type for which

$$\log \mu(r) < \cos \pi\rho \{1 - h(r)\} \log M(r), \qquad r \geq r_0.$$

For a set E on the positive real axis, we denote the lower logarithmic density of E by $\underline{\Lambda}(E)$ (see e.g. [3] or [7]).

Theorem 3. *Let $f(z)$ be a nonconstant entire function of order ρ $(0 \leq \rho < \frac{1}{2})$ and let $\rho < \alpha < \frac{1}{2}$. Then for $k > 1$,*

$$\underline{\Lambda}\{r \mid \log \mu(r) > (k^{-\alpha} \cos \pi\alpha) \log M(kr)\} \geq 1 - \rho/\alpha.$$

The estimate (1.6) is of interest when it is better than (1.5). For this we need

$$\frac{\cos \pi\lambda}{k^{\lambda}} \geq \frac{\cos \pi\rho}{k^{\rho}}, \qquad \text{for} \quad \lambda < \rho < \frac{1}{2},$$

i.e.

$$\log k \geq \frac{\log \cos \pi\rho - \log \cos \pi\lambda}{\rho - \lambda},$$

and making $\rho \to \lambda$, we get that we need

$$(1.10) \qquad \log k \geq -\pi \tan \pi\lambda$$

which is implied with inequality by $k^{\lambda} \geq \cos \pi\lambda$ and which implies that $k^{-\rho} \cos \pi\rho$ is decreasing on $(\lambda, \frac{1}{2})$. Another factor is that in our proof it is the hypothesis that $f(z)$ has lower order at most λ that is used; and, as there are functions (see [4, pp. 55–56] or [11]) of order and lower order ρ' for which

$$\limsup_{r \to \infty} \frac{\log \mu(r)}{\log M(kr)} = \frac{\cos \pi \rho'}{k^{\rho'}},$$

and since we may take ρ' arbitrarily close to 0, we see that under this hypothesis the restriction $(\cos \pi \lambda)/k^{\lambda} \leq 1$ is necessary. However, it may be that in Theorem 1(b), (1.10) is sufficient instead of $k^{-\lambda} \cos \pi \lambda \leq 1$. When $\frac{1}{2} < \lambda < 1$, the condition $1 + k^{1-\lambda} \cos \pi \lambda > 0$ guarantees that

$$\frac{\cos \pi \lambda}{k^{\lambda}} \geq \frac{\cos \pi \rho}{k^{\rho}} \qquad \text{for} \quad \lambda < \rho < 1.$$

We refer to a function $h(r)$ which is positive and continuous for $r \geq r_0$ and satisfies (1.8) as slowly varying. An example of a function satisfying our conditions and (1.9) is

$$h(r) = 1/\log r (\log \log r)^{\delta}$$

where $\delta \leq 1$. Wiman [11] gave an example of a function of order $\rho < 1$ and finite type for which

$$\frac{\log \mu(r)}{\log M(r)} < \cos \pi \rho - K \frac{\log r}{r^{\rho}} \qquad (r \geq r_0),$$

where K is a positive constant.

A proof of Theorem 3 in the case $k = 1$ is given in Barry [3].

2. Proof of Theorem 1

Since our proofs are modifications of proofs for the case $k = 1$ we just give them in outline. Since $\rho < 1$, we know that

$$f(z) = cz^p \prod_{n=1}^{\infty} \left(1 - \frac{z}{a_n}\right),$$

where a_n's are the nonzero zeros of $f(z)$ arranged in order of increasing magnitude. Set $r_n = |a_n|$ and

$$F(z) = \prod_{n=1}^{\infty} \left(1 + \frac{z}{r_n}\right).$$

Let $r^{\rho(r)}$ be a proximate order for $F(z)$. By Cartwright [5, pp. 55, 63] for $\varepsilon > 0$, $0 < \rho < \frac{1}{2}$, there is a sequence of $r \to \infty$ on which

$$\log \mu(r) > (\cos \pi \rho - \varepsilon) r^{\rho(r)}$$

$$> \frac{\cos \pi \rho - 2\varepsilon}{k^{\rho}} (kr)^{\rho(kr)}$$

uniformly for $0 < k_1 \leqq k \leqq k_2 < \infty$,

$$> \frac{\cos \pi\rho - 3\varepsilon}{k^\rho} \log F(kr)$$

$$> \frac{\cos \pi\rho - 4\varepsilon}{k^\rho} \log M(kr)$$

so that

$$\limsup_{r \to \infty} \frac{\log \mu(r)}{\log M(kr)} \geqq \frac{\cos \pi\rho}{k^\rho}$$

and, in fact

$$\log \mu(r) > \frac{\cos \pi\rho - \varepsilon}{k^\rho} \log M(kr)$$

on a sequence of $r \to \infty$, uniformly for $k_1 \leqq k \leqq k_2$, where $k_1 > 0$.
When $\frac{1}{2} \leqq \rho < 1$, the corresponding argument shows that

$$\log |F(-r)F(r)| > \frac{(\cos \frac{1}{2}\pi\rho - \varepsilon)2 \cos \frac{1}{2}\pi\rho}{k^\rho} (kr)^{\rho(kr)}$$

from which we deduce that

$$\log \mu(r) > \frac{\cos \pi\rho + 1 - \varepsilon}{k^\rho} \log M(kr) - \log M(r)$$

on a sequence of $r \to \infty$.

To use the argument of Beurling in [7, p. 15] or [4, pp 4–5] we suppose that $f(z)$ is nonconstant and has order at most ρ, minimal type, where $\rho < \frac{1}{2}$. For $\varepsilon > 0$, we choose $a(\varepsilon)$ so that $\log |F(-r)| - \varepsilon r^\rho \cos \pi\rho \leqq a(\varepsilon)$ for $r \geqq 0$ with equality at $r(\varepsilon)$. Then

$$\log |F(z)e^{-\varepsilon z^\rho - a(\varepsilon)}| \leqq 0$$

on $z = -r$ and so throughout the plane, giving for any R,

$$\log F(R) \leqq \varepsilon R^\rho + a(\varepsilon).$$

Then

$$\log |F\{-r(\varepsilon)\}| = \varepsilon r(\varepsilon)^\rho \cos \pi\rho + a(\varepsilon)$$
$$\geqq \cos \pi\rho (r(\varepsilon)/R)^\rho \log F(R)$$
$$+ \{1 - \cos \pi\rho (r(\varepsilon)/R)^\rho\}a(\varepsilon)$$

and as $a(\varepsilon)$ and $r(\varepsilon) \to \infty$ $(\varepsilon \to 0)$, we see that on a sequence of $r \to \infty$,

$$\log |F(-r)| \geqq (\cos \pi\rho)(r/R)^\rho \log F(R)$$

for all R such that

$$R^\rho \geqq r^\rho \cos \pi\rho.$$

If we put $R = kr$, this gives that on a sequence of $r \to \infty$,

$$\log |F(-r)| \geqq \frac{\cos \pi\rho}{k^\rho} \log F(kr)$$

for all k such that $k^{-\rho} \cos \pi\rho \leqq 1$, the sequence of r being independent of k.

To turn now to Theorem 1(b), we recall from [4, p. 41] that for $0 < \alpha < 1$,

$$\int_0^\infty \frac{\log |1 - t|}{t^{1+\alpha}} \, dt = \frac{\pi}{\alpha} \cot \pi\alpha, \qquad \int_0^\infty \frac{\log(1 + t)}{t^{1+\alpha}} \, dt = \frac{\pi}{\alpha \sin \pi\alpha}.$$

In the second of these, the change of variable $t = kx$ $(k > 0)$ yields

$$\int_0^\infty \frac{\log(1 + kx)}{x^{1+\alpha}} \, dx = \frac{\pi k^\alpha}{\alpha \sin \pi\alpha}$$

and so

$$\int_0^\infty \left\{ \log |1 - t| - \frac{\cos \pi\alpha}{k^\alpha} \log(1 + kt) \right\} \frac{dt}{t^{1+\alpha}} = 0.$$

Let

$$\theta(t) = \log |1 - t| - \frac{\cos \pi\alpha}{k^\alpha} \log(1 + kt), \qquad t \geqq 0, 0 < \alpha < 1,$$

and

$$\psi(x) = \int_x^\infty \theta(t) t^{-1-\alpha} \, dt.$$

Then $\psi(0) = \psi(\infty) = 0$. As $x \to \infty$,

$$\psi(x) \sim \alpha^{-1}(1 - k^{-\alpha} \cos \pi\alpha) \frac{\log(1 + x)}{x^\alpha}$$

provided

(2.1) $1 - k^{-\alpha} \cos \pi\alpha > 0.$

As $x \to 0$,

$$\psi(x) \sim \frac{1 + k^{1-\alpha} \cos \pi\alpha}{1 - \alpha} \frac{\log(1 + x)}{x^\alpha}$$

provided

(2.2) $1 + k^{1-\alpha} \cos \pi\alpha > 0.$

An examination of $\theta'(t)$, using (2.1) and (2.2), shows that it has no positive zeros so $\theta'(t) < 0$ for $0 < t < 1$, $\theta'(t) > 0$, $t > 1$. It follows that $\theta(t)$ is negative from 0 to a point and subsequently positive, which means in turn that $\psi(t)$ is strictly increasing to that point and then strictly decreasing, so that $\psi(t) > 0$ for $t > 0$. Then $\psi(x)x^\alpha/\log(1 + x)$ has a positive minimum $\delta(\alpha, k)$ and finite maximum $K(\alpha, k)$. It follows that, for $0 < r < R$,

$$(2.3) \qquad \int_r^R \{\log |1 - t| - k^{-\alpha} \cos \pi\alpha \, \log(1 + kt)\} \frac{dt}{t^{1+\alpha}}$$

$$> \delta(\alpha, k) \frac{\log(1 + r)}{r^\alpha} - K(\alpha, k) \frac{\log(1 + R)}{R^\alpha},$$

and if we denote $\log |F(z)|$ by $U(z)$,

$$\int_r^R \{U(-t) - (k^{-\alpha} \cos \pi\alpha)U(kt)\} \frac{dt}{t^{1+\alpha}} > \delta(\alpha, k) \frac{U(r)}{r^\alpha} - K(\alpha, k) \frac{U(R)}{R^\alpha}.$$

We now proceed as in Kjellberg [8]. Let $f(0) = 1$. If $\{a_n\}$ are the zeros of f, we set

$$f_1(z) = \prod_{|a_n| \leq R} \left(1 - \frac{z}{a_n}\right), \qquad f_2(z) = \prod_{|a_n| \leq R} \left(1 + \frac{z}{|a_n|}\right),$$

$$\text{and} \quad f(z) = f_1(z)f_3(z).$$

Let $\mu_2(r) = \min_{|z|=r} |f_2(z)|$, etc. Then for $0 < R_1 < R_2$,

$$(2.4) \qquad \int_{R_1}^{R_2} \{\log \mu_2(r) - k^{-\lambda} \cos \pi\lambda \, \log M_2(kr)\} \frac{dr}{r^{1+\lambda}}$$

$$> \delta(\lambda, k) \frac{\log M_2(R_1)}{R_1^\lambda} - K(\lambda, k) \frac{\log M_2(R_2)}{R_2^\lambda}.$$

As $\lambda < \frac{1}{2}$, we add to each side

$$I(R_1, R_2) = \int_{R_1}^{R_2} \{\log \mu_3(r) - k^{-\lambda} \cos \pi\lambda \, \log M_3(kr)\} \frac{dr}{r^{1+\lambda}},$$

denoting the resulting expression on the left by $A(R_1, R_2)$.

We choose R as in [8], i.e., so that

$$\log M(2R) < (\alpha + \varepsilon)(2R)^\lambda,$$

where

$$\alpha = \liminf_{r \to \infty} \frac{\log M(r)}{r^\lambda}, \qquad \beta = \limsup_{r \to \infty} \frac{\log M(r)}{r^\lambda},$$

but set $R_2 = R/2k'$, where $k' = \max(1, k)$. For $r \leq R/2k'$,

$$\log \mu_3(r) - k^{-\lambda} \cos \pi\lambda \log M_3(kr) \geq -4(1 + k'k^{-\lambda} \cos \pi\lambda)rR^{-1} \log M(2R)$$
$$> -(\alpha + \varepsilon)4(1 + k'k^{-\lambda} \cos \pi\lambda)2^{\lambda}R^{\lambda-1}r,$$

so

$$I(R_1, R_2) > \frac{-(\alpha + \varepsilon)2^{2\lambda+1}}{1 - \lambda}(1 + k'k^{-\lambda} \cos \pi\lambda)k'^{\lambda-1}.$$

Kjellberg's other estimations remain valid and we deduce that

$$A(R_1, R/2k') > \delta(\lambda, k)\frac{\log M(R_1)}{R_1^{\lambda}}$$

$$- \delta(\lambda, k)(\alpha + \varepsilon)2^{\lambda+2}(R_1/R)^{1-\lambda}$$

$$- K(\lambda, k)(\alpha + \varepsilon)2^{1+2\lambda}k'^{\lambda}$$

$$- \frac{(\alpha + \varepsilon)2^{2\lambda+1}}{1 - \lambda}(1 + k'k^{-\lambda} \cos \pi\lambda)k'^{\lambda-1}.$$

Now assuming that $\alpha = 0$, or else that $\alpha < +\infty$, $\beta = +\infty$, we infer that for arbitrarily large R_1 and R, $A(R_1, R/2k') > 0$, and so on a sequence of $r \to \infty$,

$$\log \mu_2(r)\mu_3(r) > \frac{\cos \pi\lambda}{k^{\lambda}} \log M_2(kr)M_3(kr)$$

from which

$$\log \mu(r) > \frac{\cos \pi\lambda}{k^{\lambda}} \log M(kr).$$

The extension to when $f(0) \neq 1$ is handled as in [8].

When $\frac{1}{2} < \lambda < 1$, our assumption that $a_n < 0$ for all n implies that $f_1(z) \equiv f_2(z)$. This time we add

$$\int_{R_1}^{R_2} \{\log \mu_3(r) - k^{-\lambda} \cos \pi\lambda \log \mu_3(kr)\} \frac{dr}{r^{1+\lambda}}$$

to each side of (2.4) and by a similar argument deduce that, on a sequence of $r \to \infty$

$$\log \mu_1(r)\mu_3(r) > \frac{\cos \pi\lambda}{k^{\lambda}} \log M_1(kr)\mu_3(kr).$$

As $\mu(r) \geq \mu_1(r)\mu_3(r)$, $M(kr) \geq M_1(kr)\mu_3(kr)$, and $\cos \pi\lambda < 0$ we deduce that,

$$\log \mu(r) > \frac{\cos \pi\lambda}{k^\lambda} \log M(kr),$$

again on the assumption that $\alpha = 0$, or $\alpha < +\infty$, $\beta = +\infty$.

It remains to show that if (1.7) holds and $\alpha < \infty$, then $\alpha = \beta$, and for this we modify the proofs of Kjellberg [9] and Essén [6]. Since α and β are both finite, $\rho = \lambda$. As in [9] we may suppose (1.7) holds for all $r \geqq 0$. Then, as in [9],

$$F(R) = \frac{2R}{\pi^2} \int_0^\infty \frac{\log |F(-r)| + \log F(r)}{R^2 - r^2} \log \frac{R}{r} \, dr,$$

so

$$\frac{\log M(R)}{R^\lambda} \leqq \frac{2}{\pi^2} \int_0^\infty \frac{\log \mu(r) + \log M(r)}{r^\lambda} \left(\frac{r}{R}\right)^\lambda \frac{R \log R/r}{R^2 - r^2} \, dr$$

$$\leqq \frac{2}{\pi^2} \int_0^\infty \frac{k^{-\lambda} \cos \pi\lambda \log M(kr) + \log M(r)}{r^\lambda} \left(\frac{r}{R}\right)^\lambda \frac{R \log R/r}{R^2 - r^2} \, dr$$

$$= \int_0^\infty \frac{\log M(r)}{r^\lambda} K(r, R) \, dr,$$

where

$$K(r, R) = \frac{2}{\pi^2} \left\{ \left(\frac{r}{R}\right)^\lambda \frac{R \log R/r}{R^2 - r^2} + \cos \pi\lambda \left(\frac{r}{kR}\right)^\lambda \frac{kR \log(kR/r)}{k^2R^2 - r^2} \right\}.$$

Then as in Essén [6], we set $R = e^x$, $r = e^y$, $\phi_0(x) = e^{-\lambda x} \log^+ M(e^x)$, and deduce that

$$\phi_0(x) \leqq \int_{-\infty}^\infty K_0(x - y)\phi_0(y) \, dy,$$

where

$$K_0(x) = \frac{2}{\pi^2} \left\{ \frac{e^{x(1-\lambda)}x}{e^{2x} - 1} + \cos \pi\lambda \frac{k^{1-\lambda}e^{x(1-\lambda)}(x + \log k)}{k^2e^{2x} - 1} \right\}$$

which is nonnegative for all x, as we suppose $\lambda < \frac{1}{2}$.

Now

$$\int_{-\infty}^\infty K_0(x) \, dx = \frac{2}{\pi^2} (1 + \cos \pi\lambda) \int_{-\infty}^\infty \frac{e^{x(1-\lambda)}x}{e^{2x} - 1} \, dx$$

$$= 1,$$

by [9]. Also,

$$\int_{-\infty}^{\infty} xK_0(x)\,dx = \frac{2}{\pi^2}(1+\cos\pi\lambda)\int_{-\infty}^{\infty}\frac{x^2 e^{x(1-\lambda)}}{e^{2x}-1}\,dx$$

$$-\frac{2\log k}{\pi^2}\cos\pi\lambda\int_{-\infty}^{\infty}\frac{xe^{x(1-\lambda)}}{e^{2x}-1}\,dx$$

$$= -\pi\tan\tfrac{1}{2}\pi\lambda - \frac{\cos\pi\lambda}{1+\cos\pi\lambda}\log k \qquad \text{(by Essén's calculations)},$$

$$= -\tfrac{1}{2}\sec^2\tfrac{1}{2}\pi\lambda\{\pi\sin\pi\lambda + \cos\pi\lambda\log k\}$$

$$< 0$$

as $k > e^{-\pi\tan\pi\lambda}$, since $k^{-\lambda}\cos\pi\lambda \leq 1$. The conditions on ϕ_0 are unchanged, so by Essen's argument,

$$\lim_{x\to\infty}\phi_0(x)$$

exists.

3. Proof of Theorem 2

We set $h(t) = (t/r_0)h(r_0)$ for $0 \leq t \leq r_0$ and

$$\phi(r) = \exp\left\{\delta\int_1^r \frac{h(t)}{t}\,dt\right\},$$

where $\delta > 0$. Since $h(t) \to 0$ $(t \to \infty)$, $\phi(r) \sim \phi(sr)$ $(r \to \infty)$ for each $s > 0$. It follows (see e.g. [1, p. 86]) that

$$\phi(r) = o(r^\varepsilon) \qquad (r \to \infty)$$

for each $\varepsilon > 0$, and (1.9) implies that $\phi(r) \to \infty$.

Set $\psi(r) = r^\rho\phi(r)$. Then $\psi(r)$ has order ρ and as $\log M(r)$ is of mean type, $\log M(r) = o\{\psi(r)\}$ $(r \to \infty)$.

Further

$$r\psi'(r) = r^\rho\phi(r)\{\rho + \delta h(r)\}$$

so that

$$r\frac{d}{dr}\{r\psi'(r)\} = r^\rho\phi(r)[\{\rho + \delta h(r)\}^2 + \delta r h'(r)]$$

$$\geq r^\rho\phi(r)\{\rho^2 + \delta r h'(r)\}$$

$$\geq 0,$$

if $rh'(r) \geq -\rho^2/\delta$, which is true if δ is small enough. This is the sole use we make of the condition $rh'(r) \geq -O(1)$, i.e., to show that $r\psi'(r)$ is non-decreasing.

We now set

$$H(re^{i\theta}) = \frac{1}{\pi} \int_0^\infty \frac{r^{1/2}(r+s)\psi(s)\cos\frac{1}{2}\theta}{s^{1/2}\{(r+s)^2 - 4rs\sin^2\frac{1}{2}\theta\}} \, ds$$

as in [2, p. 461], which provides a solution of the Dirichlet problem with boundary values $H(-r) = \psi(r)$ in the plane slit along the real axis from 0 to $-\infty$. Since $r\psi'(r)$ is nondecreasing, the argument in [2, pp. 461–462] shows that $\max_{0 \leq \theta \leq 2\pi} H(re^{i\theta}) = H(r)$. Now

$$H(r) = \frac{r^{1/2}}{\pi} \int_0^\infty \frac{(r+s)s^\rho\phi(s)}{s^{1/2}(r+s)^2} \, ds$$

$$= \frac{r^\rho}{\pi} \int_0^\infty \frac{t^\rho\phi(rt)}{t^{1/2}(1+t)} \, dt$$

$$\sim \phi(r)\frac{r^\rho}{\pi} \int_0^\infty \frac{t^\rho}{t^{1/2}(1+t)} \, dt,$$

since ϕ is slowly varying (see [1, p. 82], quoted in [2, p. 462]), and so

$$H(r) \sim \phi(r)r^\rho \sec\pi\rho \qquad (r \to \infty).$$

Since $H(-r) = \psi(r) = r^\rho\phi(r)$, we see that

$$\frac{H(-r)}{H(r)} \geq \{1 + o(1)\}\cos\pi\rho \qquad (r \to \infty).$$

The argument of Beurling, applied with $H(re^{i\theta})$ instead of Re z^ρ, now gives (see [2, Lemma 5]) that

$$\frac{\log\mu(r)}{\log M(r)} \geq \frac{H(-r)}{H(r)} \geq \frac{r^\rho\phi(r)}{H(r)}$$

on a sequence $r \to \infty$. It remains to estimate this last term. Now

$$H(r) = \frac{r^{1/2}}{\pi} \int_0^\infty \frac{t^\rho\phi(t)}{t^{1/2}(t+r)} \, dt$$

$$= \frac{r^{1/2}}{\pi} \int_0^\infty \frac{t^\rho\phi(r)}{t^{1/2}(t+r)} \, dt + \frac{r^{1/2}}{\pi} \int_0^\infty \frac{t^\rho\{\phi(t) - \phi(r)\}}{t^{1/2}(t+r)} \, dt$$

$$= r^\rho\phi(r)\sec\pi\rho + I(r), \quad \text{say.}$$

But

$$I(r) = \frac{r^{1/2}}{\pi} \sum_{n=0}^{\infty} (-1)^n \left[\frac{1}{r} \int_0^r t^{\rho - 1/2 + n} r^{-n} \{\phi(t) - \phi(r)\} \, dt \right.$$

$$\left. + \int_r^{\infty} t^{\rho - 1/2 - 1} t^{-n} r^n \{\phi(t) - \phi(r)\} \, dt \right]$$

$$= \frac{r^{1/2}}{\pi} \sum_{n=0}^{\infty} (-1)^n \left\{ -\frac{1}{n + \frac{1}{2} + \rho} r^{-n-1} \int_0^r t^{n + \rho - 1/2} t\phi'(t) \, dt \right.$$

$$\left. + \frac{1}{n + \frac{1}{2} - \rho} r^n \int_r^{\infty} t^{\rho - n - 3/2} t\phi'(t) \, dt \right\}$$

$$= \frac{r^{\rho}}{\pi} \sum_{n=0}^{\infty} (-1)^n \left\{ -\frac{1}{n + \frac{1}{2} + \rho} \int_0^1 s^{n + \rho - 1/2} rs\phi'(rs) \, ds \right.$$

$$\left. + \frac{1}{n + \frac{1}{2} + \rho} \int_1^{\infty} s^{\rho - n - 3/2} rs\phi'(rs) \, ds \right\}$$

$$\sim \delta \pi^{-1} r^{\rho} \phi(r) h(r) \sum_{n=0}^{\infty} (-1)^n \{-(n + \tfrac{1}{2} + \rho)^{-2} + (n + \tfrac{1}{2} - \rho)^{-2}\}$$

$$= \delta \pi \sin \pi\rho \sec^2 \pi\rho \; r^{\rho} \phi(r) h(r),$$

since $r\phi'(r) = \delta h(r)\phi(r) \sim \delta h(sr)\phi(sr)$ $(r \to \infty)$, where $s > 0$, so $r\phi'(r)$ is slowly varying. Thus

$$\frac{\log \mu(r)}{\log M(r)} \geq r^{\rho}\phi(r) / [r^{\rho}\phi(r)\sec \pi\rho + \{1 + o(1)\} \delta C(\rho) r^{\rho} h(r)\phi(r)]$$

$$= \cos \pi\rho / [1 + \{1 + o(1)\} h(r) \delta C(\rho)\cos \pi\rho]$$
$$> \cos \pi\rho \{1 - (1 + \varepsilon) h(r) \delta C(\rho)\cos \pi\rho\}, \quad \text{since} \quad h(r) \to 0,$$
$$> \cos \pi\rho \{1 - h(r)\},$$

if δ is chosen sufficiently small. This completes the proof.

The use we made of (1.9) was to ensure that $\log M(r) = o\{r^{\rho}\phi(r)\}$ $(r \to \infty)$. The method could clearly be adapted for hypotheses such as

$$\log M(r) = O\{r^{\rho}(\log r)^q\},$$

where $q > 0$, or

$$\log M(r) = o\left[r^{\rho} \exp\left\{ \int_1^r h(t) t^{-1} \, dt \right\} \right],$$

where $h(t)$ is a sufficiently smooth slowly varying function.

To construct our counterexample, set

$$\log f(z) = \int_0^\infty \log(1 + z/t)\, d[\phi(t)t^\rho]$$

$$= z \int_0^\infty \frac{[\phi(t)t^\rho]}{t(t+z)}\, dt = z \int_1^\infty \frac{[\phi(t)t^\rho]}{t(t+z)}\, dt.$$

Then

$$\log f(z) = z \int_0^\infty \frac{\phi(t)t^\rho}{t(t+z)}\, dt + I_1 + I_2,$$

where

$$|I_1| \leq \frac{r}{r-1} \int_0^1 t^{\rho-1}\, dt < 2/\rho \qquad (r \to \infty),$$

and

$$|I_2| \leq \int_1^\infty \frac{|z|}{t\,|t+z|}\, dt.$$

We set $z = re^{i\theta}$, $|\theta| < \pi$. Then

$$|I_2| \leq \int_1^{r/2} \frac{r}{t\,|t+z|}\, dt + \int_{r/2}^{2r} \frac{r}{t\,|t+z|}\, dt + \int_{2r}^\infty \frac{r}{t\,|t+z|}\, dt$$

$$\leq 1 + 2\log r + 2 \int_{r/2}^{2r} \frac{dt}{|t+z|}.$$

Now for $|\theta| \leq \frac{1}{2}\pi$, $|t+z|^2 \geq t^2 + r^2$, so the last integral is at most $2\log 4$; otherwise if $z = x + iy$ $(x < 0)$, $|t+z|^2 = (t+x)^2 + y^2$ and

$$\int_{r/2}^{2r} \{(t - |x|)^2 + y^2\}^{-1/2}\, dt < \log(8r/|y|),$$

as we see by integrating over $(\frac{1}{2}r, 2r)$ when $|x| \leq \frac{1}{2}r$, and over $(\frac{1}{2}r, |x|)$ and $(|x|, 2r)$ otherwise. Thus

$$|I_2| \leq 1 + 2\log r + 2\log 4, \qquad\qquad \text{for} \quad x \geq 0,$$
$$|I_2| \leq 1 + 2\log r + 2\log(8r/|y|), \qquad \text{for } x < 0,\ y \neq 0,$$
$$= 1 + 2\log r + 2\log 8\,|\mathrm{cosec}\,\theta|.$$

Thus

$$\log f(r) \geq r \int_0^\infty \frac{\phi(t)t^\rho}{t(t+r)}\, dt - (1 + 2\log 4 + 2/\rho + 2\log r)$$

$$= \pi r^\rho \phi(r)\mathrm{cosec}\,\pi\rho + r \int_0^\infty \{\phi(t) - \phi(r)\}t^{\rho-1}(t+r)^{-1}\, dt - O(\log r).$$

Now

$$r \int_0^\infty \{\phi(t) - \phi(r)\} t^{\rho-1}(t + r)^{-1} \, dt$$

$$= \sum_{n=0}^\infty (-1)^n \left[-(\rho + n)^{-1} r^{-n} \int_0^r t^{\rho+n-1} t\phi'(t) \, dt \right.$$

$$\left. + (n + 1 - \rho)^{-1} r^{n+1} \int_r^\infty t^{\rho-n-2} t\phi'(t) \, dt \right]$$

$$\sim \delta r^\rho \phi(r) h(r) \sum_{n=0}^\infty (-1)^n \{(n + 1 - \rho)^{-2} - (\rho + n)^{-2}\}$$

$$= \delta r^\rho \phi(r) h(r) C_1(\rho), \qquad \text{say,}$$

since $\phi(r)h(r)$ is slowly varying. Hence

$$\log f(r) \geqq r^\rho \phi(r)[\pi \operatorname{cosec} \pi\rho + \{1 - o(1)\}\delta h(r) C_1(\rho)] - O (\log r) \quad (r \to \infty).$$

Also

$$\log f(r) \leqq r^\rho \phi(r)[\pi \operatorname{cosec} \pi\rho + \{1 + o(1)\}\delta h(r) C_1(\rho)] + O(\log r),$$

and as $\phi(r) = O(1)$ by hypothesis, since (1.9) is not satisfied, f has order ρ, mean type.

Now

$$\log \mu(r) \leqq \operatorname{Re} \int_0^\infty \phi(t) t^{\rho-1} z(t + z)^{-1} \, dt + |I_2| + 2/\rho$$

$$= \operatorname{Re} \int_0^\infty \phi(r) t^{\rho-1} z(t + z)^{-1} \, dt$$

$$+ \operatorname{Re} \int_0^\infty \{\phi(t) - \phi(r)\} t^{\rho-1} z(t + z)^{-1} \, dt + |I_2| + 2/\rho$$

$$= \phi(r)\pi r^\rho \cos \rho\theta \operatorname{cosec} \pi\rho$$

$$+ \operatorname{Re} \int_0^\infty \{\phi(t) - \phi(r)\} t^{\rho-1} z(t + z)^{-1} \, dt + |I_2| + 2/\rho.$$

If we take $\theta(r) = \pi - r^{-K}$, with $K > 1$, then $\cos \rho\theta(r) = \cos \pi\rho + O(r^{-K})$, so

$$\phi(r)\pi r^\rho \cos \rho\theta(r)\operatorname{cosec} \pi\rho = \phi(r)\pi r^\rho \cos \pi\rho \operatorname{cosec} \pi\rho + o(1).$$

Also

$$|I_2| \leqq 1 + 2 \log r + 2 \log(8 \operatorname{cosec} r^{-K}) \leqq 2(1 + K + \varepsilon)\log r \qquad (r \geqq r_0).$$

The remaining integral is

$$\int_0^\infty \{\phi(t) - \phi(r)\} t^{\rho-1} \operatorname{Re} z(t+z)^{-1} \, dt$$

$$= \int_0^r \left[t^{\rho-1}\{\phi(t) - \phi(r)\} \sum_{n=0}^\infty (-1)^n (t/r)^n \cos n\theta \right] dt$$

$$+ \int_r^\infty \left[t^{\rho-1}\{\phi(t) - \phi(r)\} \sum_{n=0}^\infty (-1)^n (r/t)^{n+1} \cos(n+1)\theta \right] dt$$

$$= -\delta \sum_{n=0}^\infty \left\{ r^{-n}(n+\rho)^{-1} \int_0^r t^{n+\rho-1} h(t)\phi(t) \, dt \right.$$

$$\left. + r^{n+1}(n+1-\rho)^{-1} \int_r^\infty t^{\rho-n-2} h(t)\phi(t) \, dt \right\}$$

$$+ \delta \sum_{n=0}^\infty \left[(1 - \cos n\pi \cos n\theta) r^{-n}(n+p)^{-1} \int_0^r t^{n+\rho-1} h(t)\phi(t) \, dt \right.$$

$$\left. + \{1 + \cos n\pi \cos(n+1)\theta\} r^{n+1}(n+1-\rho)^{-1} \int_r^\infty t^{\rho-n-2} h(t)\phi(t) \, dt \right]$$

$$= -\delta r^\rho \sum_{n=0}^\infty \left\{ (n+\rho)^{-1} \int_0^1 s^{n+\rho-1} h(rs)\phi(rs) \, ds \right.$$

$$\left. + (n+1-\rho)^{-1} \int_1^\infty s^{\rho-n-2} h(rs)\phi(rs) \, ds \right\}$$

$$+ \delta r^\rho \sum_{n=0}^\infty \left[(1 - \cos n\pi \cos n\theta)(n+\rho)^{-1} \int_0^1 s^{n+\rho-1} h(rs)\phi(rs) \, ds \right.$$

$$\left. + \{1 + \cos n\pi \cos(n+1)\theta\}(n+1-\rho)^{-1} \int_1^\infty s^{\rho-n-2} h(rs)\phi(rs) \, ds \right]$$

$$= \Sigma_1 + \Sigma_2, \text{ say.}$$

We are supposing that $\theta = \theta(r)$ in Σ_2.

For Σ_2 we use the fact that if $L(t)$ is slowly varying with $r_0 = 0$, and $\alpha > 0$, then (see [1, p. 86])

$$L_1(x) = x^{-\alpha} \max_{0 \leq t \leq x} \{t^\alpha L(t)\}, \qquad L_2(x) = x^\alpha \max_{x \leq t < \infty} \{t^{-\alpha} L(t)\}$$

are both slowly varying and $L_1(x) \sim L_2(x) \sim L(x)$ $(x \to \infty)$. For the integral over $(0, 1)$ in Σ_2 we use $L_1(x)$ with $L(x) = h(x)\phi(x)$ and $\alpha = \frac{1}{2}\rho$. Then

$$r^{-\rho/2} \int_0^1 \left\{ \sum_{n=0}^\infty \frac{1 - \cos n\pi \cos n\theta}{n+\rho} s^{n+\rho/2-1} (rs)^{\rho/2} h(rs)\phi(rs) \right\} ds$$

$$\leq L_1(r) \sum_{n=0}^\infty \frac{1 - \cos n\pi \cos n\theta}{(n+\rho)(n+\frac{1}{2}\rho)}.$$

Now

$$|1 - \cos n\pi \cos(n\pi - nr^{-K})| = |1 - \cos nr^{-K}| \leq nr^{-K} \leq r^{-1}$$

for $n \leq r^{K-1}$, and so

$$\sum_{n=0}^{\infty} \frac{(1 - \cos n\pi \cos n\theta)}{(n + \rho)(n + \frac{1}{2}\rho)} \leq \frac{1}{r} \sum_{n=0}^{r^{K-1}} \frac{1}{(n + \rho)(n + \frac{1}{2}\rho)} + 2 \sum_{1 + r^{K-1}}^{\infty} \frac{1}{(n + \rho)(n + \frac{1}{2}\rho)}$$

$$\to 0 \qquad (r \to \infty), \qquad \text{since} \quad K > 1,$$

and as $L_1(r) \sim h(r)\phi(r) \, (r \to \infty)$ we see that the first term in Σ_2 is $o(1)h(r)\phi(r)r^\rho$ $(r \to \infty)$. Similarly with the integral over $(1, \infty)$, and so

$$\Sigma_2 = o(1)h(r)\phi(r)r^\rho \qquad (r \to \infty).$$

As in the estimation of $\log f(r)$,

$$\Sigma_1 \sim -\delta r^\rho h(r)\phi(r) \sum_{n=0}^{\infty} \{(n + \rho)^{-2} + (n + 1 - \rho)^{-2}\},$$

since $h(r)\phi(r)$ is slowly varying.

Thus

$$\frac{\log \mu(r)}{\log M(r)} < \frac{\pi \cot \pi\rho - \delta h(r) \sum\limits_{n=0}^{\infty} \{(n + \rho)^{-2} + (n + 1 - \rho)^{-2}\} + o(1)h(r) + O\{\log r/r^\rho \phi(r)\}}{\pi \operatorname{cosec} \pi\rho + \delta h(r) \sum\limits_{n=0}^{\infty} (-1)^n \{(n + 1 - \rho)^{-2} - (\rho + n)^{-2}\} + o(1)h(r) - O\{\log r/r^\rho \phi(r)\}}.$$

Since $h(r)\phi(r)$ is slowly varying, $r^{-\rho} \log r = o(1)h(r)\phi(r)$, and so, since $h(r) \to 0$,

$$\frac{\log \mu(r)}{\log M(r)} < \cos \pi\rho - \{\delta C_2(\rho) + o(1)\}h(r) < \cos \pi\rho - h(r), \qquad r \geq r_0,$$

where δ is chosen sufficiently small.

This completes the proof.

4. Proof of Theorem 3

We suppose that $\rho < \alpha < \frac{1}{2}$ and $k > 1$. Now

$$\int_r^{\infty} \{\log |1 - t| - k^{-\alpha} \cos \pi\alpha \log(1 + kt)\} \frac{dt}{t^{1+\alpha}}$$

$$= \int_r^{\infty} \{\log |1 - t| - \cos \pi\alpha \log(1 + t)\} \frac{dt}{t^{1+\alpha}}$$

$$+ \cos \pi\alpha \int_r^\infty \{\log(1 + t) - k^{-\alpha} \log(1 + kt)\} \frac{dt}{t^{1+\alpha}}$$

$$= I_1 + I_2, \text{ say.}$$

But

$$I_1 > \frac{1 - \cos \pi\alpha}{\alpha} \frac{\log(1 + r)}{r^\alpha}$$

for all $r > 0$, by [7, pp. 17–18]. Further,

$$\int_r^\infty \frac{\log(1 + kt)}{t^{1+\alpha}} dt = k^\alpha \int_{kr}^\infty \frac{\log(1 + s)}{s^{1+\alpha}} ds$$

and so

$$I_2 = \cos \pi\alpha \int_r^{kr} \frac{\log(1 + s)}{s^{1+\alpha}} ds \geq \cos \pi\alpha \frac{(1 - k^{-\alpha}) \log(1 + r)}{\alpha r^\alpha},$$

so by addition

(4.1) $$I_1 + I_2 > \alpha^{-1}(1 - k^{-\alpha} \cos \pi\alpha) r^{-\alpha} \log(1 + r).$$

For $\alpha > \rho$, set

$$G(r) = r^\alpha \int_r^\infty \{U(-t) - (k^{-\alpha} \cos \pi\alpha)U(kt)\} t^{-1-\alpha} dt$$

so that for $r > 0$,

$$G(r) > \alpha^{-1}(1 - k^{-\alpha} \cos \pi\alpha) r^{-\alpha} \log(1 + r).$$

Also

$$tG'(t) = \alpha G(t) + k^{-\alpha} \cos \pi\alpha U(kt) - U(-t)$$

almost everywhere

$$= \alpha G(t) - (1 - k^{-\alpha} \cos \pi\alpha)U(t)$$
$$+ k^{-\alpha} \cos \pi\alpha\{U(kt) - U(t)\} + U(t) - U(-t)$$
$$> 0,$$

by (4.1) and the fact that $k > 1$. Thus, for $r \geq r_1$, where $U(r_1) > 0$

$$\int_{r_1}^r \frac{G'(t)}{G(t)} dt = \int_{r_1}^r \left\{\alpha + \frac{k^{-\alpha} \cos \pi\alpha U(kt) - U(-t)}{G(t)}\right\} \frac{dt}{t}$$

$$\leq \log \frac{G(r)}{G(r_1)}$$

$$\leq \rho\{1 + o(1)\}\log r \qquad (r \to \infty)$$

since $U(-t) - k^{-\alpha}\cos \pi\alpha U(kt) \leqq (1 - k^{-\alpha}\cos \pi\alpha)U(kt)$ and $U(t)$ has order ρ.
Since the integrand is nonnegative, we see that if E is the set in which $U(-t) \leqq k^{-\alpha}\cos \pi\alpha U(kt)$, then

$$\alpha \int_{E \cap (r_1, r)} t^{-1}\, dt \leqq \rho\{1 + o(1)\}\log r \qquad (r \to \infty)$$

and so

$$\limsup_{r \to \infty} \frac{1}{\log r}\int_{E \cap (1, r)} t^{-1}\, dt \leqq \frac{\rho}{\alpha}$$

as required.

REFERENCES

1. S. Aljančic, R. Bojanic, and M. Tomić, Sur la valeur asymptotique d'une classe des intégrales définies, *Publ. Inst. Math. Acad. Serbe Sci.* **7** (1954), 81–94.
2. P. D. Barry, The minimum modulus of small integral and subharmonic functions, *Proc. London Math. Soc.* (3) **12** (1962), 445–495.
3. ———, On a theorem of Besicovitch, *Quart. J. Math. Oxford Ser.* 2 **14** (1963), 293–302.
4. R. P. Boas, Jr., *Entire Functions*, Academic Press, New York, 1954.
5. M. L. Cartwright, *Integral Functions*, Cambridge University Press, Cambridge, England, 1956.
6. M. Essén, Note on "A theorem on the minimum modulus of entire functions" by Kjellberg, *Math. Scand.* **12** (1963), 12–14.
7. B. Kjellberg, *On certain integral and harmonic functions*, Dissertation, Uppsala, 1948.
8. ———, On the minimum modulus of entire functions of lower order less than one, *Math. Scand.* **8** (1960), 189–197.
9. ———, A theorem on the minimum modulus of entire functions, *Math. Scand.* **12** (1963), 5–11.
10. G. Valiron, *Lectures on the general theory of integral functions*, Imprimerie et Librairie Édouard Privat, Librarie de l'Universite, Toulouse, 1923; Chelsea, New York, 1949.
11. A. Wiman, Über die angenäherte Darstellung von ganzen Funktionen, *Ark. Mat. Astronom. Fys.* **1**, 1903.

(Received December 28, 1968)

Integers—
No Three in
Arithmetic Progression

P. S. Chiang

Department of Mathematics
Valdosta State College
Valdosta, Ga. 31601
U.S.A.

H. Shankar

Department of Mathematics
Ohio University
Athens, Ohio, 45701
U.S.A.

1. Introduction

Following Erdős and Turan [1] a sequence of integers $0 < a_1 < a_2 < \cdots \leq N$ which does not have any of its three terms in arithmetic progression will be called an A-sequence belonging to N, and in short, as an $A(N)$-sequence. For example, $1 < 2 < 4$; $1 < 2 < 4 < 5$; $1 < 2 < 4 < 5 < 10$ are $A(10)$-sequences. From now on an $A(N)$-sequence will be written in set notation as $\{a_1, a_2, \ldots, a_n\}_N$ or simply as $\{a_1, a_2, \ldots, a_n\}$ if there is no ambiguity about the value of N. We consider those $A(N)$-sequences which contain maximum number of terms, and we shall call them *maximal $A(N)$-sequences*. Let us denote by $r(N)$ the number of terms in a maximal $A(N)$-sequence. Erdős and Turan [1] have given the exact values of $r(N)$ for $N = 1, 2, \ldots, 23$. In general they have proved

Theorem A. *If* $N \geqq 8$, *then* $r(2N) \leqq N$.

Theorem B. *For* $\varepsilon > 0$ *and* $N > N_0(\varepsilon)$, $r(N) < (4/9 + \varepsilon)N$.

Also, they have predicted that $r(N) = o(N)$. G. Szekeres has conjectured [1] that $r\{(3^k + 1)/2\} = 2^k$. This has been proved for $k = 1, 2, 3, 4$.

It is interesting to note that Theorem A is true for $N = 4, 5, 6$, but not for $N = 7$ (see (2.1)). In a recent paper, Chiang and Macintyre [2], and Makowski [3] proved that $r(20) = 9$ contrary to the claim made by Erdős and Turan [1] that $r(20) = 8$. Our aim in the present paper is to complete the list of exact values of $r(N)$ up to $N = 41$; and according to the traditions in number theory, we make some conjectures beyond this point in Section 4.

2. Observations and definitions

The precise values of $r(N)$ for $N = 1, 2, \ldots, 20$, are listed in the following table which will be used frequently in the sequel.

(2.1)

N	1	2	3	4	5	6	7	8	9	10	11	12	13	14	15	16	17	18	19	20
$r(N)$	1	2	2	3	4	4	4	4	5	5	5	6	6	7	8	8	8	8	8	9.

It should be noted that if

(2.2) $$\{a_1, a_2, \ldots, a_n\}$$

forms an $A(N)$-sequence, then

(2.3) $$\{N + 1 - a_n, N + 1 - a_{n-1}, \ldots, N + 1 - a_1\}$$

and

(2.4) $$\{a_1 - k, a_2 - k, \ldots, a_n - k\},$$

for any integer $k < a_1$, also form $A(N)$-sequences. Consequently, for any pair M, N of positive integers, we have

(2.5) $$r(M + N) \leqq r(M) + r(N).$$

The set (2.3) will be said to be in *reverse order* to the set (2.2); and the term $N + 1 - a_i$ will be called *inverse* to the term a_i. We further note that (cf. [2, p. 128]) there are only the following two maximal $A(20)$-sequences and that $r(20) = 9$:

(2.6)
$$\{1, 2, 6, 7, 9, 14, 15, 18, 20\},$$
$$\{1, 3, 6, 7, 12, 14, 15, 19, 20\}.$$

3. Statement of result

We now give our main

Theorem. *For $N = 21, 22, \ldots, 41$, the values of $r(N)$ are given by*

N	21	22	23	24	25	26	27	28	29	30	31	32	33	34	35	36	37	38	39	40	41
r(N)	9	9	9	10	10	11	11	11	11	12	12	13	13	13	13	14	14	14	14	15	16.

The proof of this theorem is contained in the following series of Lemmas. The techniques used to prove these Lemmas are almost the same, yet their proofs could not be omitted since the arguments depend on tables of maximal sequences which should be available to the reader.

Lemma 3.1. $r(21) = r(22) = 9$.

Suppose $r(22) \geq 10$. Then maximal $A(22)$-sequence of ten terms from 1 to 22 must include both the integers 21 and 22; for neither 21 nor 22 alone can be adjoined to either of the maximal $A(20)$-sequences of nine terms given by (2.6). Therefore, in consequence of (2.3) the maximal $A(22)$-sequences must also include both the integers 1 and 2 which are the inverse terms of 21 and 22, respectively. Further, since the sets $\{1, 11, 21\}$ and $\{2, 12, 22\}$ form arithmetic progressions, the integers 11 and 12 must be omitted from maximal $A(22)$-sequences.

Next, we note that $A(9)$-sequences with $r(9) = 5$ (only three with integers 1 and 2) are

(3.1)
$$\{1, 2, 4, 8, 9\} \qquad \{1, 2, 6, 7, 9\}$$
$$\{1, 2, 6, 8, 9\} \qquad \{1, 3, 4, 8, 9\}.$$

By arguments similar to those used above, one can easily verify that none of the sequences listed in (3.1) permit extension to an $A(22)$-sequence with $r(22) \geq 10$. Hence the integer 10 must be included in a maximal $A(22)$-sequence. Consequently its inverse term 13 must also be included in such a sequence.

Thus from our foregoing analysis we conclude that the integers 3, 4, 5, 6, 7 and 16, 17, 18, 19, 20 must be excluded from the maximal $A(22)$-sequences. Also one can not have three terms from the integers 8, 9, 10 and 13, 14, 15. Therefore the assumption $r(22) \geq 10$ is impossible; and since $9 = r(20) \leq r(21) \leq r(22) < 10$, the lemma follows.

Lemma 3.2. $r(23) = 9$ *and* $r(24) = 10$.

Again, suppose $r(23) \geq 10$. Then, since $r(22) = 9$, by arguments similar to those used in Lemma 3.1, we conclude that the integers 1 and 23 should be included in a maximal $A(23)$-sequence; and the integer 12 must be excluded. But we observe that the integer 23 can not be adjoined to either of the maximal $A(20)$-sequences of nine terms given by (2.6). Hence the sequence must contain any two of the integers 21, 22, 23. This implies that two of the corresponding inverse terms 3, 2, 1 should also occur in a maximal $A(23)$-sequence.

Furthermore, there are seven maximal $A(11)$-sequences with $r(11) = 6$. These are

$$\{1, 2, 4, 5, 10, 11\} \qquad \{1, 2, 7, 8, 10, 11\}$$
$$\{1, 2, 4, 8, 9, 11\} \qquad \{1, 3, 4, 8, 9, 11\}$$
$$\{1, 2, 4, 8, 10, 11\} \qquad \{1, 3, 4, 8, 10, 11\}$$
$$\{1, 2, 5, 7, 10, 11\}.$$

But it is evident that none of these sequences can lead to $r(23) = 10$. Consequently, we can not adjoin to these sequences, six integers from 13 to 23. Therefore, in a maximal $A(23)$-sequence we must have five integers from 1 to 11 and five from 13 to 23.

Now, we check $A(9)$-sequences with $r(9) = 5$ listed in (3.1), and the following $A(10)$-sequences with $r(10) = 5$.

$$\{1, 2, 4, 5, 10\} \quad \{1, 2, 7, 8, 10\} \quad \{1, 3, 4, 9, 10\} \quad \{1, 4, 5, 8, 10\}$$
$$\{1, 2, 4, 8, 10\} \quad \{1, 2, 7, 9, 10\} \quad \{1, 3, 6, 7, 10\} \quad \{1, 4, 6, 9, 10\}$$
$$\{1, 2, 4, 9, 10\} \quad \{1, 3, 4, 6, 10\} \quad \{1, 3, 7, 8, 10\} \quad \{1, 5, 7, 8, 10\}$$
$$\{1, 2, 5, 7, 10\} \quad \{1, 3, 4, 8, 10\} \quad \{1, 3, 7, 9, 10\} \quad \{1, 6, 7, 9, 10\}.$$

Again we find that none of these can be extended to $A(23)$ sequence with $r(23) = 10$. Hence the integer 11, and therefore, in consequence of (2.3), the integer 13 must be used in $A(23)$-sequences. Thus, our choice, as indicated earlier, of any two from the integers 21, 22, 23 must fall on 22 and 23; and the integers 3, 4, 6, 7, and 9 must be ruled out from our choice. So there is no possibility of having five integers from 1 to 11 in an $A(23)$-sequence. This contradicts the assumption and we conclude that $r(23) = 9$.

Finally, by inequality (2.5), we have $r(24) \leq r(23) + r(1) = 10$. But there are at least two $A(24)$-sequences with $r(24) = 10$, namely

$$\{1, 2, 6, 7, 9, 14, 18, 20, 23, 24\}$$

and

$$\{1, 2, 5, 7, 11, 16, 18, 19, 23, 24\}.$$

Hence we have $r(24) = 10$.

Lemma 3.3. $r(25) = 10$ and $r(26) = 11$.

Suppose that $r(25) \geq 11$. Then by usual arguments we conclude that integers 24 and 25 must be used in $A(25)$-sequences, and hence also 2 and 1. Now, we may suppose that $A(25)$-sequences with $r(25) \geq 11$ can have six terms out of integers from 1 to 12 and five terms from 13 to 25. The following are $A(12)$-sequences with $r(12) = 6$, and which begin with 1 and 2.

$\{1, 2, 4, 5, 10, 11\}$ $\{1, 2, 4, 9, 10, 12\}$
$\{1, 2, 4, 5, 10, 12\}$ $\{1, 2, 4, 9, 11, 12\}$
$\{1, 2, 4, 5, 11, 12\}$ $\{1, 2, 5, 7, 10, 11\}$
$\{1, 2, 4, 8, 9, 11\}$ $\{1, 2, 7, 8, 10, 11\}$
$\{1, 2, 4, 8, 10, 11\}$ $\{1, 2, 8, 9, 11, 12\}$.

The integer 13 can not be adjoined to these and none of them lead to $A(25)$-sequence with $r(25) = 11$. It follows by inversion (2.3) that a combination of five terms from 1 to 12 and six terms from 14 to 25 is also impossible. Thus we conclude that $r(25) = 10$.

Now we consider $r(26)$ and by (2.5), $r(26) \leq r(25) + r(1) = 11$. But $\{1, 3, 4, 8, 9, 11, 16, 20, 22, 25, 26\}$ and $\{1, 2, 5, 7, 11, 16, 18, 19, 23, 24, 26\}$ are maximal $A(26)$-sequences with $r(26) = 11$ and the lemma follows.

Lemma 3.4. $r(27) = r(28) = r(29) = 11$ *but* $r(30) = 12$.

We consider $r(29)$ and suppose that $r(29) \geq 12$. There are two types of $A(29)$-sequences with $r(29) = 12$:

CASE (*a*). nine integers from 1 to 22 and three integers from 23 to 29.

CASE (*b*). eight integers from 1 to 22 and four from 23 to 29.

Consider Case (*a*). There are forty-four $A(22)$-sequences with $r(22) = 9$. These are (for the sake of convenience commas and brackets are dropped)

1 2 4 5 10 14 17 21 22	1 2 4 10 14 15 17 21 22	1 2 6 7 9 15 18 19 22
1 2 4 5 12 14 15 17 21	1 2 5 6 12 14 15 17 21	1 2 6 7 9 15 19 20 22
1 2 4 5 12 14 17 18 21	1 2 5 6 12 14 15 20 21	1 2 6 7 9 18 19 21 22
1 2 4 5 14 15 17 21 22	1 2 5 6 12 15 17 20 21	1 2 6 8 9 13 19 21 22
1 2 4 5 14 15 19 20 22	1 2 5 6 13 15 16 18 22	1 2 6 8 9 18 19 21 22
1 2 4 5 14 15 19 21 22	1 2 5 6 13 15 18 19 22	1 2 6 9 13 15 18 19 22
1 2 4 5 14 16 17 21 22	1 2 5 7 10 16 17 20 21	1 2 6 9 13 18 19 21 22
1 2 4 8 9 18 19 21 22	1 2 5 7 14 15 19 20 22	1 2 7 8 10 16 17 20 21
1 2 4 9 12 13 18 19 21	1 2 6 7 9 14 15 18 20	1 3 4 6 13 14 18 19 21
1 3 4 8 9 11 16 20 22	1 3 6 7 12 14 15 19 20	1 4 5 8 14 16 17 21 22
1 3 4 8 9 16 18 19 21	1 3 6 7 12 15 19 20 22	
1 3 4 8 9 16 18 21 22	1 3 7 12 14 15 19 20 22	
1 3 4 8 9 18 19 21 22		1 5 7 8 10 16 17 20 21
1 3 4 8 11 16 17 20 22	1 4 5 8 10 14 17 21 22	1 5 7 8 10 17 18 20 21
1 3 4 8 14 16 17 21 22	1 4 5 8 10 17 18 20 21	1 5 7 8 10 17 18 21 22
1 3 4 9 10 13 18 20 21	1 4 5 8 10 17 18 21 22.	

None of these, however, can be extended to form sequences with $r(29) = 12$.

Regarding Case (b), we note that there are ten sets listed below of four terms from 23 to 29 and none of these can be extended to form sequences with $r(29) = 12$.

$$\{23,\ 24,\ 26,\ 27\} \qquad \{23,\ 25,\ 26,\ 28\} \qquad \{24,\ 25,\ 28,\ 29\}$$
$$\{23,\ 24,\ 27,\ 28\} \qquad \{23,\ 25,\ 28,\ 29\} \qquad \{24,\ 26,\ 27,\ 29\}$$
$$\{23,\ 24,\ 27,\ 29\} \qquad \{24,\ 25,\ 27,\ 28\} \qquad \{25,\ 26,\ 28,\ 29\}$$
$$\{23,\ 24,\ 28,\ 29\}.$$

Therefore $r(29) = 11$, and hence $r(27) = r(28) = 11$.

Since $r(30) \leq 12$, and there is at least one $A(30)$-sequence of twelve terms, namely 1, 3, 4, 8, 9, 11, 20, 22, 23, 27, 28, 30 we have $r(30) = 12$. This completes the proof of the lemma.

Lemma 3.5. $r(31) = 12$ *and* $r(32) = 13$.

Suppose $r(31) \geq 13$, then an $A(31)$-sequence with $r(31) = 13$ must be composed of either

CASE (a). nine terms from integers 1 to 22 and four from integers 23 to 31; or

CASE (b). eight terms from integers 1 to 22 and five terms from integers 23 to 31.

It is easy to dispose of Case (a) since it is a matter of simple verification that none of the forty-four sets listed in Lemma 3.4 Case (a) permit extensions by four terms from integers 23 to 31. Now we discuss Case (b). Since $r(31) = 13$ and $r(29) = 11$, the $A(31)$-sequence must contain the integers 30 and 31. Further, note that there are only three A-sequences of five terms picked out from integers 23 to 31. These are $\{23, 24, 26, 30, 31\}$, $\{23, 24, 28, 30, 31\}$, and $\{23, 25, 26, 30, 31\}$. None of these three allow us to adjoin eight terms from integers 1 to 22. Thus we have $r(31) = 12$. Since $r(32) \leq 13$ and

$$\{1,\ 2,\ 4,\ 8,\ 9,\ 11,\ 19,\ 22,\ 23,\ 26,\ 28,\ 31,\ 32\},$$
$$\{1,\ 2,\ 5,\ 7,\ 10,\ 11,\ 14,\ 22,\ 24,\ 25,\ 29,\ 31,\ 32\}$$

form $A(32)$-sequences of thirteen terms, we have $r(32) = 13$. This completes the proof of the lemma.

Lemma 3.6. $r(33) = r(34) = r(35) = 13$ *and* $r(36) = 14$.

If $r(35) \geq 14$, then the three possibilities of composing $A(35)$-sequences with $r(35) = 14$, are as follows.

CASE (a). pick nine terms from integers 1 to 22 and five terms from 23 to 35;

CASE (b). pick eight terms from integers 1 to 22 and six terms from 23 to 35;

CASE (c). pick seven terms from integers 1 to 22 and seven terms from 23 to 35.

Since all forty-four sets listed in Lemma 3.4 Case (a) do not allow us to adjoin five terms from integers 23 to 35, the Case (a) can not occur. We also find that the Case (b) can not occur since none of the seventy-eight sets of six terms (see Table A) from integers 23 to 35 permit extension by eight terms chosen from integers 1 to 22. To discuss Case (c), we first observe that there are only six sets of seven terms from integers 23 to 35 listed as follows:

$$
\begin{array}{ll}
23\ 24\ 26\ 27\ 32\ 33\ 35 & 23\ 24\ 26\ 31\ 32\ 34\ 35 \\
23\ 24\ 26\ 27\ 32\ 34\ 35 & 23\ 25\ 26\ 28\ 32\ 34\ 35 \\
23\ 24\ 26\ 30\ 32\ 33\ 35 & 23\ 25\ 26\ 31\ 32\ 34\ 35.
\end{array}
$$

It is easy to check that none of these permit seven terms from integers 1 to 22 to be added. Thus we conclude that $r(35) = 13$, and obviously, then $r(33) = r(34) = 13$. Finally, since $r(36) \leq 14$, and there are at least two $A(36)$-sequences of fourteen terms, namely

$$\{1,\ 2,\ 5,\ 7,\ 10,\ 11,\ 14,\ 16,\ 24,\ 28,\ 29,\ 33,\ 35,\ 36\}$$

and

$$\{1,\ 2,\ 4,\ 8,\ 9,\ 13,\ 21,\ 23,\ 26,\ 27,\ 30,\ 32,\ 35,\ 36\},$$

we have $r(36) = 14$.

Lemma 3.7. $r(37) = r(38) = r(39) = 14$, $r(40) = 15$, *and* $r(41) = 16$.

We consider $r(39)$ and suppose that $r(39) \geq 15$. Then $A(39)$-sequensce with $r(39) = 15$ can be formed as follows.

CASE (a). nine terms from integers 1 to 22 and six terms from 23 to 39.

CASE (b). eight terms from integers 1 to 22 and six terms from 23 to 39.

CASE (c). seven terms from integers 1 to 22 and eight terms from 23 to 39.

It is easy to verify that Case (a) can not occur since none of the forty-four sets of nine terms, listed in Lemma 3.4 Case (a), permit us to adjoin six more terms from integers 23 to 39. Nor does Case (c) occur, since none of the forty-three sets of eight terms (see Table B) from integers 23 to 39, allow us to add seven more terms from integers 1 to 22.

We now discuss Case (b). Since $r(35) = 13$, the $A(39)$-sequences with $r(35) = 15$ must be one of the following types:

$$
\begin{array}{lllll}
(b_1) & 1, \ldots, 36, & 37, & & 39. \\
(b_2) & 1, \ldots, 36, & & 38, & 39. \\
(b_3) & 1, \ldots, 36, & 37, & & \\
(b_4) & 1, \ldots, 36, & & 38. & \\
(b_5) & 1, \ldots, 36, & & & 39. \\
(b_6) & 1, \ldots, & 37, & 38. & \\
(b_7) & 1, \ldots, & 37, & & 39. \\
(b_8) & 1, \ldots, & & 38, & 39.
\end{array}
$$

There are thirty-four, fifty-seven, twenty-three, thirty-two, thirty, fifty-one, twenty, and fifty-six sets, respectively, for the types (b_1)–(b_8). These various sets are listed in Tables B_1, B_2, $\ldots$, B_8, respectively. We can check by a similar but very much longer argument that none of these sets permit extension by eight terms from integers 1 to 22. Thus $r(39) = 14$. Therefore, since $r(36) = 14$, $r(37) = r(38) = 14$.

Finally, since $r(40) \leq 15$ and there are at least two $A(40)$-sequences of fifteen terms, namely $\{1, 2, 4, 5, 10, 11, 13, 14, 28, 29, 31, 32, 37, 38, 40\}$ and $\{1, 3, 4, 9, 10, 12, 13, 27, 28, 30, 31, 36, 37, 39, 40\}$, we have $r(40) = 15$. Similarly $r(41) \leq 16$ and $\{1, 2, 4, 5, 10, 11, 13, 14, 28, 29, 31, 32, 37, 38, 40, 41\}$ is an $A(41)$-sequence with $r(41) = 16$, so $r(41) = 16$. This proves Lemma (3.7).

4. Conjectures

The following estimates follow immediately from the preceding lemmas.

$$
\begin{aligned}
r(59) &\geq r(20) + r(20)^* = 18 \\
r(71) &\geq r(24) + r(24)^* = 20 \\
r(77) &\geq r(26) + r(26)^* = 22 \\
r(89) &\geq r(30) + r(30)^* = 24 \\
r(98) &\geq r(33) + r(33)^* = 26 \\
r(107) &\geq r(36) + r(36)^* = 28 \\
r(119) &\geq r(40) + r(40)^* = 30 \\
r(122) &\geq r(41) + r(41)^* = 32
\end{aligned}
$$

where $r(20)^*$, $r(24)^*$, $r(26)^*$, $r(30)^*$, $r(33)^*$, $r(36)^*$, $r(40)^*$ and $r(41)^*$ denote, respectively, the number of terms of an A-sequence formed out of integers from 40 to 59, 48 to 71, 52 to 77, 60 to 89, 66 to 98, 72 to 107, 80 to 119, and 82 to 122.

We conjecture that the following hold: $r(51) = 17$, $r(54) = 18$, $r(59) = 19$, $r(61) = r(62) = \cdots = r(71) = 20$.

5. Acknowledgment

The authors are indebted to the late Professor A. J. Macintyre who pointed out this problem to them and made valuable suggestions in the initial stage of this paper during his last days of life. Also the authors wish to thank the referee for his comments, and Mr. R. E. Menninger, Xavier University, Cincinnati, Ohio, who obtained the tables of *A*-sequences for this paper.

REFERENCES

1. P. Erdős and P. Turan, On some sequences of integers, *J. London Math. Soc.* **11** (1936), 261–264.
2. P. S. Chiang and A. J. Macintyre, Integers, no three in arithmetic progression, *Math. Mag.* **41** (1968), 128–130.
3. A. Makowski, Remark on a paper of Erdős and Turan, *J. London Math. Soc.* **34** (1959), 480.

(*Received February 9, 1968 and, in revised form, September 13, 1968*)

TABLE A

23	24	26	27	32	33	23	24	31	32	34	35	24	25	27	28	33 34
23	24	26	27	32	34	23	25	26	28	32	34	24	25	27	28	33 35
23	24	26	27	32	35	23	25	26	28	32	35	24	25	27	28	34 35
23	24	26	27	33	34	23	25	26	28	34	35	24	25	27	31	32 34
23	24	26	27	33	35	23	25	26	30	31	33	24	25	27	31	33 34
23	24	26	27	34	35	23	25	26	30	32	33	24	25	27	32	33 35
23	24	26	30	31	33	23	25	26	31	32	34	24	25	27	32	34 35
23	24	26	30	31	35	23	25	26	31	32	35	24	25	28	30	33 34
23	24	26	30	32	33	23	25	26	31	33	34	24	25	30	31	33 34
23	24	26	30	32	35	23	25	26	31	34	35	24	25	31	32	34 35
23	24	26	30	33	35	23	25	26	32	33	35	24	26	27	29	33 35
23	24	26	31	32	34	23	25	26	32	34	35	24	26	27	31	32 34
23	24	26	31	32	35	23	25	28	29	32	34	24	26	27	31	33 34
23	24	26	31	33	34	23	25	28	32	34	35	24	26	27	32	33 35
23	24	26	31	34	35	23	25	29	31	32	34	24	26	27	32	34 35
23	24	26	32	33	35	23	25	30	31	33	34	24	26	29	30	33 35
23	24	26	32	34	35	23	25	31	32	34	35	24	26	30	32	33 35
23	24	27	28	34	35	23	26	27	30	32	35	24	26	31	32	34 35
23	24	27	29	32	33	23	26	27	32	33	35					
23	24	27	32	33	35	23	26	27	32	34	35	25	26	28	29	34 35
23	24	27	32	34	35	23	26	28	31	32	35	25	26	28	32	33 35
23	24	28	30	31	35	23	26	28	32	34	35	25	26	28	32	34 35
23	24	28	30	34	35	23	26	30	32	33	35	25	26	29	31	34 35
23	24	29	30	32	33	23	26	31	32	34	35	25	26	31	32	34 35
23	24	30	31	33	34	23	27	28	30	34	35	25	27	28	32	33 35
23	24	30	31	34	35	23	27	28	32	34	35	25	27	28	32	34 35
23	24	30	32	33	35											

TABLE B

23	24	26	27	32	33	35	36	23	26	27	30	35	36	38	39
23	24	26	27	32	34	35	39	23	27	28	30	35	36	38	39
23	24	26	27	32	35	36	39	23	27	29	30	32	36	38	39
23	24	26	27	33	34	36	37								
23	24	26	27	33	34	37	38	24	25	27	28	33	34	36	37
23	24	26	27	33	35	36	38	24	25	27	28	34	35	37	38
23	24	26	27	34	35	37	38	24	25	27	28	34	35	38	39
23	24	26	27	34	35	38	39	24	25	27	28	34	36	37	39
23	24	26	27	34	36	37	39	24	25	27	28	35	36	38	39
23	24	26	27	35	36	38	39	24	25	28	29	35	36	38	39
23	24	26	30	32	33	35	39	24	25	28	30	33	34	37	39
23	24	26	30	33	35	38	39	24	26	27	29	33	35	36	38
23	24	27	28	34	35	37	38	24	26	27	29	35	36	38	39
23	24	27	28	34	35	38	39	24	26	29	30	33	35	38	39
23	24	27	28	34	36	37	39								
23	24	27	28	35	36	38	39	25	26	28	29	34	35	37	38
23	24	27	29	32	33	36	38	25	26	28	29	35	36	38	39
23	24	27	29	32	36	38	39	25	27	28	30	34	36	37	39
23	24	29	30	32	33	38	39								
23	25	26	28	32	34	35	37	26	27	29	30	35	36	38	39
23	25	26	28	34	35	37	38								
23	25	26	28	34	35	38	39								
23	25	26	28	34	36	37	39								
23	25	26	28	35	36	38	39								
23	25	28	29	32	34	37	38								
23	26	27	30	32	35	36	39								

TABLE B$_1$

23	24	26	27	36	37	39	25	26	28	29	36	37	39
23	24	26	32	36	37	39	25	26	28	34	36	37	39
23	24	26	34	36	37	39	25	26	29	30	36	37	39
23	24	27	28	36	37	39	25	27	28	30	36	37	39
23	24	27	29	36	37	39	25	27	28	34	36	37	39
23	24	27	34	36	37	39	25	27	30	34	36	37	39
23	24	28	29	36	37	39	25	28	30	34	36	37	39
23	24	28	34	36	37	39							
23	24	29	32	36	37	39	26	27	29	30	36	37	39
23	25	26	28	36	37	39							
23	25	26	34	36	37	39	27	28	30	31	36	37	39
23	25	28	29	36	37	39	27	28	30	34	36	37	39
23	25	28	34	36	37	39							
23	26	27	34	36	37	39							
23	27	28	34	36	37	39							
24	25	27	28	36	37	39							
24	25	27	34	36	37	39							
24	25	28	29	36	37	39							
24	25	28	34	36	37	39							
24	26	27	29	36	37	39							
24	26	27	34	36	37	39							
24	27	28	31	36	37	39							
24	27	28	34	36	37	39							
24	28	29	31	36	37	39							

TABLE B$_2$

23	24	26	27	36	38	39	23	27	29	30	36	38	39	25	26	29	30	36	38	39
23	24	26	35	36	38	39	23	27	29	32	36	38	39	25	26	29	35	36	38	39
23	24	27	28	36	38	39	23	27	30	32	36	38	39	25	27	28	30	36	38	39
23	24	27	29	36	38	39	23	27	30	35	36	38	39	25	27	28	35	36	38	39
23	24	27	32	36	38	39	23	28	30	35	36	38	39	25	27	30	31	36	38	39
							23	29	30	32	36	38	39	25	28	29	35	36	38	39
23	24	27	35	36	38	39														
23	24	28	29	36	38	39	24	25	27	28	36	38	39	26	27	29	30	36	38	39
23	24	28	35	36	38	39	24	25	27	35	36	38	39	26	27	29	35	36	38	39
23	24	29	32	36	38	39	24	25	28	29	36	38	39	26	27	30	35	36	38	39
23	25	26	28	36	38	39	24	25	28	35	36	38	39	26	28	29	35	36	38	39
23	25	26	30	36	38	39	24	25	29	35	36	38	39	26	29	30	35	36	38	39
23	25	26	35	36	38	39	24	26	27	29	36	38	39							
23	25	28	29	36	38	39	24	26	27	35	36	38	39	27	28	30	31	36	38	39
23	25	28	30	36	38	39	24	26	29	35	36	38	39	27	28	30	35	36	38	39
23	25	28	35	36	38	39	24	27	28	35	36	38	39	27	29	30	32	36	38	39
23	25	29	30	36	38	39	24	27	29	32	36	38	39	27	29	30	35	36	38	39
23	26	27	30	36	38	39	24	27	29	35	36	38	39							
23	26	27	35	36	38	39	24	28	29	35	36	38	39							
23	26	28	35	36	38	39														
23	26	30	35	36	38	39	25	26	28	29	36	38	39							
23	27	28	30	36	38	39	25	26	28	35	36	38	39							
23	27	28	35	36	38	39														

TABLE B$_3$

23	24	26	27	33	36	37	24	25	27	28	33	36	37
23	24	26	27	34	36	37	24	25	27	28	34	36	37
23	24	26	32	33	36	37	24	25	27	33	34	36	37
23	24	26	33	34	36	37	24	25	28	33	34	36	37
23	24	27	28	34	36	37	24	26	27	33	34	36	37
23	24	27	33	34	36	37	24	27	28	31	33	36	37
23	24	28	29	31	36	37	24	27	28	33	34	36	37
23	24	29	31	32	36	37							
23	25	26	28	34	36	37	25	26	28	29	34	36	37
23	25	26	32	33	36	37	25	26	28	33	34	36	37
23	25	26	33	34	36	37	25	27	28	30	34	36	37
23	25	28	29	34	36	37							
23	26	27	33	34	36	37							

TABLE B₄

23	24	26	27	33	36	38		24	25	27	28	35	36	38
23	24	26	27	35	36	38		24	25	27	32	33	36	38
23	24	26	33	35	36	38		24	25	27	33	35	36	38
23	24	27	28	35	36	38		24	25	28	29	35	36	38
23	24	27	29	32	36	38		24	26	27	29	33	36	38
23	24	27	29	33	36	38		24	26	27	29	35	36	38
23	24	27	32	33	36	38		24	26	27	33	35	36	38
23	24	27	33	35	36	38		24	26	29	33	35	36	38
23	24	29	32	33	36	38		24	27	29	32	33	36	38
23	25	26	28	35	36	38		24	27	29	33	35	36	38
23	25	26	33	35	36	38								
23	25	29	30	32	36	38		25	26	28	29	35	36	38
23	25	29	31	32	36	38								
23	26	27	30	35	36	38		26	27	29	30	35	36	38
23	26	27	33	35	36	38		26	27	29	33	35	36	38
23	27	28	30	35	36	38								
23	27	29	30	32	36	38								
23	27	29	32	33	36	38								
23	28	30	31	35	36	38								

TABLE B₅

23	24	26	27	32	36	39		24	25	27	28	34	36	39
23	24	26	27	34	36	39		24	25	27	28	35	36	39
23	24	26	27	35	36	39		24	25	27	31	34	36	39
23	24	26	32	35	36	39		24	25	28	29	35	36	39
23	24	27	28	34	36	39		24	26	27	29	35	36	39
23	24	27	28	35	36	39		24	26	27	32	35	36	39
23	24	27	29	32	36	39								
23	24	27	32	35	36	39		25	26	28	29	35	36	39
23	25	26	28	34	36	39		25	27	28	30	34	36	39
23	25	26	28	35	36	39		25	27	30	31	34	36	39
23	25	28	30	34	36	39								
23	26	27	30	32	36	39		26	27	29	30	35	36	39
23	26	27	30	35	36	39		26	27	30	32	35	36	39
23	26	27	32	35	36	39								
23	26	30	32	35	36	39								
23	27	28	30	34	36	39								
23	27	28	30	35	36	39								
23	27	29	30	32	36	39								
23	27	30	32	35	36	39								

TABLE B₆

23	24	26	27	33	37	38	24	25	27	28	34	37	38
23	24	26	27	34	37	38	24	25	27	28	35	37	38
23	24	26	27	35	37	38	24	25	27	33	34	37	38
23	24	26	33	34	37	38	24	25	27	34	35	37	38
23	24	26	34	35	37	38	24	25	28	29	35	37	38
23	24	27	28	34	37	38	24	25	28	34	35	37	38
23	24	27	28	35	37	38	24	25	29	30	32	37	38
23	24	27	33	34	37	38	24	25	30	32	33	37	38
23	24	27	34	35	37	38	24	26	27	29	35	37	38
23	24	28	34	35	37	38	24	26	27	33	34	37	38
23	25	26	28	34	37	38	24	26	27	34	35	37	38
23	25	26	28	35	37	38	24	26	29	30	35	37	38
23	25	26	29	34	37	38	24	27	28	34	35	37	38
23	25	26	33	34	37	38							
23	25	26	34	35	37	38	25	26	28	29	34	37	38
23	25	28	29	32	37	38	25	26	28	29	35	37	38
23	25	28	29	34	37	38	25	26	28	34	35	37	38
23	25	28	32	34	37	38	25	26	29	34	35	37	38
23	25	28	34	35	37	38	25	27	28	34	35	37	38
23	25	29	32	34	37	38	25	28	29	32	34	37	38
23	26	27	33	34	37	38	25	28	29	34	35	37	38
23	26	27	34	35	37	38							
23	26	28	31	35	37	38	26	27	29	30	35	37	38
23	26	28	34	35	37	38	26	27	29	34	35	37	38
23	27	28	34	35	37	38	26	28	29	31	35	37	38
23	28	29	31	32	37	38	26	28	29	34	35	37	38
23	28	29	32	34	37	38							

TABLE B₇

23	24	26	27	34	37	39	25	26	28	33	34	37	39
23	24	26	32	33	37	39	25	27	28	30	34	37	39
23	24	26	32	34	37	39	25	28	30	33	34	37	39
23	24	26	33	34	37	39							
23	24	27	28	34	37	39							
23	25	26	28	34	37	39							
23	25	26	33	34	37	39							
23	26	28	32	34	37	39							
24	25	27	28	34	37	39							
24	25	28	30	33	37	39							
24	25	28	30	34	37	39							
24	25	28	33	34	37	39							
24	25	30	33	34	37	39							
24	26	30	31	33	37	39							
24	26	30	32	33	37	39							
24	28	30	31	33	37	39							
24	28	30	33	34	37	39							

TABLE B$_8$

23 24 26 27 34 38 39	23 25 28 34 35 38 39	25 26 28 29 35 38 39
23 24 26 27 35 38 39	23 26 27 30 35 38 39	25 26 28 34 35 38 39
23 24 26 30 33 38 39	23 26 27 34 35 38 39	25 26 30 31 33 38 39
23 24 26 30 35 38 39	23 26 28 34 35 38 39	25 26 31 33 34 38 39
23 24 26 33 34 38 39	23 26 30 33 35 38 39	25 27 28 34 35 38 39
23 24 26 33 35 38 39	23 27 28 30 35 38 39	
23 24 26 34 35 38 39	23 27 28 32 34 38 39	26 27 29 30 35 38 39
23 24 27 28 34 38 39	23 27 28 34 35 38 39	26 29 30 33 35 38 39
23 24 27 28 35 38 39	23 27 29 30 32 38 39	
23 24 27 29 32 38 39	23 29 30 32 33 38 39	
23 24 27 32 34 39 39		
23 24 27 34 35 38 39	24 25 27 28 34 38 39	
23 24 28 30 35 38 39	24 25 27 28 35 38 39	
23 24 28 34 35 38 39	24 25 27 34 35 38 39	
23 24 29 30 32 38 39	24 25 28 29 35 38 39	
23 24 29 30 33 38 39	24 25 28 34 35 38 39	
23 24 29 32 33 38 39	24 26 27 29 35 38 39	
23 24 30 32 33 38 39	24 26 27 34 35 38 39	
23 24 30 33 35 38 39	24 26 29 30 33 38 39	
23 25 26 28 34 38 39	24 26 29 30 35 38 39	
23 25 26 28 35 38 39	24 26 29 33 35 38 39	
23 25 26 30 33 38 39	24 26 30 33 35 38 39	
23 25 26 33 34 38 39	24 27 28 34 35 38 39	
23 25 26 33 35 38 39	24 29 30 32 33 38 39	
23 25 26 34 35 38 39	24 29 30 33 35 38 39	

The Composition of Entire and Meromorphic Functions

J. Clunie

Department of Mathematics
Imperial College
London, S.W.7
England

1. Introduction

If $f(z)$ is entire or meromorphic (in the plane) and $g(z)$ is entire then $f \circ g(z) = f(g(z))$ is entire or meromorphic, respectively. In this paper we shall be concerned with the behavior as $r \to \infty$ of the ratios

$$\frac{\log M(r, f \circ g)}{\log M(r, f)}, \quad \frac{\log M(r, f \circ g)}{\log M(r, g)}$$

when $f(z)$ and $g(z)$ are both entire; and with the behavior as $r \to \infty$ of the ratios

$$\frac{T(r, f \circ g)}{T(r, f)}, \quad \frac{T(r, f \circ g)}{T(r, g)}$$

when $f(z)$ is meromorphic and $g(z)$ is entire. The terms $M(r, \ldots)$ and $T(r, \ldots)$ denote, as usual, the maximum modulus and Nevanlinna characteristic, respectively. We always assume that $f(z)$ and $g(z)$ are transcendental. When one of $f(z)$ or $g(z)$ is not transcendental the appropriate form of any of our results will be immediately apparent.

A number of results related to those of this paper have been published. The first such result seems to be one due to Pólya [7], who showed that if $f(z)$ and $g(z)$ are entire and $g(0) = 0$ then there is an absolute constant $A > 0$ such that

(1.1) $$M(r, f \circ g) \geq M(AM(r/2, g), f).$$

From this inequality and its proof as given by Pólya it follows easily that if $f(z)$ and $g(z)$ are entire and transcendental, even if $g(0) \neq 0$, then

$$\frac{\log M(r, f \circ g)}{\log M(r, f)} \to \infty \qquad (r \to \infty).$$

Pólya points out that given any ρ $(0 < \rho < 1)$ there are results similar to (1.1) with $r/2$ replaced by ρr and the value of A modified. Later we shall prove that given ρ $(0 < \rho < 1)$ and $f(z)$, $g(z)$ entire with $g(0) = 0$, then

(1.2) $$M(r, f \circ g) \geq M(c(\rho)M(\rho r, g), f),$$

where $c(\rho) = (1 - \rho)^2/4\rho$. Pólya's proof depended on a result of Bohr and the sharp form of this result has been obtained by Hayman [5]. The proof of Pólya when this result of Hayman is used gives (1.2). When $\rho = \frac{1}{2}$, then $A = \frac{1}{8} = c(\frac{1}{2})$ and this result is stated by Hayman on p. 50 of his book *Meromorphic Functions* [3], which will subsequently be referred to as M.F. Hence (1.2) is known, but the proof we give does not use the idea of Pólya and in fact depends on the "principle of subordination" and some of the most elementary distortion theorems for univalent functions. It is independent of Hayman's sharp form of Bohr's result.

Inequalities (1.1) and (1.2) are universal results in the sense that given $r \geq 0$ and any entire functions $f(z)$ and $g(z)$ with $g(0) = 0$, then these inequalities are satisfied. Later we shall show that, as such, a universal result (1.2) is best possible for each $r > 0$ and ρ $(0 < \rho < 1)$, i.e., given $r > 0$ and ρ $(0 < \rho < 1)$ and any number $K > c(\rho)$ there are entire functions $f(z)$, $g(z)$ with $g(0) = 0$ such that

$$M(r, f \circ g) < M(KM(\rho r, g), f).$$

If one is concerned with a given pair of entire functions $f(z)$, $g(z)$ and the behavior of $M(r, f \circ g)$ as $r \to \infty$, then one can find a much better result than any that can be derived from (1.2). I have shown [1] in fact that

$$M(r(1 + o(1)), f \circ g) \geq M(M(r, g), f) \qquad (r \to \infty),$$

where $o(1)$ depends on $g(z)$. By the same method I use in [1] one can also show that

(1.3) $$M(r, f \circ g) \geq M((1 + o(1))M(r, g), f)$$

as $r \to \infty$ outside a set of r of finite logarithmic measure which depends, as does $o(1)$, on $g(z)$.

An exercise in M.F. (p. 54) is to show that

$$\frac{T(r, f \circ g)}{T(r, g)} \to \infty \qquad (r \to \infty),$$

when $f(z)$, $g(z)$ are entire and transcendental. In M.F. a proof of this is outlined and one sees that it is equally applicable to the case of $f(z)$ meromorphic and $g(z)$ entire, both being transcendental. Later I shall give a short proof of this result by a method different from that indicated in M.F. For completeness I shall also prove that there is an entire function $g(z)$ such that

$$\liminf_{r \to \infty} \frac{\log M(r, \exp \circ g)}{\log M(r, g)} = 0.$$

In M.F. this result is stated, without proof, in a footnote following the exercise referred to above.

Finally, there is a paper of Edrei and Fuchs [2] which deals with the zeros of $f(g(z))$ when $f(z)$, $g(z)$ are entire. Their results are of a different kind from ours, but they clearly fall into a similar context of ideas. As a consequence of this paper Hayman [6] constructed "unusual" entire functions which will be described at the appropriate place in our discussion. We shall show that there is a meromorphic function $f(z)$ and an "unusual" entire function $g(z)$ such that

$$\liminf_{r \to \infty} \frac{T(r, f \circ g)}{T(r, f)} = 0.$$

2. Statement of Results

Our first two theorems contains those results which are true in general.

Theorem 1. *Let $f(z)$ and $g(z)$ be entire and transcendental. Then*

(i) $\displaystyle \lim_{r \to \infty} \frac{\log M(r, f \circ g)}{\log M(r, f)} = \infty,$

(ii) $\displaystyle \lim_{r \to \infty} \frac{T(r, f \circ g)}{T(r, f)} = \infty,$

(iii) $\displaystyle \limsup_{r \to \infty} \frac{\log M(r, f \circ g)}{\log M(r, g)} = \infty.$

Theorem 2. *Let $f(z)$ be meromorphic and $g(z)$ be entire and suppose that $f(z)$ and $g(z)$ are transcendental. Then*

(i) $\displaystyle \lim_{r \to \infty} \frac{T(r, f \circ g)}{T(r, g)} = \infty,$

(ii) $\displaystyle \limsup_{r \to \infty} \frac{T(r, f \circ g)}{T(r, f)} = \infty.$

The next theorem supplements Theorems 1 and 2 and deals with the case when one or other of $f(z)$ and $g(z)$ is of finite order.

Theorem 3. (i) *Let $f(z)$ be meromorphic and $g(z)$ be entire. Suppose that $f(z)$ and $g(z)$ are transcendental and at least one of them is of finite order. Then*

$$\lim_{r \to \infty} \frac{T(r, f \circ g)}{T(r, f)} = \infty.$$

(ii) *Let $f(z)$ and $g(z)$ be entire and transcendental and suppose that $g(z)$ is of finite order. Then*

$$\lim_{r \to \infty} \frac{\log M(r, f \circ g)}{\log M(r, g)} = \infty.$$

The next two theorems deal with the possibilities still left open in the preceding theorems.

Theorem 4. *There is an entire function $g(z)$ such that*

$$\liminf_{r \to \infty} \frac{\log M(r, \exp \circ g)}{\log M(r, g)} = 0.$$

Theorem 5. *There is a meromorphic function $f(z)$ and an entire function $g(z)$ such that*

$$\liminf_{r \to \infty} \frac{T(r, f \circ g)}{T(r, f)} = 0.$$

Finally we state formally the result referred to in Section 1 which is related to Pólya's theorem.

Theorem 6. *Let $f(z)$ and $g(z)$ be entire with $g(0) = 0$. Let ρ satisfy $0 < \rho < 1$ and let $c(\rho) = (1 - \rho)^2 / 4\rho$. Then for $R \geq 0$,*

$$M(R, f \circ g) \geq M(c(\rho)M(\rho R, g), f).$$

Furthermore, this result is best possible.

3. Proofs of Theorems

Proof of Theorem 6

Let $g(z)$ be entire with $g(0) = 0$ and let ρ satisfy $0 < \rho < 1$. For $R > 0$ fixed, the complement of the set $\{\omega: \omega = g(z), |z| = R\}$ in the ω-plane has a finite number of components each of which is a domain. Let D_R be the unbounded component and let Γ_R be its boundary. It is easy to see that Γ_R is a Jordan curve and we denote by Δ_R the domain other than D_R having Γ_R as its boundary.

For all small $\eta > 0$ the circle $|\omega| = \eta$ lies in Δ_R, since $g(0) = 0$. Let $K > 0$ be the largest number such that $|\omega| = K$ lies in $\Delta_R \cup \Gamma_R$. From the maximum modulus principle,

$$M(K, f) \leq \max_{\omega \in \Gamma_R} |f(\omega)| \leq M(R, f \circ g)$$

for any entire function $f(z)$. Hence the first part of Theorem 6 follows if $c(\rho)M(\rho R, g) \leq K$, and this we now proceed to prove.

In the following discussion we shall use subordination and certain properties of univalent functions without individual references. All the results used will be found in [4]. In the above notation there is a Jordan path in the ω-plane joining a point on $|\omega| = K$ to $\omega = \infty$ which lies apart from its end points in D_R. We now change the variables to $W = \omega/K$ and $\zeta = z/R$ and let γ be the path in the W-plane corresponding to the preceding path in the ω-plane. Let $h(\zeta)$ be analytic and univalent in $|\zeta| < 1$ and let $W = h(\zeta)$ map $|\zeta| < 1$ onto the complement of γ and satisfy $h(0) = 0$. Then $g(\zeta R)/K$ is subordinate to $h(\zeta)$ in $|\zeta| < 1$ and so

$$(3.1) \qquad\qquad \max_{|\zeta| = \rho} \frac{|g(\zeta R)|}{K} \leq \max_{|\zeta| = \rho} |h(\zeta)|.$$

If $W = k(\zeta) = \zeta + \cdots$ is analytic and univalent in $|\zeta| < 1$, then the image of $|\zeta| < 1$ by $k(\zeta)$ covers the disk $|W| < \frac{1}{4}$ and $\max_{|\zeta| = \rho} |k(\zeta)| \leq \rho/(1 - \rho)^2$. Setting $k(\zeta) = h(\zeta)/h'(0)$ we see that if $W = h(\zeta)$ then the image of $|\zeta| < 1$ by $h(\zeta)$ covers the disk $|W| < |h'(0)|/4$. This image omits a point on $|W| = 1$ and so $|h'(0)| \leq 4$. Consequently,

$$\max_{|\zeta| = \rho} |h(\zeta)| \leq \frac{|h'(0)| \rho}{(1 - \rho)^2} \leq \frac{4\rho}{(1 - \rho)^2}.$$

From (3.1) and the above it follows that

$$\frac{M(\rho R, g)}{K} \leq \frac{4\rho}{(1 - \rho)^2} = \frac{1}{c(\rho)},$$

and so $c(\rho)M(\rho R, g) \leq K$. This is the inequality we wished to prove.

In proving Theorem 6 best possible for a given $R > 0$ and a given ρ $(0 < \rho < 1)$ it is sufficient to deal with the case $R = 1$. If in fact we have shown that given ρ $(0 < \rho < 1)$ and $K > c(\rho)$ there are entire functions $f(z)$, $g(z)$ such that

(3.2) $$M(1, f \circ g) < M(KM(\rho, g), f),$$

then for $R > 0$ we can take $G(z) = g(z/R)$ and obtain

$$M(R, f \circ G) < M(KM(\rho R, G), f).$$

Suppose that ρ $(0 < \rho < 1)$ and $K > c(\rho)$ are given. Choose t so that $0 < t < 1$ and

(3.3) $$c(\rho) \cdot \frac{\rho}{(1 - \rho)^2} \cdot \frac{(1 - \rho t)^2}{\rho t} < K.$$

If $h(z) = zt/(1 - zt)^2$, then $\omega = h(z)$ maps $|z| < 1$ onto a domain in the ω-plane which is disjoint from the strip $|\operatorname{Im} \omega| \leqq \varepsilon$, $\operatorname{Re} \omega \leqq -\frac{1}{4} + \varepsilon$, provided $\varepsilon > 0$ is small enough. There is an entire function $f(z)$ such that $|f(z)| \leqq 1$ if z lies outside the strip $|\operatorname{Im} z| \leqq \varepsilon$, $\operatorname{Re} z \leqq 0$ and $M(\frac{1}{4}, f) > 1$. Such a function can be constructed from $E_0(z)$ given in Lemma 4.1 of M.F. If ε is such that $(\frac{1}{4} - \varepsilon)^2 + \varepsilon^2 < (\frac{1}{4})^2$ then one sees that

$$M(1, f \circ h) < M(\tfrac{1}{4}, f).$$

Now, from (3.3),

$$\frac{1}{4} = c(\rho) \cdot \frac{\rho}{(1 - \rho)^2} \cdot \frac{(1 - \rho t)^2}{\rho t} \cdot M(\rho, h) < KM(\rho, h),$$

and so

$$M(1, f \circ h) < M(KM(\rho, h), f).$$

The Maclaurin series for $f(z)$ converges in $|z| < 1/t$ and as $t < 1$ we can take a partial sum $g(z)$ of this series of sufficiently high order and so obtain (3.2). To obtain a transcendental $g(z)$ in (3.2) one adds to the polynomial $g(z)$ just constructed infinitely many terms with nonzero but sufficiently small coefficients.

Proof of Theorem 1

In order to prove (i) we first of all note that if $f(z)$ is entire and transcendental then for all large r, depending on $f(z)$, $[\log M(r, f)]/\log r$ is an increasing function of r; and furthermore $[\log M(r, f)]/\log r \to \infty$ $(r \to \infty)$. The latter result is well known and the former is an immediate consequence of the fact that $\log M(r, f)$ is a convex function of $\log r$.

From the proof of Theorem 6 we see that if r is large enough then

$$\log M(r, f \circ g) \geqq M(\tfrac{1}{8}M(r/2, g) - |g(0)|, f)$$

and hence, from the preceding observations, for large r

$$\frac{\log M(r, f \circ g)}{\log M(r, f)} \geqq \frac{\log M(r/2, g)}{\log r} + o(1)$$

$$\to \infty \qquad (r \to \infty),$$

since $g(z)$ is transcendental.

In proving (ii) we use the following well-known inequalities [3, p. 18]: if $f(z)$ is entire then

$$T(r, f) \leq \log M(r, f) \leq \frac{R + r}{R - r} T(R, f) \qquad (R > r \geqq 0).$$

For large r it follows that

$$T(r, f \circ g) \geqq \tfrac{1}{3} \log M(r/2, f \circ g),$$

and

$$T(r, f) \leqq \log M(r, f).$$

From the proof of (i) it follows that for large r,

$$\frac{\log M(r/2, f \circ g)}{\log M(r, f)} \geqq \frac{\log M(r/4, g)}{\log r} + o(1)$$

$$\to \infty \qquad (r \to \infty),$$

and so, together with the preceding inequalities, we obtain the desired result.

To prove (iii) we recall the result quoted in (1.3), viz., if $f(z)$ and $g(z)$ are entire then

$$M(r, f \circ g) \geqq M((1 + o(1))M(r, g), f)$$

for arbitrarily large r. As $f(z)$ is transcendental, given any $K > 0$ the above inequality gives

$$M(r, f \circ g) \geqq (1 + o(1))^K (M(r, g))^K$$

and so

$$\frac{\log M(r, f \circ g)}{\log M(r, g)} \geqq K + o(1).$$

Therefore

$$\limsup_{r \to \infty} \frac{\log M(r, f \circ g)}{\log M(r, g)} \geq K,$$

and as K can be chosen arbitrarily large the result follows.

Proof of Theorem 2

In the following discussion many of the standard results of Nevanlinna theory will be used without particular references. These will be found in the first two chapters of M.F. As well as the more usual results of Nevanlinna theory we require the following [3, p. 9]: if $f(z)$ is meromorphic then for all complex ω outside a set of zero capacity, depending on $f(z)$,

$$N\left(r, \frac{1}{f - \omega}\right) \sim T(r, f) \qquad (r \to \infty).$$

In proving (i) we first of all choose α so that $f(z) - \alpha$ has infinitely many zeros $\zeta_1, \zeta_2, \ldots$ and

$$N\left(r, \frac{1}{f \circ g - \alpha}\right) \sim T(r, f \circ g) \qquad (r \to \infty),$$

$$N\left(r, \frac{1}{g - \zeta_n}\right) \sim T(r, g) \qquad (r \to \infty, n = 1, 2, \ldots).$$

That α can be so chosen is an immediate consequence of the preceding result. Given any integer $N > 0$ one sees that

$$N\left(r, \frac{1}{f \circ g - \alpha}\right) \geq \sum_{n=1}^{N} N\left(r, \frac{1}{g - \zeta_n}\right)$$

and so

$$\liminf_{r \to \infty} \frac{T(r, f \circ g)}{T(r, g)} \geq N.$$

As N can be chosen arbitrarily large the result follows.

The proof of (ii) requires the following result.

Lemma 1. *Let $g(z)$ be entire and transcendental. Given $K > 0$ there is a number $R_0 > 0$ and an increasing sequence $(R_n)_1^{\infty}$ with $R_1 > R_0$ and $R_n \to \infty$ $(n \to \infty)$ such that for $n \geq 1$ and all r in $R_n \leq r \leq R_n^2$ and all ω satisfying $R_0 \leq |\omega| \leq r$ we have*

$$n\left(r, \frac{1}{g - \omega}\right) > K.$$

Proof. Suppose at first that there is an $R_0 > 0$ such that for all $r \geq R_0$ and all ω satisfying $|\omega| = r$ we have

$$n\left(r, \frac{1}{g - \omega}\right) > K.$$

In this case, if $R \geq R_0$ and $|\omega| = r$, where $R_0 \leq r \leq R$, then

$$n\left(R, \frac{1}{g - \omega}\right) \geq n\left(r, \frac{1}{g - \omega}\right) > K.$$

This is clearly a stronger result than that given in the lemma.

Suppose now that the above is false, i.e., for arbitrarily large r there exists ω satisfying $|\omega| = r$ such that

$$n\left(r, \frac{1}{g - \omega}\right) \leq K.$$

First of all we choose β so that $|\beta| > |g(0)|$ and

$$N\left(r, \frac{1}{g - \beta}\right) \sim T(r, g) \qquad (r \to \infty).$$

Such a choice of β is possible from the result quoted earlier. Now take $R_0 = |g(0)| + |\beta| + 1$. Let $(R_n)_1^\infty$ be an increasing sequence with $R_1 > R_0$ and $R_n \to \infty$ $(n \to \infty)$ such that for $n \geq 1$ there is a ω satisfying $|\omega| = R_n^4$ and

(3.4)
$$n\left(R_n^4, \frac{1}{g - \omega}\right) \leq K.$$

Asymptotic relations in what follows will refer to $n \to \infty$, and hence $R_n \to \infty$.

Assume that for all large n the statement of Lemma 1 does not hold where R_0 and $(R_n)_1^\infty$ are defined above. Then for all large n there is an Ω, depending on n, such that $R_0 \leq |\Omega| \leq R_n^2$ and

(3.5)
$$n\left(\rho, \frac{1}{g - \Omega}\right) \leq K \qquad (\rho \leq R_n).$$

Choose ρ to satisfy $R_n/2 \leq \rho \leq R_n$ such that

(3.6)
$$\begin{cases} m\left(\rho, \dfrac{g'}{g - \omega}\right) = o(T(\rho, g)), \\[2mm] m\left(\rho, \dfrac{g'}{g - \Omega}\right) = o(T(\rho, g)), \end{cases}$$

where ω with $|\omega| = R_n^4$ satisfies (3.4). The only difference between (3.6) and similar results in M.F. is that in (3.6), ω and Ω depend on ρ. However, since $g(z)$ is transcendental and Ω, ω do not exceed $(2\rho)^4$ in modulus the results in M.F. remain valid. The poles of $g'(z)/[g(z) - \omega]$ and $g'(z)/[g(z) - \Omega]$ are the zeros of $g(z) - \omega$ and $g(z) - \Omega$ and as these satisfy (3.4) and (3.5) it follows that

(3.7)
$$\begin{cases} T\left(\rho, \dfrac{g'}{g - \omega}\right) = o(T(\rho, g)), \\[3mm] T\left(\rho, \dfrac{g'}{g - \Omega}\right) = o(T(\rho, g)). \end{cases}$$

Choose $\alpha = -\left(\dfrac{\beta - \omega}{\beta - \Omega}\right)$ and consider

$$h(z) = \frac{g'(z)}{g(z) - \omega} + \alpha \frac{g'(z)}{g(z) - \Omega} = \frac{(\omega - \Omega)g'(z)(g(z) - \beta)}{(g(z) - \omega)(g(z) - \Omega)}.$$

The zeros of $h(z)$ include those of $g(z) - \beta$, there being no cancellation, and so

$$N\left(\rho, \frac{1}{h}\right) \geq N\left(\rho, \frac{1}{g - \beta}\right) = (1 + o(1))T(\rho, g),$$

from the choice of β. The only thing that would upset the conclusion that

(3.8) $$T(\rho, h) \geq (1 + o(1))T(\rho, g)$$

is that $(\omega - \Omega)(g(0) - \beta)/(g(0) - \omega)(g(0) - \Omega)$ is very small, much smaller than ρ^{-2} since $g(z)$ is transcendental. But from the above discussion it is clear that this is not the case and so (3.8) is satisfied. The really important aspect of what we have done is that $\omega - \Omega$ has been prevented from becoming very small. On the other hand, from (3.7) and the value of α it follows that

(3.9) $$T(\rho, h) = o(T(\rho, g)).$$

If n, and hence ρ, is large enough then (3.8) and (3.9) are mutually incompatible and so the assumption from which this contradiction followed is false. Hence one sees that there is a sequence $(R_n)_1^\infty$ with the properties given in the statement of Lemma 1.

In proving (ii) one can assume that

$$N\left(r, \frac{1}{f}\right) \sim T(r, f) \qquad (r \to \infty),$$

since otherwise one can consider $f(z) - \gamma$, with a suitable γ, in place of $f(z)$. Let $K > 0$ be given and let R_0, $(R_n)_1^\infty$ be chosen as in Lemma 1. From Lemma 1 it follows that if $n \geq 1$ then to each zero ζ of $f(z)$ satisfying $R_0 \leq |\zeta| \leq r$

with $R_n \leqq r \leqq R_n^2$ corresponds more than K zeros of $f \circ g(z)$ in $|z| \leqq r$. Therefore, in the above notation,

$$n\left(r, \frac{1}{f \circ g}\right) \geqq K\{n(r, 1/f) - n(R_0, 1/f)\},$$

and so

$$N\left(R_n^2, \frac{1}{f \circ g}\right) \geqq K \int_{R_n}^{R_n^2} \frac{n(r, 1/f)}{r} \, dr + O(\log R_n),$$

where O depends on K. Consequently,

(3.10) $T(R_n^2, f \circ g) \geqq K\{N(R_n^2, 1/f) - N(R_n, 1/f)\} + O(\log R_n).$

Now

(3.11) $\begin{cases} N(R_n^2, 1/f) = (1 + o(1))T(R_n^2, f), \\ N(R_n, 1/f) = (1 + o(1))T(R_n, f). \end{cases}$

Also, since $f(z)$ is transcendental and $T(r, f)$ is convex in $\log r$, we have for large n that

$$\frac{T(R_n^2, f)}{\log R_n^2} \geqq \frac{T(R_n, f)}{\log R_n} \quad \text{or} \quad T(R_n^2, f) \geqq 2T(R_n, f),$$

and

$$\log R_n = o(T(R_n^2, f)).$$

Hence, from the above and (3.10) and (3.11), we see that

$$\liminf_{n \to \infty} \frac{T(R_n^2, f \circ g)}{T(R_n^2, f)} \geqq \frac{K}{2}.$$

As K can be chosen arbitrarily large it follows that

$$\limsup_{r \to \infty} \frac{T(r, f \circ g)}{T(r, f)} = \infty.$$

Proof of Theorem 3

In the proof of (i) we first of all assume that $g(z)$ is of finite order. Suppose that given $K > 0$ there is an $R_0 > 0$ such that for $r \geqq R_0$ and all ω satisfying $|\omega| = r$ we have

$$n\left(r, \frac{1}{g - \omega}\right) > K.$$

Then, as in the proof of Theorem 2, for any $r \geq R_0$ and any ω satisfying $R_0 \leq |\omega| \leq r$ we have

$$n\left(r, \frac{1}{g - \omega}\right) > K.$$

In this case the desired result follows as in the proof of the second part of Theorem 2.

Suppose now that for arbitrarily large R there exists ω satisfying $|\omega| = R$ such that

(3.12) $$n\left(R, \frac{1}{g - \omega}\right) \leq K.$$

In what follows asymptotic relations refer to R as above tending to ∞. There is a ρ in $R/2 \leq \rho \leq R$ such that

(3.13) $$m\left(\rho, \frac{g'}{g - \omega}\right) = O(\log \rho),$$

where ω is the value in (3.12). In the proof of Lemma 1 a similar result was used with $O(\log \rho)$ replaced by $o(T(\rho, g))$. The estimate of Nevanlinna theory in fact is $O(\log \rho\, T(\rho, g))$ and the weaker result involving $o(T(\rho, g))$ was sufficient in Lemma 1. If $g(z)$ is of finite order, then $O(\log \rho\, T(\rho, g))$ is $O(\log \rho)$ and this is what we require here. From (3.12) and (3.13) it follows that

$$T\left(\rho, \frac{g'}{g - \omega}\right) = O(\log \rho).$$

Now, from the above,

$$N\left(\rho, \frac{1}{g'}\right) \leq T\left(\rho, \frac{g'}{g - \omega}\right) + O(\log \rho) = O(\log \rho),$$

and as this holds for a sequence $\rho \to \infty$ it follows that $g'(z)$ has only finitely many zeros. Since $g(z)$, and hence $g'(z)$, is of finite order it follows that $g'(z) = p(z)e^{q(z)}$ for polynomials $p(z)$ and $q(z)$. If $q(z) = az^n + $ lower terms, where $n \geq 1$ and $a \neq 0$ since $g(z)$ is transcendental, there is an interval I in $[0, 2\pi]$ such that if $z = re^{i\theta}$ and $\theta \in I$ then $|g'(z)| \leq Ar^m e^{-|a|r^n/2}$ for large r. If $z = re^{i\theta}$ with $\theta \in I$, then

$$g(z) = \int_0^z g'(\zeta)\, d\zeta + g(0) = O(1) \qquad (r \to \infty),$$

uniformly for $\theta \in I$. If ρ is as above and $z = \rho e^{i\theta}$ with $\theta \in I$ then for all large ρ,

$$\left|\frac{g(z) - \omega}{g'(z)}\right| \geq \frac{\rho^{1-m}}{2A} e^{|a|\rho^n/2}$$

and so, if the length of I is δ,

$$(3.14) \qquad m\left(\rho, \frac{g-\omega}{g'}\right) \geq \frac{|a|\,\delta}{4\pi}\rho^n(1+o(1)).$$

But if

$$T\left(\rho, \frac{g'}{g-\omega}\right) = O(\log \rho),$$

then (3.14) cannot also be true provided ρ is sufficiently large. Consequently for all large r and all ω satisfying $|\omega| = r$ we have

$$n\left(r, \frac{1}{g-\omega}\right) > K$$

and so the result follows, as indicated earlier.

We now prove (i) of Theorem 3 when $f(z)$ is of finite order. Given α $(0 < \alpha < \infty)$ consider those R such that $T(R, g) \leq R^\alpha$. We assume such R form an unbounded set S. Then given $K > 0$ there is an R_0 such that if ω satisfies $R_0 \leq |\omega| \leq R$ and $R \in S$, then

$$(3.15) \qquad n\left(R, \frac{1}{g-\omega}\right) > K.$$

In this case we see that

$$(3.16) \qquad \frac{T(r, f \circ g)}{T(r, f)} \to \infty \qquad (R \to \infty, R \in S).$$

If in fact the above result were false then taking $R_0 > |g(0)|$ there would be a sequence $R \to \infty$, where $R \in S$, and corresponding values ω satisfying $R_0 \leq |\omega| \leq R$ such that

$$n\left(R, \frac{1}{g-\omega}\right) \leq K.$$

Then for such R and suitable ρ satisfying $R/2 \leq \rho \leq R$ we could argue as in the earlier part of the proof and obtain that $g'(z) = p(z)e^{k(z)}$, where $p(z)$ is a polynomial and $k(z)$ is an entire function. Since we have dealt with the case of $g(z)$ of finite order we need only consider the case when $k(z)$ is transcendental. However, in this case it is known that $g'(z)$ is of infinite lower order. By considering

$$g'(z) = \frac{g'(z)}{g(z)-\omega}(g(z)-\omega),$$

with $|z| = \rho$ and ρ and ω as above, we find that

$$T(\rho, g') \leq T(\rho, g) + O(\log \rho).$$

Hence given any number $H > 0$, for large ρ we have $T(\rho, g) > \rho^H$ and so

$$T(R, g) \geq (\rho/R)^H \cdot R^H \geq 2^{-H} R^H.$$

But if $H > \alpha$ and R is large enough we find that $T(R, g) > R^\alpha$, and this is impossible. Consequently we have that (3.16) is valid.

Since $f(z)$ is of finite order we can choose α so that for all large r, $T(r, f) \leq r^\alpha$. Now we have shown in Theorem 2 that

$$\frac{T(r, f \circ g)}{T(r, g)} \to \infty \qquad (r \to \infty),$$

and considering

$$\frac{T(r, f \circ g)}{T(r, f)} = \frac{T(r, f \circ g)}{T(r, g)} \cdot \frac{T(r, g)}{T(r, f)}$$

we see that $T(r, f \circ g)/T(r, f) \to \infty$ for those r such that $T(r, g)/T(r, f) \geq 1$. On the other hand $T(r, f \circ g)/T(r, f) \to \infty$ for those r such $T(r, g)/T(r, f) \leq 1$, by (3.16). Thus the result is proved.

In proving (ii) of Theorem 3 we require the following result. Let $g(z)$ be an entire function of finite order ρ and let $K > \rho - 1$. Then

$$\log M(r, g) \sim \log M(r - r^{-K}, g) \qquad (r \to \infty).$$

This is a simple consequence of standard results for entire functions which can be found in [8].

Let $f(z)$ and $g(z)$ be entire and transcendental with $g(z)$ of finite order ρ. Let $K > \rho - 1$ and assume, for convenience, that $g(0) = 0$. The case $g(0) \neq 0$ can be dealt with as in the proof of Theorem 1. For $r > 1$ take ρ of Theorem 6 to be $(r - r^{-K})/r$ so that $c(\rho) > r^{-2(1+K)}/4$. Then from Theorem 6, for $r > 1$,

$$M(r, f \circ g) \geq M\left(\frac{r^{-2(1+K)}}{4} M(r - r^{-K}, g), f\right)$$

and so given $N > 0$ for all large r, since $f(z)$ is transcendental,

$$M(r, f \circ g) \geq \left\{\frac{r^{-2(1+K)}}{4} M(r - r^{-K}, g)\right\}^N.$$

Hence for such r,

$$\frac{\log M(r, f \circ g)}{\log M(r, g)} \geq N \frac{\log M(r - r^{-K}, g)}{\log M(r, g)} + o(1),$$

since $g(z)$ is transcendental. Using the result stated at the start of the proof we find that

$$\liminf_{r \to \infty} \frac{\log M(r, f \circ g)}{\log M(r, g)} \geqq N,$$

and as N can be arbitrarily large we obtain the desired conclusion.

Proof of Theorem 4

The proof of Theorem 4 requires the following lemma.

Lemma 2. *Given any number $M > 0$ there is a polynomial $p(z)$ such that*

$$\max_{|z|=1} \text{Re } p(z) > M, \quad \min_{|z|=1} \text{Re } p(z) > 0, \quad \max_{|z|=1/2} |p(z)| < 3.$$

Proof. The function $\omega = h(z) = (1 + z)/(1 - z)$ maps $|z| < 1$ onto Re $\omega > 0$. Hence one can choose $\rho < 1$, but sufficiently near to 1, to ensure that

$$\max_{|z|=\rho} \text{Re } h(z) > M, \quad \min_{|z|=\rho} \text{Re } h(z) > 0, \quad \max_{|z|=\rho/2} |h(z)| < 3.$$

Thus the function $h(\rho z)$ has the properties of $p(z)$ listed in Lemma 2. The series for $h(\rho z)$ converges in $|z| < 1/\rho$ and $\rho < 1$ and hence $p(z)$ can be obtained by taking a partial sum of this series of sufficiently high order.

The function $g(z)$ of Theorem 4 will be defined as a sum $-\sum_{n=0}^{\infty} p_n(z2^{-n})/2^n$, where $p_n(z)$ is a polynomial of the kind given in Lemma 2. Take $p_0(z) \equiv 1$ and suppose that $p_1(z), \ldots, p_n(z)$ have been defined. Let

$$M_{n+1} = \max_{|z|=2^{n+1}} \left| \sum_{v=0}^{n} \frac{p_v(z2^{-v})}{2^v} \right|$$

Choose $p_{n+1}(z)$ to be $p(z)$ of Lemma 2 with $M = 2^{n+1} e^{M_{n+1}^2}$.

In the first place $g(z)$ as defined above is entire. This follows at once from the observation that for $|z| \leqq 2^n$ and $N > n$,

$$\left| \sum_{v=N}^{\infty} \frac{p_v(z2^{-v})}{2^v} \right| < 3 \sum_{v=N}^{\infty} \frac{1}{2^v} = \frac{3}{2^{N-1}}.$$

We also have

$$\max_{|z|=2^{n+1}} \text{Re } \frac{p_{n+1}(z2^{-n-1})}{2^{n+1}} > e^{M_{n+1}^2}$$

and so

$$M(2^{n+1}, g) > e^{M_{n+1}^2} - M_{n+1} - 3 \cdot 2^{-n-1}.$$

On the other hand,

$$\max_{|z| = 2^{n+1}} \text{Re } g(z) < M_{n+1} + 3 \cdot 2^{-n-1}.$$

From the above it follows that

$$\frac{\log M(2^{n+1}, \exp \circ g)}{\log M(2^{n+1}, g)} \lesssim \frac{M_{n+1} + 3 \cdot 2^{-n-1}}{\log(e^{M_{n+1}^2} - M_{n+1} - 3 \cdot 2^{-n-1})}$$

$$= \frac{1 + o(1)}{M_{n+1}} \to 0 \qquad (n \to \infty).$$

Hence, for the $g(z)$ constructed above,

$$\liminf_{r \to \infty} \frac{\log M(r, \exp \circ g)}{\log M(r, g)} = 0.$$

Proof of Theorem 5

The result of Hayman referred to in Section 1 is the following [6]:

Let $(R_n)_1^\infty$ be an increasing sequence of positive numbers such that $R_n \to \infty$ $(n \to \infty)$, and let $(\omega_n)_1^\infty$ be any sequence of complex numbers such that

$$|\omega_n| > K_n \qquad (n \geq 2),$$

where K_n depends on $\omega_1, \ldots, \omega_{n-1}$ and $R_1, \ldots, R_{n-1}$ only. Then there is a transcendental entire function $g(z)$ such that $g(z) \neq \omega_n$ for $|z| \leq R_n$.

Note, as Hayman points out, R_n can be chosen after ω_n has been specified.

We also require the following result whose proof is obvious and will be omitted. Given any sequence $(\zeta_n)_1^\infty$ of complex numbers such that $\zeta_n \to \infty$ $(n \to \infty)$ and any sequence $(k_n)_1^\infty$ of integers if $(\phi_n)_1^\infty$ is a sequence of sufficiently small real numbers then

$$\prod_{n=1}^{\infty} \left\{ \frac{1 - z/\zeta_n}{1 - ze^{i\phi_n}/\zeta_n} \right\}^{k_n}$$

defines a meromorphic function $f(z)$.

Suppose that $g(z)$ is an entire function according to Hayman's result with $R_n \geq 2|\omega_n|$. We shall define a function $f(z)$ as above with $\zeta_n = \omega_n$. Let $k_1 = 1$, $\phi_1 = 0$ and suppose that $k_1, \ldots, k_n$ and $\phi_1, \ldots, \phi_n$ have been specified. Let

$$p_n(z) = \prod_{v=1}^{n} \left\{ \frac{1 - z/\omega_v}{1 - ze^{i\phi_v}/\omega_v} \right\}^{k_v}.$$

In addition to any other requirements the ϕ_n will always be assumed to be chosen so that $f(z)$ is a meromorphic function. Let

$$T(R_{n+1}, p_n \circ g) = M_{n+1}.$$

Choose $k_{n+1} > M_{n+1}^2$ and choose ϕ_{n+1} so that $g(z) \neq e^{-i\phi_{n+1}}\omega_{n+1}$ $(|z| \leq R_{n+1})$ and,

$$(3.17) \qquad 1 - \frac{1}{2^n} < \left| \frac{1 - g(z)/\omega_{n+1}}{1 - g(z)e^{i\phi_{n+1}}/\omega_{n+1}} \right|^{k_{n+1}} < 1 + \frac{1}{2^n} \qquad (|z| = R_{n+1}).$$

As

$$\left(\frac{1 - g(z)/\omega_{n+1}}{1 - g(z)e^{i\phi_{n+1}}/\omega_{n+1}} \right)^{k_{n+1}}$$

is regular in $|z| \leq R_{n+1}$ the above inequalities are also satisfied for $|z| < R_{n+1}$.

Consider now $T(R_{n+1}, f \circ g)$. Then the poles of $f \circ g(z)$ in $|z| \leq R_{n+1}$ are the same as those of $p_n \circ g(z)$ and, from (3.17) and the observation that follows, the other factors in $f \circ g(z)$ apart from those in $p_n \circ g(z)$ do not significantly alter $m(R_{n+1}, f \circ g)$ from the value of $m(R_{n+1}, p_n \circ g)$. Hence

$$T(R_{n+1}, f \circ g) = M_{n+1}(1 + o(1)).$$

But $f(z)$ has a pole of order k_{n+1} on $|z| = \omega_{n+1}$ and so

$$T(R_{n+1}, f) \geq N(R_{n+1}, f) \geq \int_{|\omega_{n+1}|}^{R_{n+1}} \frac{k_{n+1}}{t} \, dt$$

$$\geq k_{n+1} \log 2$$

$$\geq M_{n+1}^2 \log 2.$$

Therefore

$$\frac{T(R_{n+1}, f \circ g)}{T(R_{n+1}, f)} \leq \frac{1 + o(1)}{M_{n+1} \log 2} \to 0 \qquad (n \to \infty),$$

and so

$$\liminf_{r \to \infty} \frac{T(r, f \circ g)}{T(r, f)} = 0.$$

In conclusion I should like to thank Fred Gross for stimulating my interest in the topics of this paper.

REFERENCES

1. J. Clunie, The maximum modulus of an integral function of an integral function *Quart. J. Math. Oxford Ser.* 2 **6** (1955), 176–178.
2. A. Edrei and W. H. J. Fuchs, On the zeros of $f(g(z))$ where f and g are entire functions . *J. Analyse Math.* **12** (1964), 243–255.
3. W. K. Hayman, *Meromorphic functions*, Clarendon Press, Oxford, 1964.
4. ———, *Multivalent functions*, Cambridge University Press, Cambridge, England, 1958
5. ———, Some applications of the transfinite diameter to the theory of functions, *J Analyse Math.* **1** (1951), 155–179.
6. ———, Some integral functions of infinite order, *Math. Notae* **XX** (1965), 1–5.
7. G. Pólya, On an integral function of an integral function, *J. London Math. Soc.* **1** (1926) 12–15.
8. G. Valiron, *Lectures on the general theory of integral functions*, Imprimerie et Librairie Édouard Privat, Librairie de l'University, Toulouse, 1923; Chelsea, New York, 1949.

(Received March 25, 1968)

Interpolation by Entire Functions of Exponential Type

Richard F. DeMar

Department of Mathematics
University of Cincinnati
Cincinnati, Ohio 45221
U.S.A.

Let K denote the class of entire functions of exponential type, i.e., entire functions satisfying, for some A and τ,

$$|f(z)| \leqq Ae^{\tau|z|}$$

for all z. This is precisely the class of entire functions f having a Laplace transform analytic and zero at infinity. We shall denote the Laplace transform of f by F throughout. If Ω is a simply connected domain, let $K[\Omega]$ denote the class of all f in K whose Laplace transforms are analytic in Ω', the complement of Ω. If $\{g_n\}$ is a sequence of functions each analytic on Ω, then

$$(1) \qquad L_n(f) = \frac{1}{2\pi i} \int_\Gamma g_n(\zeta)F(\zeta)\,d\zeta$$

where $\Gamma \subset \Omega$ is a simple closed curve such that F is analytic outside and on Γ, defines a sequence of linear functionals on the class $K[\Omega]$. Given such a sequence of functionals and a simply connected domain $\Omega_1 \subseteq \Omega$, the interpolation problem is to determine conditions on a sequence $\{b_n\}$ of complex numbers in order that there exists a function f in $K[\Omega_1]$ such that $L_n(f) = b_n$; $n = 0, 1, \ldots$. Such a function f is said to *interpolate* the sequence $\{b_n\}$ relative to $\{L_n\}$.

93

In two earlier papers [1, 2] the author gave a solution to the interpolation problem for the case $g_n(\zeta) = [W(\zeta)]^n$ where W is univalent on Ω. In a recent paper [4], he gave a condition for $K[\Omega_1]$ to be a uniqueness class (i.e., $f, g \in K[\Omega_1]$ and $L_n(f) = L_n(g)$; $n = 0, 1, 2, \ldots$ implies $f = g$) for the more general case $g_{pn+k}(\zeta) = h_k(\zeta)[W(\zeta)]^{pn}$; $k = 0, 1, \ldots, p - 1$; $n = 0, 1, 2, \ldots$. In this paper, we obtain the generalization of the results on the interpolation problem to this more general class of sequences of linear functionals. Throughout the paper, p will denote a given positive integer and α a primitive pth root of 1. A set X will be said to be p-symmetric (with respect to the origin) if $\alpha\zeta \in X$ for all ζ in X. This condition implies that $\alpha^q\zeta \in X$ for each ζ in X; $q = 1, 2, \ldots, p - 1$.

We start out by proving a lemma which will enable us to treat the general case by dealing only with a special case.

Lemma 1. *Let $\{L_n\}$ be a sequence of linear functionals defined on $K[\Omega]$ by (1). Let W be analytic and univalent on Ω_ζ where $\Omega = W(\Omega_\zeta)$. Let $\{L_n{}^*\}$ be defined on $K[\Omega_\zeta]$ by*

$$L_n{}^*(f) = \frac{1}{2\pi i} \int_{\Gamma_1} g_n(W(\zeta))F(\zeta)\, d\zeta$$

where $\Gamma_1 \subset \Omega_\zeta$ is a simple closed curve with F analytic outside and on Γ_1 and let $\{b_n\}$ be a given sequence of complex numbers. Then there exists a function f in $K[\Omega]$ such that $L_n(f) = b_n$ if and only if there exists a function g in $K[\Omega_\zeta]$ such that $L_n{}^*(g) = b_n$; $n = 0, 1, 2, \ldots$.

Proof. Let $f \in K[\Omega]$ and suppose $L_n(f) = b_n$; $n = 0, 1, 2, \ldots$. Let $Z(w)$ be the inverse of $W(\zeta)$ which maps Ω one-to-one onto Ω_ζ. Then we have

$$b_n = \frac{1}{2\pi i} \int_\Gamma g_n(w)F(w)\, dw$$

$$= \frac{1}{2\pi i} \int_{\Gamma_\zeta} g_n(W(\zeta))F(W(\zeta))W'(\zeta)\, d\zeta,$$

where $\Gamma_\zeta = Z(\Gamma)$ is a simple closed curve contained in Ω_ζ. Now W and W' are analytic inside and on Γ_ζ and F is analytic outside and on Γ, so that $F(W(\zeta))W'(\zeta)$ is analytic on Γ_ζ. Therefore, we can write

$$F(W(\zeta))W'(\zeta) = F_1(\zeta) + F_2(\zeta),$$

where F_2 is analytic inside and on Γ_ζ and F_1 is analytic outside and on Γ_ζ with $F_1(\infty) = 0$. Then we have

$$b_n = \frac{1}{2\pi i} \int_{\Gamma_\zeta} g_n(W(\zeta))[F_1(\zeta) + F_2(\zeta)]\, d\zeta$$

$$= \frac{1}{2\pi i} \int_{\Gamma_\zeta} g_n(W(\zeta))F_1(\zeta)\, d\zeta,$$

since $g_n(W(\zeta))F_2(\zeta)$ is analytic inside and on Γ_ζ. Now since F_1 is analytic outside and on Γ_ζ and zero at infinity, it is the Laplace transform of a function f_1 in $K[\Omega_\zeta]$. Thus, we have $L_n*(f_1) = b_n$; $n = 0, 1, 2, \dots$.

The same proof with the roles of $W(\zeta)$ and $Z(w)$ interchanged proves the converse.

Because of this lemma, we need deal only with the case $g_{pn+k}(\zeta) = h_k(\zeta)\zeta^{pn}$ instead of the general case $g_{pn+k}(\zeta) = h_k(\zeta)[W(\zeta)]^{pn}$ with W univalent.

We shall need the following lemma in the proof of the main theorem.

Lemma 2. *Let Ω be a simply connected domain and suppose that for each t on a rectifiable curve γ, $f(z, t) \in K[\Omega]$ as a function of z and that for each z, f is analytic on γ as a function of t. Let L_n be defined on $K[\Omega]$ by (1). Let $u(z) = \int_\gamma f(z, t)\, dt$. Then*

(A) $u \in K[\Omega]$,
(B) $L_n(u) = \int_\gamma L_n(f)(t)\, dt.$

Proof. (A) Since differentiation under the integral is justified, u is entire. For each t on γ, let $\tau(t)$ be the type of $f(z, t)$; i.e., $\tau(t) = \mathrm{glb}\{\sigma\,|\,\text{for some } A,$ $|f(z, t)| \le Ae^{\sigma|z|}$ for all $z\}$. Then τ is continuous on the compact set γ, so it has a maximum τ_0. Let $\tau_1 > \tau_0$. Then for each t on γ, there exists $A(t)$ such that $|f(z, t)| \le A(t)e^{\tau_1|z|}$ for all z. If we take $A(t)$ to be the minimum such number, then it is a continuous function; so it has a maximum on γ. Thus, $u \in K$ and u is of type at most τ_1. Therefore, for Re $\zeta \ge \tau_1$ and U and F, the Laplace transforms of u and f, respectively,

$$U(\zeta) = \int_0^\infty e^{-s\zeta}u(s)\, ds$$

$$= \int_0^\infty e^{-s\zeta} \int_\gamma f(s, t)\, dt\, ds$$

$$= \int_\gamma \int_0^\infty e^{-s\zeta}f(s, t)\, ds\, dt$$

$$= \int_\gamma F(\zeta, t)\, dt,$$

where the interchange of the order of integration is justified by the existence of

$$\int_0^\infty \int_\gamma |e^{-s\zeta}|\,|F(s, t)|\,|dt|\,ds.$$

But since for each t on γ, $F(\zeta, t)$ is analytic on Ω', we conclude U is analytic on Ω'; i.e., $u \in K[\Omega]$.

Proof of (B):

$$\int_\gamma L_n(f)(t)\,dt = \int_\gamma \frac{1}{2\pi i} \int_\Gamma g_n(\zeta)F(\zeta, t)\,d\zeta\,dt$$

$$= \frac{1}{2\pi i} \int_\Gamma g_n(\zeta) \int_\gamma F(\zeta, t)\,dt\,d\zeta$$

$$= \frac{1}{2\pi i} \int_\Gamma g_n(\zeta)U(\zeta)\,d\zeta$$

$$= L_n(u)$$

which completes the proof.

For the theorem, we shall need a function which we shall denote by $\Delta(\zeta)$ defined as follows: Let $h_{jk}(\zeta) = h_k(\alpha^j \zeta)$; $j, k = 0, 1, \ldots, p - 1$. Then $\Delta(\zeta)$ is the determinant of the matrix $(h_{jk}(\zeta))$.

Theorem. *Let Ω be a simply connected p-symmetric domain. Let h_k be analytic on Ω; $k = 0, 1, \ldots, p - 1$, and have the property that $\Delta(\zeta) \not\equiv 0$. Let $\{L_n\}$ be the sequence of linear functionals defined on $K[\Omega]$ by (1) with $g_{pn+k}(\zeta) = \zeta^{pn} h_k(\zeta)$; i.e.,*

$$(2) \quad L_{pn+k}(f) = \frac{1}{2\pi i} \int_\Gamma \zeta^{pn} h_k(\zeta)F(\zeta)\,d\zeta; \ k = 0, 1, \ldots, p - 1; n = 0, 1, \ldots.$$

Given a sequence $\{b_n\}$ of complex numbers such that $b(z) = \sum_{n=0}^\infty b_n z^n$ is analytic at the origin, there exists a function f in $K[\Omega]$ such that $L_n(f) = b_n$; $n = 0, 1, 2, \ldots$ if and only if the function b can be continued analytically to the set $(\Omega^{-1})'$, the complement of $\Omega^{-1} = \{t \mid t^{-1} \in \Omega\}$.

Proof. Suppose there exists a function f in $K[\Omega]$ such that $L_n(f) = b_n$; $n = 0, 1, 2, \ldots$. We show that the function $b(z) = \sum_{n=0}^\infty b_n z^n$ has the representation

$$(3) \qquad b(z) = \frac{1}{2\pi i} \int_\Gamma \frac{\sum_{k=0}^{p-1} h_k(\zeta)z^k}{1 - \zeta^p z^p}\,F(\zeta)\,d\zeta,$$

where $\Gamma \subset \Omega$ encloses the origin and all singularities of F. (Since Ω is a simply connected p-symmetric domain, it contains the origin). Let $B(z, \zeta) =$

$(\sum_{k=0}^{p-1} h_k(\zeta)z^k)/(1 - \zeta^p z^p)$. Let $\sigma = 1/\max|\zeta|$ for all ζ on Γ. Then for all $|z| < \sigma$ and $\zeta \in \Gamma$, $B(z, \zeta) = \sum_{n=0}^{\infty} \sum_{k=0}^{p-1} h_k(\zeta)\zeta^{pn}z^{pn+k}$, the series being convergent. If we let $0 < \sigma_1 < \sigma$, then for $|z| \leq \sigma_1$, the series converges uniformly for $\zeta \in \Gamma$. For such z, we have

$$\frac{1}{2\pi i} \int_\Gamma \frac{\sum_{k=0}^{p-1} h_k(\zeta)z^k}{1 - \zeta^p z^p} F(\zeta)\, d\zeta$$

$$= \frac{1}{2\pi i} \int_\Gamma \sum_{n=0}^{\infty} \sum_{k=0}^{p-1} h_k(\zeta)\zeta^{pn}z^{pn+k}F(\zeta)\, d\zeta$$

$$= \sum_{n=0}^{\infty} \sum_{k=0}^{p-1} L_{pn+k}(f)z^{pn+k}$$

$$= \sum_{n=0}^{\infty} b_n z^n = b(z).$$

Thus $b(z)$ has representation (3).

From this representation, b is analytic for all z in any domain containing the origin such that $z^{-p} \neq \zeta^p$ for all ζ on Γ. Since Ω is p-symmetric, so is $(\Omega^{-1})'$. Since Γ encloses the origin, so does Γ^{-1} and since $(\Omega^{-1})'$ contains the origin and $\Gamma^{-1} \subset \Omega^{-1}$, this implies Γ^{-1} encloses $(\Omega^{-1})'$. Now $z \in (\Omega^{-1})'$ implies $(\alpha^q z) \in (\Omega^{-1})'$; $q = 0, 1, \ldots, p-1$; so that $\alpha^q z \notin \Gamma^{-1}$. This implies $(\alpha^q z)^{-1} \notin \Gamma$ which implies $z^{-p} \neq \zeta^p$ for all ζ on Γ. Thus $b(z)$ is analytic on $(\Omega^{-1})'$.

To prove the converse, let $\{b_n\}$ be a sequence such that $b(z) = \sum_{n=0}^{\infty} b_n z^n$ is analytic on $(\Omega^{-1})'$. We shall construct a function f in $K[\Omega]$ such that $L_n(f) = b_n$; $n = 0, 1, 2, \ldots$. To do this, we first construct p functions $u_k(z, \zeta)$; $k = 0, 1, \ldots, p-1$, satisfying the following conditions.

(a) For all ζ in Ω except those for which $\Delta(\zeta)$ is zero, as a function of z, $u_k(z, \zeta) \in K[\Omega]$.

(b) $u_k(z, \alpha^q \zeta) = u_k(z, \zeta)$; $q = 1, 2, \ldots, p-1$.

(c) If $\zeta \in \Omega$ with $\Delta(\zeta) \neq 0$, for u_k as a function of z,

$$L_{pn+m}(u_k) = \delta_{mk} \zeta^{pn} \qquad \text{(Kronecker } \delta\text{)}.$$

To construct these functions, we write

$$e^{z\zeta} = h_0(\zeta)u_0(z, \zeta) + \cdots + h_{p-1}(\zeta)u_{p-1}(z, \zeta).$$

Then, if the u_k can be defined to satisfy (b), we shall have

$$\exp(\alpha^q z\zeta) = h_0(\alpha^j \zeta)u_0(z, \zeta) + h_1(\alpha^j \zeta)u_1(z, \zeta) + \cdots + h_{p-1}(\alpha^j \zeta)u_{p-1}(z, \zeta).$$

We use this linear system of equations to define the u_k's. Let $\Lambda_k(z, \zeta)$ be the determinant of the matrix whose jth row is

$$(h_0(\alpha^j \zeta), \ldots, h_{k-1}(\alpha^j \zeta), \exp(\alpha^j z\zeta), h_{k+1}(\alpha^j \zeta), \ldots, h_{p-1}(\alpha^j \zeta)).$$

Define $u_k(z, \zeta) = \Lambda_k(z, \zeta)/\Delta(\zeta)$. Then for each ζ in Ω such that $\Delta(\zeta) \neq 0$, $u_k(z, \zeta)$ is a linear combination of functions $\exp(\alpha^j z \zeta)$. Since Ω is p-symmetric, $\zeta \in \Omega$ implies $\alpha^j \zeta \in \Omega$; $j = 1, 2, \ldots, p-1$; so each u_k satisfies (a). We also have $\Lambda_k(z, \alpha^j \zeta) = (-1)^j \Lambda_k(z, \zeta)$ and $\Delta(\alpha^j \zeta) = (-1)^j \Delta(\zeta)$; so $u_k(z, \alpha^j \zeta) = u_k(z, \zeta)$. To show that (c) is satisfied, we use the fact that for L_n defined by (1) and $f(z) = e^{az}$, $L_n(f) = g_n(a)$; so we have

$$L_{pn+m}(u_k)$$

$$= \frac{\det(h_0(\alpha^j\zeta), \ldots, h_{k-1}(\alpha^j\zeta), L_{pn+m}(\exp(\alpha^j z\zeta)), h_{u+1}(\alpha^j\zeta), \ldots, h_{p-1}(\alpha^j\zeta)}{\det(h_0(\alpha^j\zeta), \ldots, h_{k-1}(\alpha^j\zeta), h_k(\alpha^j\zeta) h_{k+1}(\alpha^j\zeta), \ldots, h_{p-1}(\alpha^j\zeta))},$$

$$= \frac{\det(h_0(\alpha^j\zeta), \ldots, h_{k-1}(\alpha^j\zeta), \zeta^{pn}h_m(\alpha^j\zeta), h_{k+1}(\alpha^j\zeta), \ldots, h_{p-1}(\alpha^j\zeta))}{\det(h_0(\alpha^j\zeta), \ldots, h_{k-1}(\alpha^j\zeta), h_k(\alpha^j\zeta), h_{k+1}(\alpha^j\zeta), \ldots, h_{p-1}(\alpha^j\zeta))}$$

$$= \delta_{mk} \zeta^{pn}.$$

Thus, the functions u_k have properties (a), (b), and (c).

We define f in terms of these functions u_k. Let

$$f(z) = \frac{1}{2\pi i} \int_\gamma b(t) \sum_{k=0}^{p-1} u_k(z, t^{-1}) t^{-k-1} \, dt,$$

where γ is a simple closed curve enclosing $(\Omega^{-1})'$ such that b is analytic inside and on γ and such that $\Delta(t^{-1}) \neq 0$ for all t on γ. This is possible since b is analytic on $(\Omega^{-1})'$ which is compact. Then for each t on γ, $t^{-1} \in \Omega$ and $\Delta(t^{-1}) \neq 0$, so $u_k \in K[\Omega]$. Therefore, by Lemma 2(A), $f \in K[\Omega]$. By Lemma 2(B), we have

$$L_{pn+m}(f) = \frac{1}{2\pi i} \int_\gamma b(t) \sum_{k=0}^{p-1} t^{-k-1} L_{pn+m}(u_k)(t^{-1}) \, dt$$

$$= \frac{1}{2\pi i} \int_\gamma b(t) t^{-pn-m-1} \, dt$$

$$= \frac{b^{(pn+m)}(0)}{(pn+m)!} = b_{pn+m}, \quad m = 0, 1, \ldots, p-1; n = 0, 1, \ldots.$$

Thus, f has the required properties, and this completes the proof.

From Lemma 1, we have immediately the following corollary.

Corollary 1. *Let W be analytic and univalent on a simply connected domain Ω such that $W(\Omega)$ is p-symmetric. Let Z be the inverse of W which maps $W(\Omega)$ one-to-one onto Ω. Let $h_k(\zeta)$ be analytic on Ω; $k = 0, 1, \ldots, p-1$ and denote the function $h_k(Z(\alpha^j W(\zeta)))$ by $h_{jk}(\zeta)$ and the determinant of the matrix $(h_{jk}(\zeta))$ by $\Delta(\zeta)$. Assume $\Delta(\zeta) \not\equiv 0$. Let $\{L_n\}$ be defined by (1) with $g_{pn+k}(\zeta) = h_k(\zeta)[W(\zeta)]^{pn}$. If $b(z) = \sum_{n=0}^\infty b_n z^n$ is analytic at the origin, then there exists* *a function f in $K[\Omega]$ such that $L_n(f) = b_n$; $n = 0, 1, \ldots$ if and only if b can be continued analytically to the set $(W(\Omega)^{-1})'$, the complement of*

$$\{[W(\zeta)]^{-1} \mid \zeta \in \Omega\}.$$

Except for an added restriction on the domain Ω dealt with and the condition that $\Delta(\zeta) \neq 0$, the result is the same as that obtained in [2] for functionals defined by (1) with $g_n(\zeta) = [W(\zeta)]^n$. Therefore, we can state the following corollary.

Corollary 2. *Let $\{L_n\}$ be defined by (1) with $g_{pn+k}(\zeta) = h_k(\zeta)[W(\zeta)]^{pn}$ and $\{L_n^*\}$ be defined by (1) with $g_n(\zeta) = [W(\zeta)]^n$ and let Ω, W, and the functions h_k satisfy the hypotheses of Corollary 1. Given a sequence $\{b_n\}$ of complex numbers, there exists a function f in $K[\Omega]$ such that $L_n(f) = b_n$; $n = 0, 1, 2, \ldots$ if and only if there exists a function f^* in $K[\Omega]$ such that $L_n^*(f^*) = b_n$; $n = 0, 1, 2, \ldots$.*

In [4], the author showed that zeros of $\Delta(\zeta)$ give rise to nonzero functions g such that $L_n(g) = 0$; $n = 0, 1, 2, \ldots$. Thus, if g is such a function, then for any function f in $K[\Omega]$ such that $L_n(f) = b_n$, it is also true that for any complex number a, $L_n(f + ag) = b_n$, so that if there is one function f in $K[\Omega]$ which interpolates $\{b_n\}$ relative to $\{L_n\}$, then there are infinitely many such functions. In the case of $\{L_n^*\}$ of Corollary 2, $K[\Omega]$ is a uniqueness class [3], so there can be at most one function f^* such that $L_n^*(f^*) = b_n$; $n = 0, 1, \ldots$.

In the case of $W(\Omega)$ a disk: $|\zeta| < R$, $0 < R \leq \infty$, Corollary 1 implies that if each h_k is analytic on Ω and $\Delta(\zeta) \neq 0$, then given a sequence $\{b_n\}$, there exists a function f in $K[\Omega]$ such that $L_n(f) = b_n$; $n = 0, 1, \ldots$ if and only if $\limsup_{n \to \infty} |b_n|^{1/n} < R$.

Example 1. $W(\zeta) = \zeta$; p any positive integer; Ω any simply connected p-symmetric domain on which each h_k is analytic. If $\Delta(\zeta) \neq 0$, then $\{b_n\}$ can be interpolated by a function in $K[\Omega]$ relative to $\{L_n\}$ if b is analytic on $(\Omega^{-1})'$. Specific examples: Let $\{a_k\}_{k=0}^{p-1}$ be any finite sequence of complex numbers

(1) $h_k(\zeta) = e^{a_k \zeta} \leftrightarrow L_{pn+k}(f) = f^{(pn)}(a_k)$

(2) $h_k(\zeta) = \zeta^k e^{a_k \zeta} \leftrightarrow L_{pn+k}(f) = f^{(pn+k)}(a_k)$

(3) $h_k(\zeta) = e^{a_k \zeta}(e^\zeta - 1)^k \leftrightarrow L_{pn+k}(f) = \Delta^k f^{(pn)}(a_k)$

(4) $h_k(\zeta) = (\zeta - a_k)^{-1} \leftrightarrow L_{pn+k}(f) = g_k^{(pn)}(0)$,

where

$$g_k(z) = \int_0^z e^{a_k(z-t)} f(t)\, dt.$$

Example 2. $W(\zeta) = e^{\zeta} - 1$; $p = 2$; Ω the strip $\{x + iy \,|\, |y| < \pi\}$ slit along the positive real axis from $\log 2$ to ∞. Then $(W(\Omega)^{-1})'$ is the interval $[-1, 1]$ of the real axis, so $\{b_n\}$ can be interpolated by a function f in $K[\Omega]$ relative to $\{L_n\}$ iff b is analytic on this interval. Specific examples: $k = 0, 1$.

(1) $h_k(\zeta) = \zeta^k \leftrightarrow L_{2n+k}(f) = \Delta^{2n}f^{(k)}(0)$

(2) $h_k(\zeta) = e^{a_k\zeta} \leftrightarrow L_{2n+k}(f) = \Delta^{2n}f(a_k)$

(3) $h_k(\zeta) = \zeta^k e^{a_k\zeta}(e^{\zeta} - 1)^k \leftrightarrow L_{2n+k}(f) = \Delta^{2n+k}f^{(k)}(a_k)$

(4) $h_k(\zeta) = (\zeta - (-1)^k \pi i)^{-1} \leftrightarrow L_{2n+k}(f) = \Delta^{2n}g_k(0)$,

where

$$g_k(z) = \int_0^z e^{(-1)^k \pi i(z-t)} f(t)\, dt$$

Example 3. $W(\zeta) = e^{\zeta/2} - e^{-\zeta/2} = 2\sinh(\zeta/2)$; $p = 4$; Ω is the strip

$$\{x + iy \,|\, |y| < \pi\}$$

slit along the real axis from $2\sinh^{-1} 1 \approx 1.76 \cdots$ to ∞ and from $-2\sinh^{-1} 1$ to $-\infty$. Then $(W(\Omega)^{-1})'$ is the union of the intervals $[-\frac{1}{2}, \frac{1}{2}]$ and $[-i/2, i/2]$ of the real and imaginary axes, respectively. Then $\{b_n\}$ can be interpolated by a function f in $K[\Omega]$ relative to $\{L_n\}$ iff b is analytic on this set. Specific examples: $k = 0, 1, 2, 3$:

(1) $h_k(\zeta) = \zeta^k \leftrightarrow L_{4n+k}(f) = \Delta^{4n}f^{(k)}(-2n)$

(2) $h_k(\zeta) = e^{a_k\zeta} \leftrightarrow L_{4n+k}(f) = \Delta^{4n}f(-2n + a_k)$.

(3) $h_k(\zeta) = e^{a_k\zeta}(e^{\zeta} - 1)^k \leftrightarrow L_{4n+k}(f) = \Delta^{4n+k}f(-2n + a_k)$

(4) $h_k(\zeta) = (\zeta - i^k \pi)^{-1} \leftrightarrow L_{n+k}(f) = \Delta^{4n}g_k(-2n)$,

where

$$g_k(z) = \int_0^z e^{ik\pi(z-t)} f(t)\, dt.$$

REFERENCES

1. R. F. DeMar, Existence of interpolating functions of exponential type, *Trans. Amer. Math. Soc.* **105** (1962), 359–371.

2. ———, On a theorem concerning existence of interpolating functions, *Trans. Amer. Math. Soc.* **114** (1965), 23–29.

3. ———, A uniqueness theorem for entire functions, *Proc. Amer. Math. Soc.* **16** (1965), 69–71.

4. ———, Uniqueness classes for periodic-type functionals, *Duke Math. J.* **37** (1970), 29–40.

(*Received February 23, 1968*)

Complements to Some Theorems of Bowen and Macintyre on the Radial Growth of Entire Functions with Negative Zeros

David Drasin

Department of Mathematics
Purdue University
Lafayette, Indiana 47907

Daniel F. Shea*

Department of Mathematics
University of Wisconsin
Madison, Wisconsin 53706

Some years ago, N. A. Bowen and A. J. Macintyre [5] proved the following theorem in answer to a question raised by E. C. Titchmarsh [19] concerning the growth of entire functions satisfying certain regularity conditions.

Theorem A. *Let*

(1) $$f(z) = \prod_{n=1}^{\infty} \left(1 + \frac{z}{a_n}\right) \qquad (0 < a_n \leqq a_{n+1})$$

be an entire function. Assume there is a constant ρ $(0 < \rho < 1)$ and a slowly varying function $L(x)$ such that, for each $\varepsilon > 0$,

* Research of this author supported in part by NSF grant GP 5728.

(2) $$\log |f(-x)| < (\cos \pi\rho + \varepsilon)x^\rho L(x) \qquad (x > x_0(\varepsilon)),$$

and such that

(3) $$\log |f(-x_n)| > (\cos \pi\rho - \varepsilon)x_n{}^\rho L(x_n)$$

holds for sufficiently large x_n of a sequence satisfying

(4) $$x_n \to +\infty, \; x_{n+1}/x_n \to 1 \qquad (n \to \infty).$$

Then

(5, 0) $$\log f(x) \sim x^\rho L(x) \qquad (x \to +\infty)$$

holds if either

(i) $0 < \rho < \frac{1}{2}$ *and*

(6) $$\log f(x) = o(x^{1/2}) \qquad (x \to +\infty),$$

or

(ii) $\frac{1}{2} < \rho < 1.$

The term "slowly varying" here is used in the sense of Karamata [13], and means that $L(x)$ is defined and positive for $x > x_0$, and satisfies

$$\lim_{x \to \infty} \frac{L(\sigma x)}{L(x)} = 1 \qquad (0 < \sigma < \infty).$$

Bowen and Macintyre actually stated Theorem A only for the special case $L(x) \equiv$ constant, but it is not difficult to deduce the above statement from their methods, modified in a standard way. (See, for example, the discussions in [1, 16] for all the facts needed here and in what follows concerning slowly varying functions.)

Part (i) of Theorem A was proved by Titchmarsh himself in [19], and the truth of (ii) was left as an open question by him. (Of course, no conclusion like (5, 0) is possible in general for functions satisfying (2) and (3) with $\rho = \frac{1}{2}$: just note that functions of any order $\rho_1 < \frac{1}{2}$ satisfy (2) and (3) with $\rho = \frac{1}{2}$, but not (5, 0).)

A well-known tauberian theorem of Valiron ([20]; also cf. [3, Chapter 4]) implies that the conclusion (5, 0) is equivalent to

(5, N) $$N(x, 0) = \left[\frac{\sin \pi\rho}{\pi\rho} + o(1) \right] x^\rho L(x) \qquad (x \to \infty)$$

when* $0 \leq \rho < 1$. Also, Titchmarsh had proved in [19] the following

* When $\rho = 0$, $(\sin \pi\rho)/\pi\rho$ is to be replaced by its limit, $= 1$. In (5, N), as usual,

$$N(x, 0) = \int_0^x n(t, 0) \, dt/t$$

and $n(t, 0)$ denotes the number of zeros of $f(z)$ in $|z| \leq t$.

Theorem B. *If $f(z)$ has the form* (1), *and satisfies* (5, 0) *or* (5, N) *for some* ρ $(0 < \rho < 1)$, *then* (2)–(4) *are true.*

Hence Theorem A is the converse of the elementary (although technically quite complicated) "abelian" Theorem B.

Theorem A may be regarded in another way; it implies that, if $f(z)$ satisfies hypotheses (2)–(4) and (i) or (ii), then

$$(7) \qquad \lim_{n \to \infty} \frac{\log |f(-x_n)|}{\log f(x_n)} = \cos \pi\rho,$$

where $\{x_n\}$ is a sequence satisfying (4). Clearly, the fact that $\log f(x)$ increases yields at once that (7) is equivalent to (5, 0), when (2) and (3) are given.

It is the main purpose of this note to show that a condition like (7)— in which the regularity of $f(z)$ along the negative real axis is measured in terms of the *intrinsic* comparison function $\log f(x)$ rather than one of the form x^ρ or $x^\rho L(x)$, say—is by itself sufficient to yield the regularity conditions (5, 0) and (5, N).

Theorem 1. *Let $f(z)$ be an entire function of the form* (1), *and put*

$$(8) \qquad \alpha = \limsup_{x \to \infty} \frac{\log |f(-x)|}{\log f(x)}.$$

Assume that there is a sequence satisfying (4) *and such that*

$$(9) \qquad \log |f(-x_n)| > (\alpha - \varepsilon)\log f(x_n) \qquad (n > n_0(\varepsilon))$$

holds for each $\varepsilon > 0$.

Then $-1 \leq \alpha \leq 1$, and the regularity statements (5, 0) *and* (5, N) *as well as*

$$(5, \pi) \qquad \log |f(-x)| = [\cos \pi\rho + o(1)]x^\rho L(x) \qquad (x \to \infty, x \notin E)$$

hold for a suitable slowly varying function $L(x)$. The set E in (5, π) *is of density zero* and ρ is determined by*

$$(10) \qquad \cos \pi\rho = \alpha \qquad (0 \leq \rho \leq 1).$$

We note in passing that functions of order $\frac{1}{2}$ play no exceptional role here, as was the case in Theorem A, and that no auxiliary condition such as (6) in Theorem A need be assumed.

Theorem 1 can be described as a "ratio tauberian" theorem, and may be compared with some recent results of Edrei and Fuchs [9], and of the

---

* A set E has density zero if $\text{meas}\{E \cap [0, x]\} = o(x)$ when $x \to +\infty$.

authors [6, 18], in which conclusions like (5, 0) are deduced from hypotheses such as

$$\alpha = \lim_{x \to \infty} \frac{N(x, 0)}{\log f(x)} \quad \text{exists} \quad (\alpha > 0).$$

But the proof of Theorem 1 here is complicated by the erratic asymptotic behavior of $\log |f(-x)|$.

Analogous to Titchmarsh's Theorem B is the result (quite easily proved— cf. last part of Section 5) that condition (5, 0) for functions of the form (1) implies

$$(5, \theta) \qquad \log |f(re^{i\theta})| = [\cos \theta \rho + o(1)] r^\rho L(r) \qquad (r \to \infty)$$

for each fixed θ, $|\theta| < \pi$. In [5, Theorems II, III], Bowen and Macintyre proved the converse theorem (5, θ) $\Rightarrow$ (5, 0), and this result may be regarded as an analogue of Theorem A for $\theta \neq \pi$. The following theorem is the corresponding analogue of Theorem 1.

Theorem 2. *Let $f(z)$ be an entire function of the form* (1), *and assume that*

$$(11) \qquad\qquad A = \lim_{r \to \infty} \frac{\log |f(re^{i\phi})|}{\log f(r)}$$

exists for some fixed ϕ, $0 < |\phi| < \pi$.

Then $\cos \phi \leq A \leq 1$, and there exists a slowly varying function $L(r)$ such that (5, θ) holds for each θ ($|\theta| < \pi$), with ρ determined by

$$(12) \qquad\qquad \cos \phi \rho = A \qquad (0 \leq \rho \leq 1).$$

As an application of Theorem 2, we answer a question suggested by some recent work of Edrei and Fuchs [9] and Baernstein [1]: Let $f(z)$ be a function of the form (1), of order ρ, and consider the region

$$D_f = \{re^{i\theta} : |f(re^{i\theta})| \geq 1, 0 \leq \theta \leq \pi, r \geq 0\}.$$

It is not difficult to see that the boundary of D_f consists of the positive real axis together with a continuous curve $\Gamma = \{re^{i\beta(r)} : 0 \leq r < \infty\}$ emanating from the origin and subject to $\pi/2 < \beta(r) \leq \pi$.

Now, we have already pointed out that (5, 0) implies (5, θ) and hence also

$$(13) \qquad\qquad \beta_0 = \lim_{r \to \infty} \beta(r) \text{ exists}, \; \beta_0 = \min\{\pi, \pi/2\rho\}.$$

Thus it follows from Theorems A, B, 1, 2 that a function satisfying the conditions of those theorems also satisfies (13). (The results in [9, 1] give still other hypotheses sufficient to yield (13).) It is therefore natural to ask whether (13) is actually *equivalent* to the other conditions mentioned, and this question is answered by the following easy consequence of Theorem 2.

Corollary 2.1. *Let $f(z)$ be an entire function of the form* (1), *of order ρ ($\frac{1}{2} < \rho \leq 1$).*
If

$$\beta = \lim_{r \to \infty} \beta(r) \ \text{exists,}$$

then $\beta_0 = \pi/2\rho$, and $f(z)$ satisfies (5, θ) *for $|\theta| < \pi$.*

No such conclusions are possible with $\rho \leq \frac{1}{2}$, as follows from the examples in [11, Section 6].

Finally, we emphasize that while the restriction to functions of the special form (1) may seem artificial, it is justified by the important role such functions play in the class of all entire functions of order less than one. In fact, a form of Theorem 1 involving no assumption on the location of the zeros of $f(z)$ also holds (see [7]), but we shall have to leave these matters for another paper.

1. Proof of Theorem 1

As is well known, a function of the form (1) has order $\rho \leq 1$. In Sections 1 and 2 we shall demand that $\rho < 1$; the case $\rho = 1$ is considered in Section 3.

Because of the irregular behavior of $\log|f(z)|$ on the negative real axis, we introduce the modified functions

(1.1)
$$g(x) = \int_0^x \frac{\log f(t)}{t^\gamma} \, dt,$$

(1.2)
$$h(x) = \int_0^x \{\log f(t) + \log|f(-t)|\} \frac{dt}{t^\gamma},$$

where γ is a fixed number, $0 < \gamma < 1$.

The major portion of our argument—deferred to Section 2—is to show that the hypotheses (8) and (9) (the latter assumption involving only a discrete sequence $\{x_n\}$) are completely equivalent to the global assertion

(1.3)
$$\lim_{x \to \infty} \frac{g(x)}{h(x)} = (1 + \alpha)^{-1}.$$

(The classical "cos $\pi\rho$ theorem" [3, p. 40] yields that $\alpha \geq \cos \pi\rho > -1$, so that $(1 + \alpha)^{-1}$ is finite.)

We begin by comparing $g(z)$ and $\log f(x)$. In the first place,

$$g(x) \leq \log f(x) \int_0^x t^{-\gamma}\,dt = (1-\gamma)^{-1} x^{1-\gamma} \log f(x).$$

An inequality in the opposite direction follows from the standard relation

(1.4) $$\log f(x) \leq \frac{x}{y}\log f(y) \qquad (y < x),$$

since then

$$x^{1-\gamma}\log f(x) \leq 2x^{1-\gamma}\log f(x/2)$$

$$= B\log f(x/2) \int_{x/2}^x t^{-\gamma}\,dt$$

$$\leq B\int_{x/2}^x \log f(t)\frac{dt}{t^{\gamma}} \leq Bg(x),$$

where $B = B(\gamma)$ is a positive constant.

Hence, we have

(1.5) $$k_1 x^{1-\gamma}\log f(x) \leq g(x) \leq k_2 x^{1-\gamma}\log f(x),$$

where the positive constants k_1 and k_2 depend only on γ.

The next goal is to show that $g(x)$ and $h(x)$ are related by

(1.6) $$g(x) = \int_0^\infty h(t)k(x/t)\frac{dt}{t},$$

with

(1.7) $$k(u) = \frac{2}{\pi^2}\frac{u^{2-\gamma}}{u^2 - 1}\log u.$$

As a first step in obtaining (1.6), recall the representation recently established by Kjellberg [15]:

(1.8) $$\log f(y) = \int_0^\infty \{\log f(t) + \log |f(-t)|\} K(y/t)\frac{dt}{t},$$

with

$$K(u) = 2\pi^{-2}u(u^2 - 1)^{-1}\log u.$$

Hence

(1.9) $$g(x) = \frac{2}{\pi^2}\int_0^x dy \int_0^\infty \{\log f(t) + \log |f(-t)|\} \frac{y^{1-\gamma}t^{\gamma}}{y^2 - t^2}\log\left(\frac{y}{t}\right)\frac{dt}{t^{\gamma}}$$

$$= \frac{2}{\pi^2}\int_0^x dy\left\{\int_0^\infty \frac{y^{1-\gamma}t^{\gamma}}{y^2 - t^2}\log\left(\frac{y}{t}\right)\,dh(t)\right\}.$$

Since we are assuming $\rho < 1$ and (1.3), it follows that for some $\varepsilon > 0$

(1.10) $\quad h(t) \leqq (1 + \alpha + \varepsilon)g(t) < kt^{1-\gamma} \log f(t) < t^{2-\gamma-\varepsilon} \qquad (t > t_0)$,

and hence (1.9) becomes, after an integration by parts,

(1.11)

$$g(x) = \frac{2}{\pi^2} \int_0^x dy \int_0^\infty h(t) \left[\frac{y^{1-\gamma} t^{\gamma-1}}{y^2 - t^2} - \log\left(\frac{y}{t}\right) \frac{\gamma y^{3-\gamma} t^{\gamma-1} + (2-\gamma)t^{1+\gamma} y^{1-\gamma}}{(y^2 - t^2)^2} \right] dt.$$

Writing $y = tu$ and interchanging the order of integration (permissible because of (1.10)), (1.11) becomes

(1.12) $\quad g(x) = \dfrac{2}{\pi^2} \displaystyle\int_0^\infty h(t) \left\{ \int_0^{x/t} \dfrac{u^{1-\gamma}}{u^2 - 1} - \log u \, \dfrac{\gamma u^{3-\gamma} + (2-\gamma)u^{1-\gamma}}{(u^2 - 1)^2} \, du \right\} \dfrac{dt}{t}$

$$= \frac{2}{\pi^2} \int_0^\infty h(t) K_1(x/t) \frac{dt}{t}.$$

If we regard the integral defining K_1 in the principal value sense at $u = 1$, we may integrate the first term by parts to obtain

$$K_1(y) = \frac{y^{2-\gamma}}{y^2 - 1} \log y,$$

and this establishes the representation (1.6), (1.7).

We are now in a position to apply Theorem 1 of [6] to (1.6).* Since g increases, assumption (1.3) shows that g and h satisfy the hypotheses of that theorem, and similarly it is easy to see that the kernel k in (1.7) satisfies the required conditions, as

(1.13)

$$\int_{-\infty}^\infty k(e^{-t})e^{-st} \, dt = \frac{2}{\pi^2} \int_0^\infty \{\log \xi/(\xi^2 - 1)\} \xi^{1-\gamma-s} \, d\xi$$

$$= [1 + \cos \pi(1 - \gamma - s)]^{-1} \qquad (-\gamma < s < 2 - \gamma)$$

by a standard contour integration (cf. for example, [21, p. 118]). Hence

(1.14) $\qquad\qquad g(x) = \displaystyle\int_0^x \log f(t) \frac{dt}{t^\gamma} = x^\lambda L_1(x)$,

where L_1 varies slowly and λ is determined by

(1.15) $\quad [1 + \cos \pi(1 - \gamma - \lambda)]^{-1} = (1 + \alpha)^{-1} \qquad (-\gamma < \lambda < 2 - \gamma)$.

(We require from this point on that γ has been chosen in $0 < \gamma < 1$ so that $\lambda \neq 0$.)

---

* We remark that the method developed in [18] gives an alternate way of obtaining (1.14) from (1.3).

An elementary tauberian method [3, p. 58], when used with (1.14), shows that

(1.16) $\log f(x) \sim \lambda x^{\lambda+\gamma-1} L_1(x)$ $(x \to \infty)$

and hence f has order $\rho = \lambda + \gamma - 1$. Thus (1.15) and (1.16) yield (5, 0) and (10), with $L(x) = \lambda L_1(x)$.

Conclusion (5, N) follows from Valiron's tauberian theorem [20], or from a suitable extension (cf. Baerstein's Lemma 1 in [1]) of Wiener's general tauberian theorem.

For a function $f(z)$ of order $\rho < 1$, one can obtain (5, π) from (5, N) using a suitable modification of Titchmarsh's abelian argument [19, Theorem III]. Since we shall also need to consider the case $\rho = 1$ in Section 3, we give here a short direct proof of (5, π) valid for any ρ, $0 \leq \rho \leq 1$. First, notice that (1.3), (1.14), and (1.15) imply

(1.17)

$$\int_0^r \frac{\log |f(-t)|}{t^\gamma} dt = [\cos \pi\rho + o(1)](1 + \rho - \gamma)^{-1} r^{1+\rho-\gamma} L(r) \qquad (r \to \infty).$$

Also, hypothesis (8) together with (1.16) implies

(1.18) $\log |f(-t)| \leq [\cos \pi\rho + o(1)] t^\rho L(t)$ $(t \to \infty)$.

Hence, if (5, π) were false there would exist $\eta > 0$, $\delta > 0$, and a sequence $r_m \to \infty$ such that the set

(1.19) $\mathscr{E} = \{t : \log |f(-t)| < (\cos \pi\rho - \eta) t^\rho L(t)\}$

satisfied

(1.20) $$\int_{\mathscr{E}(0,r_m)} dt > 2\delta r_m,$$

where $\mathscr{E}(a, b)$ denotes the intersection of $\mathscr{E}$ and the interval (a, b).

Now from (1.20) follows

$$\int_{\mathscr{E}(\delta r_m, r_m)} dt > \delta r_m,$$

and from (1.17), (1.18), (1.19), and (1.14) that

$$\int_0^{r_m} \frac{\log |f(-t)|}{t^\gamma} dt \leq [\cos \pi\rho + o(1)] \int_0^{r_m} t^{\rho-\gamma} L(t) \, dt - \eta \int_{\mathscr{E}(0,r_m)} t^{\rho-\gamma} L(t) \, dt$$

$$\leq [\cos \pi\rho + o(1)](1 + \rho - \gamma)^{-1} r_m^{1+\rho-\gamma} L(r_m)$$

$$- \frac{1}{2} \eta L(r_m) \int_{\mathscr{E}(\delta r_m, r_m)} t^{\rho-\gamma} \, dt$$

$$\leq [\cos \pi\rho + o(1)] g(r_m) - B r_m^{1+\rho-\gamma} L(r_m)$$

for all sufficiently large m, where B is a positive constant. Hence, from (1.2),

$$\frac{h(r_m)}{g(r_m)} \leq 1 + \cos \pi\rho - B(1 + \rho - \gamma) + o(1) \qquad (m \to \infty)$$

which contradicts (1.3) since $\alpha = \cos \pi\rho$. This proves $(5, \pi)$ and completes the deduction of Theorem 1 from (1.3) when $0 \leq \rho < 1$.

2. Proof of (1.3) for functions of order less than one

An easy calculation using (8) shows that

$$(2.1) \qquad \liminf_{x \to \infty} \frac{g(x)}{h(x)} \geq (1 + \alpha)^{-1},$$

where g and h are defined by (1.1) and (1.2). To show that

$$(2.2) \qquad \limsup_{x \to \infty} \frac{g(x)}{h(x)} \leq (1 + \alpha)^{-1}$$

requires more effort, since (9) is assumed to hold only for a discrete set of points on $\arg z = \pi$. We first follow Titchmarsh [19, p. 196] and Bowen and Macintyre [5, p. 118] to get an inequality similar to (9) which is valid for all points on a ray $\arg z = \pi - \delta$, $\delta > 0$. That is, if one lets

$$h_\delta(x) = \int_0^x \{\log f(t) + \log |f(-te^{-i\delta})|\} \frac{dt}{t^\gamma},$$

it will be shown that, for any $\delta > 0$,

$$(2.3) \qquad \limsup_{x \to \infty} \frac{g(x)}{h_\delta(x)} \leq (1 + \alpha)^{-1}.$$

The proof of (1.3) then follows upon showing that (2.3) also holds with $\delta = 0$.
    We first notice that

$$(2.4) \qquad \log |f(-xe^{-i\delta})| \geq \log |f(-x_n)|$$

certainly holds when

$$\left| 1 - \frac{xe^{-i\delta}}{a_m} \right|^2 \geq \left| 1 - \frac{x_n}{a_m} \right|^2 \qquad (m = 1, 2, \ldots),$$

and so when x satisfies

$$(2.5) \qquad x_n \leq x \leq x_n / \cos \delta.$$

Now, if $\delta > 0$ is fixed, every $x > x_0(\delta)$ lies in an interval of the form (2.5) for some n, since $x_{n+1}/x_n \to 1$. Hence it follows from (2.4), (1.4) and hypothesis (9) that

$$(2.6) \qquad \log|f(-xe^{-i\delta})| \geq (\alpha - \varepsilon)\log f(x_n) \geq (\alpha - \varepsilon)\frac{x_n}{x}\log f(x)$$

$$\geq (\alpha - 2\varepsilon)\log f(x) \qquad (x_{n+1} > x \geq x_n > x_0),$$

and now (2.3) follows by an obvious calculation, similar to that used in obtaining (2.1).

It remains to improve (2.3) to (2.2), and this will be a direct consequence of (2.3) and the estimate

$$(2.7) \qquad\qquad h_\delta(x) \leq h(x) + B(\delta)g(x),$$

where $B(\delta) \to 0$ as $\delta \to 0$.

Consider

$$E_\delta(x) = h_\delta(x) - h(x)$$

$$= \sum \int_0^x \log\left|\frac{a_n - te^{-i\delta}}{a_n - t}\right|\frac{dt}{t^\gamma} \geq 0.$$

Set

$$(2.8) \qquad\qquad \eta = 2(1 - \cos \delta)$$

and $\xi = t/a_n$; then since

$$\int_0^x \log\left|\frac{a_n - te^{-i\delta}}{a_n - t}\right|\frac{dt}{t^\gamma} = \frac{1}{2}\int_0^x \log\left[1 + \eta\frac{a_n t}{(a_n - t)^2}\right]\frac{dt}{t^\gamma}$$

$$= \frac{1}{2}a_n^{1-\gamma}\int_0^{x/a_n} \log\left[1 + \frac{\eta\xi}{(1 - \xi)^2}\right]\frac{d\xi}{\xi^\gamma},$$

we obtain

$$E_\delta(x) = \frac{1}{2}\int_0^\infty t^{1-\gamma}\left\{\int_0^{x/t} \log\left[1 + \frac{\eta\xi}{(1 - \xi)^2}\right]\frac{d\xi}{\xi}\right\} dn(t, 0).$$

For $\eta > 0$, let

$$(2.9) \qquad\qquad g_\eta(\xi) = \xi^{-\gamma}\log\left[1 + \frac{\eta\xi}{(1 - \xi)^2}\right]$$

and

$$G_\eta(u) = \int_0^u g_\eta(\xi)\, d\xi.$$

Clearly,

$$n(t, 0)t^{1-\gamma}G_\eta(x/t) \to 0 \qquad (t \to 0, \infty),$$

so that

(2.10) $$E_\delta(x) = -\frac{1}{2}\int_0^\infty n(t, 0)[(1 - \gamma)t^{-\gamma}G_\eta(x/t) - xt^{-1-\gamma}g_\eta(x/t)]\, dt.$$

From (2.10) follows

(2.11) $$E_\delta(x) \leq \frac{x}{2}\int_0^\infty \frac{n(t, 0)}{t^{1+\gamma}}g_\eta(x/t)\, dt,$$

and we shall deduce (2.7) from this.

To estimate the integral in (2.11), first notice that

$$\int_0^{2x} n(t, 0)g_\eta(x/t)\frac{dt}{t^{1+\gamma}} \leq n(2x, 0)\int_0^{2x} g_\eta(x/t)\frac{dt}{t^{1+\gamma}}$$

$$= n(2x, 0)x^{-\gamma}\int_{1/2}^\infty g_\eta(\xi)\xi^{\gamma-1}\, d\xi,$$

and so by Jensen's inequality [10] and (1.4)

(2.12) $$\int_0^{2x} n(t, 0)g_\eta(x/t)\frac{dt}{t^{1+\gamma}} \leq x^{-\gamma}\log f(x)\left\{\frac{4}{\log 2}\int_{1/2}^\infty g_\eta(\xi)\xi^{\gamma-1}\, d\xi\right\}.$$

It is easy to see that

$$\int_{1/2}^\infty g_\eta(\xi)\xi^{\gamma-1}\, d\xi \to 0 \qquad (\eta = 2(1 - \cos \delta) \to 0),$$

by dominated convergence, and hence (2.12) and (1.5) imply

(2.13) $$\frac{x}{2}\int_0^{2x} n(t, 0)g_\eta(x/t)\frac{dt}{t^{1+\gamma}} = A(\delta)\, g(x), \qquad A(\delta) \to 0 \text{ as } \delta \to 0.$$

Now consider the integral in (2.11) for $2x < t < \infty$. First, (2.9) implies

$$g_\eta(\xi) < 4\eta\xi^{1-\gamma} \qquad (\xi = x/t \leq \tfrac{1}{2}),$$

and then

$$\int_{2x}^\infty n(t, 0)g_\eta(x/t)\frac{dt}{t^{1+\gamma}} \leq 4\eta x^{1-\gamma}\int_{2x}^\infty \frac{n(t, 0)}{t^2}\, dt$$

$$\leq 6\eta x^{1-\gamma}\int_{2x}^\infty \frac{n(t, 0)}{t(t + x)}\, dt \leq 6\eta x^{-\gamma}\log f(x).$$

This, together with (1.5), (2.11) and (2.13) shows that

$$E_\delta(x) \leq B(\delta)g(x)$$

where $B(\delta) \to 0$ $(\delta \to 0)$, and proves (2.7).

3. Functions of order $\rho = 1$

The methods of Section 1 clearly must be modified for functions of order $\rho = 1$, since then $1 + \cos \pi\rho = 0$ and the basic relation (1.3) breaks down. However, one can avoid such difficulties by using the representation

$$(3.1) \qquad \log f(x) = \frac{(-1)^p}{\pi} \int_0^\infty \left(\frac{x}{t}\right)^{p+1/2} \frac{\log|f(-t)|}{t+x} \, dt,$$

due to Bowen and Macintyre [5, p. 121], in place of Kjellberg's representation (1.8). The parameter p in (3.1) must be an integer chosen to satisfy

$$(3.2) \qquad |\log f(re^{i\theta})| = o(r^{p-1/2}) \qquad (r \to 0),$$

$$(3.3) \qquad \int_{-\pi}^{\pi} |\log f(re^{i\theta})| \, d\theta = o(r^{p+1/2}) \qquad (r \to \infty),$$

and for functions of genus zero it is not difficult to see that (3.2), (3.3) are satisfied with $p = 1$ [5, pp. 117, 119].

Assume that $f(z)$ has the form (1), and that $p = 1$; define

$$(3.4) \qquad G(x) = \int_0^x \frac{\log f(t)}{t^{1/2}} \, dt,$$

$$(3.5) \qquad H(x) = \int_0^x \frac{\log|f(-t)|}{t^{1/2}} \, dt$$

analogously to (1.1) and (1.2), and integrate (3.1) as in Section 1 to obtain

$$(3.6) \qquad G(x) = \int_0^\infty H(t) K_p(x/t) \frac{dt}{t},$$

where

$$(3.7) \qquad K_p(x) = \frac{(-1)^p}{\pi} \int_0^x t^p \frac{p+1+pt}{(t+1)^2} \, dt;$$

(3.6) is the desired analogue of (1.6).

The arguments of Section 2 apply also to $G(x)$ and $H(x)$, and lead at once to

$$(3.8) \qquad \lim_{x \to \infty} \frac{H(x)}{G(x)} = \alpha.$$

In order to see that $\alpha \neq 0$ (so that Theorem 1 of [6] may be applied), we prove a lemma which is valid for any entire function of genus zero having only negative zeros. It can be regarded as a "two-sided" form of the "$\cos \pi\rho$ theorem."

Lemma 1. *Let $f(z)$ be a function of the form* (1), *and let $G(x)$ and $H(x)$ be defined by* (3.4), (3.5). *Then*

(3.9)
$$\liminf_{x \to \infty} \frac{H(x)}{G(x)} \leq \cos \pi \rho \leq \limsup_{x \to \infty} \frac{H(x)}{G(x)},$$

where ρ denotes the order of $f(z)$.

Before proving Lemma 1, we notice that (3.9) and (3.8) together establish that $\alpha = -1$ for functions of order $\rho = 1$ which satisfy the hypotheses of Theorem 1. (This remark completes the proof of (10).) In particular $\alpha \neq 0$, so we can deduce, as in Section 1, that

$$G(x) = x^{\lambda} L_1(x)$$

where L_1 varies slowly. Hence, as before,

$$\log f(x) \sim \lambda x^{\lambda - 1/2} L_1(x) \qquad (x \to \infty)$$

so that necessarily $\lambda = \frac{3}{2}$. This proves (5, 0).

To obtain (5, N), we appeal to the representation

$$\log f(r) = r \int_0^{\infty} \frac{N(t, 0)}{(t + r)^2} \, dt$$

[3, p. 55], and rewrite it in terms of

$$Q(r) = \int_r^{\infty} \frac{N(t, 0)}{t^2} \, dt,$$

using an integration by parts; this leads to

$$\log f(r) = 2r^2 \int_0^{\infty} \frac{t Q(t)}{(t + r)^3} \, dt,$$

and hence (5, 0) implies

$$\frac{1}{L(r)} \int_0^{\infty} \left\{ \frac{2rt^2}{(t + r)^3} \right\} Q(t) \frac{dt}{t} \to 1 \qquad (r \to \infty).$$

From this it is not difficult to deduce, via a tauberian argument (cf. [1, p. 62]), that

$$Q(r) \sim L(r)$$

and then

$$N(r, 0) = o(r L(r))$$

as $r \to \infty$.

Finally, we recall that the argument at the end of Section 1, proving that (5, 0) implies (5, π) when $\rho < 1$, also applies when $\rho = 1$. This observation completes the proof of all the assertions of Theorem 1, and it remains only to verify Lemma 1.

Proof of Lemma 1

Let $0 < r < R < \infty$, $0 < \sigma < 1$, and

(3.10) $$I_\sigma(r, R) = \int_r^R \{\log |f(-t)| - \cos \pi\sigma \log f(t)\} \frac{dt}{t^{1+\sigma}}.$$

Then

(3.11) $$k(\sigma) \frac{\log f(r)}{r^\sigma} - 10 \frac{\log f(R)}{R^\sigma} \leq I_\sigma(r, R)$$

holds, with

(3.12) $$k(\sigma) = \frac{1 - \sin \pi |\frac{1}{2} - \sigma|}{\frac{1}{2} - |\frac{1}{2} - \sigma|},$$

as Kjellberg [14] has shown.

It is easy to show in the same way that the (essentially equivalent) inequality

(3.13) $$I_\sigma(r, R) \leq 10 \frac{\log f(r)}{r^\sigma} - k(\sigma) \frac{\log f(R)}{R^\sigma}$$

also holds for $0 < r < R < \infty$, $0 < \sigma < 1$.

We can write I_σ in terms of the functions G and H defined in (3.4), (3.5) by an obvious integration by parts:

(3.14) $$I_\sigma(r, R) = (\tfrac{1}{2} + \sigma) \int_r^R \{H(t) - \cos \pi\sigma G(t)\} \frac{dt}{t^{3/2+\sigma}}$$

$$+ [H(R) - \cos \pi\sigma G(R)]R^{-1/2-\sigma} - [H(r) - \cos \pi\sigma G(r)]r^{-1/2-\sigma}.$$

Hence (3.13), and (1.5) (with $\gamma = \tfrac{1}{2}$) imply

(3.15) $$(\tfrac{1}{2} + \sigma) \int_r^R \{H(t) - \cos \pi\sigma G(t)\} \frac{dt}{t^{3/2+\sigma}}$$

$$\leq B \frac{\log f(r)}{r^\sigma} - k(\sigma) \frac{\log f(R)}{R^\sigma} - \frac{H(R) - \cos \pi\sigma G(R)}{R^{1/2+\sigma}}$$

for a suitable constant $B > 0$.

Consider the first inequality in (3.9), and notice first that it is trivial when $\rho = 0$; hence we assume that $0 < \rho \leq 1$. If, then,

(3.16) $$\liminf_{x \to \infty} \frac{H(x)}{G(x)} > \cos \pi\rho,$$

there exist constants x_0 and σ, $0 < \sigma < \rho$, such that

(3.17) $$H(x) - \cos \pi\sigma G(x) > 0 \qquad (x > x_0).$$

Let $r > x_0$ be given, and use (3.17) in (3.15) to deduce

(3.18)
$$0 < (\tfrac{1}{2} + \sigma) \int_r^R \{H(t) - \cos \pi\sigma G(t)\} \frac{dt}{t^{3/2+\sigma}} \leq B \frac{\log f(r)}{r^\sigma} - k(\sigma) \frac{\log f(R)}{R^\sigma}$$

for all $R > r > x_0$. Since $\sigma < \rho$, there exists a sequence $R_\nu \to \infty$ such that

$$\log f(R_\nu) R_\nu^{-\sigma} \to \infty \qquad (\nu \to \infty)$$

and thus (3.18) yields a contradiction. Hence (3.16) is false, and the first inequality in (3.9) is true for all $\rho \leq 1$.

We consider now the right inequality in (3.9), and assume that it is false; then

(3.19) $$H(t) - \cos \pi\sigma G(t) < 0 \qquad (t > t_0)$$

holds for a suitable σ $(\rho < \sigma < 1)$ if $\rho < 1$, and for $\sigma = 1$ if $\rho = 1$. In either case,

(3.20) $$\log f(r) = o(r^\sigma) \qquad (r \to \infty),$$

since f has genus zero.

It is easy to see that (3.11), (3.14) remain valid with $\sigma = 1$, if we define

$$k(1) = \lim_{\sigma \to 1-} k(\sigma) = 0.$$

Hence, for the choice of σ made in the previous paragraph, we deduce

(3.21) $$(\tfrac{1}{2} + \sigma) \int_r^R \{H(t) - \cos \pi\sigma G(t)\} \frac{dt}{t^{3/2+\sigma}} \geq -10 \frac{\log f(R)}{R^\sigma}$$

$$+ [H(r) - \cos \pi\sigma G(r)] r^{-1/2-\sigma} - [H(R) - \cos \pi\sigma G(R)] R^{-1/2-\sigma}.$$

For the same choice of σ, consider the closed set

$$\mathscr{E}_\sigma = \{r \geq 0 : H(r) - \cos \pi\sigma G(r) \geq 0\}.$$

By our assumption that (3.9) is false, (3.19) holds and hence $\mathscr{E}_\sigma$ is bounded. Put $r' = \sup \mathscr{E}_\sigma$, and use (3.21) with $r = r'$ to deduce

(3.22) $$0 > (\tfrac{1}{2} + \sigma) \int_{r'}^R \{H(t) - \cos \pi\sigma G(t)\} \frac{dt}{t^{3/2+\sigma}} \geq -10 \frac{\log f(R)}{R^\sigma}$$

for every $R > r'$. But this is impossible since, by the definition of r', the middle term in (3.22) decreases as $R \to +\infty$, while by (3.20) the right-hand term tends to zero. Hence the second inequality in (3.9) must be true, and this completes the proof of Lemma 1.

4. Remarks on Theorem 1

There is a variant of Theorem 1, of some interest in itself, which is easy to prove using the methods developed above. An application of this modified form of Theorem 1 will be given in another paper (cf. [7]).

Theorem 1a. *Let f be an entire function of the form* (1), *and assume that there exists a set E of density zero such that*

$$(4.1) \qquad a = \lim_{\substack{x \to \infty \\ x \notin E}} \frac{\log |f(-x)|}{\log f(x)}$$

exists.

Then $-1 \leq a \leq 1$, *and* (5, 0), (5, N) *hold with ρ determined by*

$$\cos \pi\rho = a \qquad (0 \leq \rho \leq 1).$$

Proof. Notice that, by the definition of sets of density zero, there exists a sequence $\{x_n\}$ satisfying (4) and such that

$$\log |f(-x_n)| > (a - \varepsilon)\log f(x_n) \qquad (n > n_0(\varepsilon)),$$

and hence the arguments of Section 2 lead to

$$\log |f(-xe^{i\delta})| > (a - \varepsilon)\log f(x) \qquad (x > x_0(\varepsilon, \delta))$$

and, if $0 \leq \rho < 1$,

$$(4.2) \qquad \liminf_{x \to \infty} \frac{h(x)}{g(x)} \geq 1 + a$$

where g, h are defined by (1.1), (1.2). If $\rho = 1$, we deduce instead

$$(4.3) \qquad \liminf_{x \to \infty} \frac{H(x)}{G(x)} \geq a$$

by the arguments of Section 2 and 3. (That $a > -\infty$ is clear.)
    On the other hand, since

$$\log |f(-x)| < (a + \varepsilon)\log f(x) \qquad (x > x_0(\varepsilon), x \notin E)$$

and outside of E we know that $\log|f(-x)| \leq \log f(x)$, it follows easily from the basic properties of sets of density zero that

$$h(x) \leq (1 + a + o(1))g(x) \qquad (x \to \infty)$$

when $0 \leq \rho < 1$, with the corresponding inequalities for H and G if $\rho = 1$. These facts together with (4.2) and (4.3) yields (1.3), with α replaced by a, and the arguments of Sections 1, 3 complete the proof of Theorem 1a.

5. Proof of Theorem 2

The proof of Theorem 2 can be made to follow directly from the method of Section 1. We determine λ $(0 < \lambda < 1)$ by

$$(5.1) \qquad \lambda = \phi/\pi$$

(there is no loss of generality in assuming $\phi > 0$, since $|f(re^{i\phi})|$ is even in ϕ), and let

$$(5.2) \qquad f_1(z) = f(z^\lambda).$$

Then $f_1(z)$ is holomorphic and free of zeros on the plane cut along the negative axis, and $|f_1(z)|$ is continuous for $|\arg z| \leq \pi$. Also

$$(5.3) \quad \log|f_1(z)| \leq \log f_1(r) = O(r^{\rho_1 + \varepsilon}) \qquad (|\arg z| \leq \pi; r = |z| \to \infty)$$

holds for each $\varepsilon > 0$, and for no $\varepsilon < 0$, with $\rho_1 = \lambda\rho$ and with ρ $(0 \leq \rho \leq 1)$ the order of $f(z)$.

It follows that the Kjellberg representation (1.8) may be applied to $f_1(z)$, and then (5.2) yields

$$\log f_1(x) = \int_0^\infty g_1(t)K(x/t)\frac{dt}{t}$$

with

$$g_1(t) = \log f_1(t) + \log|f_1(-t)|$$
$$= \log f(t^\lambda) + \log|f(t^\lambda e^{i\pi\lambda})|,$$

and

$$K(u) = \frac{2}{\pi^2}\frac{\log u}{u - u^{-1}}$$

as in Section 1.

By hypothesis (1.1) and (5.1),

$$(5.4) \qquad \lim_{t \to \infty}\frac{g_1(t)}{\log f_1(t)} = 1 + \lim_{t \to \infty}\frac{\log|f(t^\lambda e^{i\phi})|}{\log f(t^\lambda)} = 1 + A.$$

To see that $1 + A \neq 0$, notice that the function $u(z)$ defined for $|\arg z| \leq \pi$ by

$$u(z) = \log |f(z^\lambda)|$$

is subharmonic in the whole plane, with

$$\max_{|z|=r} u(z) = \log f(r^\lambda), \quad \min_{|z|=r} u(z) = \log |f(r^\lambda e^{i\pi\lambda})|.$$

Hence the "$\cos \pi\rho$ theorem" for subharmonic functions [12, 2] implies

$$\limsup_{r\to\infty} \frac{\log |f(r^\lambda e^{i\pi\lambda})|}{\log f(r^\lambda)} \geq \cos \pi\rho_1,$$

so that by (5.1), (5.3), and (5.4)

$$A \geq \cos \pi\rho_1 = \cos \phi\rho \geq \cos \phi > -1.$$

As in Section 1, we can apply Theorem 1 of [6] (or [18]) to deduce

$$\log f_1(r) = r^\sigma L_1(r),$$

with $L_1(r)$ slowly varying and σ determined by

$$\int_0^\infty t^{\sigma-1} K(t)\, dt = (1 + A)^{-1} \qquad (0 \leq \sigma < 1).$$

But $f(z)$ has order ρ, and (5.2) and (5.1) hold, so that in fact $\sigma = \lambda\rho = \phi\rho/\pi$, and

$$(1 + A)^{-1} = \int_0^\infty t^{\lambda\rho-1} K(t)\, dt = (1 + \cos \phi\rho)^{-1} \qquad (0 \leq \rho \leq 1),$$

$$\log f(r) = r^\rho L(r),$$

where $L(r) = L_1(r^{1/\lambda})$ is again slowly varying.

We have thus shown that (11) implies (12), and also $(5, \theta)$ for $\theta = 0$, ϕ. To get $(5, \theta)$ for general θ is easy when $0 \leq \rho < 1$, since by Valiron's theorem $(5, 0)$ implies $(5, N)$, and then an abelian argument (see, for example, [9, Theorem 2]) may be used to prove

(5.5) $\log |f(re^{i\theta})| = [\cos \theta\rho + o(1)]r^\rho L(r) \qquad (r \to \infty),$

uniformly in any interval $|\theta| \leq \pi - \eta < \pi$.

When $\rho = 1$, $(5, N)$ shows that the relation between $N(r, 0)$ and $r^\rho L(r)$ breaks down. In fact, $N(r, 0)$ need not be of regular variation in this case (see [1, p. 67] for a counterexample), and although (5.5) remains true its proof is a little more difficult.

One way to prove (5.5) when $\rho = 1$ involves using formula (17) in [5] together with Baernstein's extension of Wiener's tauberian theorem (see [1, Chapter II, Lemma 1]); the details here are easy to supply, provided one knows that the function

$$g(t) = t^{-\lambda} \log |f(t^\lambda e^{i\theta})| \qquad (\lambda = \theta/\pi)$$

is slowly decreasing in the sense of [22, p. 211]. But this follows at once from the mean value theorem together with the estimate

(5.6)
$$\left| \frac{\partial}{\partial r} \log |f(re^{i\theta})| \right| < \frac{K}{r} \log f(r),$$

where $K = K(\theta) < \infty$ for $0 \leq \theta < \pi$. [The inequality (5.6) is an easy consequence of

$$\left| \frac{\partial}{\partial r} \log |f(z)| \right| \leq \left| \frac{\partial}{\partial r} \log f(z) \right| \leq \int_0^\infty \frac{n(t, 0)}{|t + z|} \frac{dt}{t} + \int_0^\infty \frac{|z|\, n(t, 0)}{|t + z|^2} \frac{dt}{t}$$

and

$$|t + z| \geq (t + |z|)\sin \tfrac{1}{2}(\pi - \theta) \qquad (\arg z = \theta).]$$

This leads to (5.5) for $|\theta| < \pi/2$ and also for $\pi/2 < |\theta| < \pi$, and then the θ-monotonicity of $\log |f(re^{i\theta})|$ implies (5.5) also for $\theta = \pi/2$, as well as the statement about uniform convergence in (5.5).

An alternative proof, depending only on function-theoretic methods, follows from Nevanlinna's "two-constant" inequality applied to the functions

$$u_t(z) = \frac{\log f(z)}{zL(t)} - 1 \qquad (t > 0)$$

in the region $\tfrac{1}{2}t \leq |z| \leq 2t$, $0 \leq \arg z \leq \pi - \tfrac{1}{2}\eta \ (\eta > 0)$. Minor modifications in an argument of Bowen and Macintyre [4] then lead to

$$\lim_{t \to \infty} \frac{\log f(te^{i\theta})}{tL(t)} = e^{i\theta},$$

uniformly for $|\theta| \leq \pi - \eta$, and hence to (5.5). For details of the proof, we refer the reader to [4, pp. 244–245].

It might be of some interest to point out that Theorem 2 remains true even if hypothesis (11) is assumed only for a sequence $r = r_n \to \infty$ satisfying $r_{n+1}/r_n \to 1$. In this case, one merely notices that (5.6) and the mean value theorem in fact imply (11) for *all* $r \to \infty$, so that the two hypotheses are actually equivalent.

6. Deduction of Corollary 2.1

By definition of $\beta(r)$, the limit β_0 defined in (12) satisfies $\pi/2 \leqq \beta_0 \leqq \pi$. If $\beta_0 = \pi$, then by a lemma of Edrei and Fuchs [8, p. 322]

$$m(r, 1/f) = o(T(r,f)) \qquad (r \to \infty),$$

and hence $N(r, 0) \sim T(r,f)$. (Here we have used the standard notation and first fundamental theorem of Nevanlinna's theory of meromorphic functions, as described, for example, in [10].) But this contradicts the result

$$\liminf_{r \to \infty} \frac{N(r, 0)}{T(r, f)} \leqq \sin \pi\rho,$$

valid for functions of the form (1) having order ρ, $\frac{1}{2} < \rho \leqq 1$ ([17], Theorem 1). Hence $\pi/2 \leqq \beta_0 < \pi$, and we can appeal to the inequality

$$(6.1) \qquad \left| \frac{\partial}{\partial \theta} \log |f(re^{i\theta})| \right| \leqq K(\delta) \log f(r) \qquad (|\theta| \leqq \pi - \delta < \pi),$$

easily established in the same way that (5.6) was. Choose $\delta > 0$ so that $\beta(r) < \beta_0 + \delta < \pi - \delta \ (r > r_0)$, and notice that by (6.1) and the mean value theorem

$$|\log |f(re^{i\beta_0})|| \leqq K(\delta) |\beta_0 - \beta(r)| \log f(r) \qquad (r > r_0).$$

Thus f satisfies the conditions of Theorem 2, with $\phi = \beta_0$ and $A = 0$, and the conclusions of the corollary then follow from those of Theorem 2.

REFERENCES

1. A. Baernstein, A nonlinear tauberian theorem in function theory and some results on tauberian oscillations, Ph.D. dissertation, University of Wisconsin, 1968.
2. P. D. Barry, On a theorem of Besicovitch, *Quart. J. Math. Oxford Ser. 2* **14** (1963), 293–302.
3. R. P. Boas, Jr., *Entire Functions*, Academic Press, New York, 1954.
4. N. A. Bowen and A. J. Macintyre, An oscillation theorem of tauberian type, *Quart. J. Math. Oxford Ser. 2* **1** (1950), 243–247.
5. ———, Some theorems on integral functions with negative zeros, *Trans. Amer. Math. Soc.* **70** (1951), 114–126.
6. D. Drasin, Tauberian theorems and slowly varying functions, *Trans. Amer. Math. Soc.* **133** (1968), 333–356.
7. D. Drasin and D. F. Shea, Asymptotic properties of entire functions extremal for the cos $\pi\rho$ theorem, *Bull. Amer. Math. Soc.* **75** (1969), 119–122.
8. A. Edrei and W. H. J. Fuchs, Bounds for the number of deficient values of certain classes of meromorphic functions, *Proc. London Math. Soc.* **12** (1962), 315–344.

9. ———, Tauberian theorems for a class of meromorphic functions with negative zeros and positive poles, *Contemporary problems of the theory of analytic functions*, "Nauka," Moscow, 1966, 339–358.

10. W. K. Hayman, *Meromorphic functions*, Clarendon Press, Oxford, 1964.

11. S. Hellerstein and J. Williamson, Entire functions with negative zeros and a problem of R. Nevanlinna, *J. Analyse Math.* **22** (1969), 233–267.

12. A. Huber, Über Wachstumseigenschaften gewisser Klassen von subharmonischen Funktionen, *Comment. Math. Helv.* **26** (1952), 81–116.

13. J. Karamata, Sur un mode de croissance régulière, *Mathematica (Cluj)* **4** (1930), 38–53.

14. B. Kjellberg, A relation between the maximum and minimum modulus of a class of entire functions, C. R. du 12. Congrès des Mathématiciens Scandinaves tenu á Lund, 1953 (Lund, 1954), 135–138.

15. ———, A theorem on the minimum modulus of entire functions, *Math. Scand.* **12** (1963), 5–11.

16. B. Ja. Levin, *Distribution of zeros of entire functions*, American Mathematical Society Translation, Providence, 1964.

17. D. F. Shea, On the Valiron deficiencies of meromorphic functions of finite order, *Trans. Amer. Math. Soc.* **124** (1966), 201–222.

18. ———, On a complement to Valiron's tauberian theorem for the Stieltjes transform *Proc. Amer. Math. Soc.*, **21** (1969), 1–9.

19. E. C. Titchmarsh, On integral functions with real negative zeros, *Proc. London Math Soc.* (2) **26** (1927), 185–200.

20. G. Valiron, Sur les fonctions entières d'ordre nul et d'ordre fini, et en particulier les fonctions a correspondance régulière, *Ann. Fac. Sci. Univ. Toulouse* (3) **5** (1913), 117–257.

21. E. T. Whittaker and G. N. Watson, *A course in modern analysis*, Cambridge University Press, Cambridge, England, 1927.

22. D. V. Widder, *The Laplace transform*, Princeton University Press, Princeton, N.J., 1946.

(Received October 1, 1968)

Some Extremal Problems in Combinatorial Number Theory

Paul Erdős

Nemetvolgyi UT 72C
Budapest XII
Hungary
and
Department of Mathematics
University of Colorado
Boulder, Colorado 80302
U.S.A.

In this note I discuss problems in number theory most of which have a combinatorial flair. Section 2 is a joint work with A. Sárkőzi and E. Szemerédi.

First we introduce some notations which will be used frequently in this paper. The sequence $a_1, a_2, \ldots$, will be denoted by A, $A(x) = \sum_{a_i \leq x} 1$. The limit, $\lim_{x \to \infty} A(x)/x$, if it exists, is called the density of A (the upper density is the lim sup of the same expression). The term $V(n)$ denotes the number of prime factors of n, and $V(n, l)$, the number of prime factors of n not exceeding l (in both cases multiple factors are counted multiply). The symbols $c, c_1, \ldots$, will denote positive absolute constants not necessarily the same at each occurrence; $\varepsilon, \delta, \eta$ denote positive numbers which can be chosen arbitrarily small. The letters $a, b, t, l, \ldots$ denote integers; p is a prime; $P(t)$ is the greatest and $p(t)$ the least prime factor of t.

1.

Denote by $f(k, x)$ the maximum number of integers $a_1 < \cdots < a_r \leq x$ so that no k of them have pairwise the same common divisor. I have proved [6] that for every k if $x > x_0(k)$

(1) $\exp\left(c_k \dfrac{\log x}{\log \log x}\right) < f(k, x) < x^{3/4 + \varepsilon},$ where $\exp z = e^z.$

It was conjectured in [6] that the lower bound seems to give the right order of magnitude for $f(k, x)$.

Denote by $F(k, x)$ the maximum number of integers $a_1 < \cdots < a_s \leqq x$ so that no k of them have pairwise the same least common multiple. I conjectured that $F(k, x) = o(x)$ for every $k \geqq 3$. Recently, I proved that for $k \geqq 4$ this conjecture is certainly false. At present I cannot disprove this conjecture for $k = 3$.

The falseness of the conjecture will easily follow from the following result which is of independent interest:

Theorem 1. *The density of integers having three relatively prime divisors satisfying $b_1 < b_2 < b_3 < 2b_1$ exists and is less than* 1.

I have proved [7] that the density of integers having two relatively prime divisors $b_1 < b_2 < 2b_1$ is 1. The proof has not been published and is quite complicated, but we will not need this result here.

Let us assume that Theorem 1 is already proved. Then consider the integers $x/2 < a_1 < \cdots < a_s < x$ no one of which has three pairwise relatively prime divisors $b_1 < b_2 < b_3 < 2b_1$. By our theorem $s > cx$. Now we show that there are no four a's, say $a_1 < a_2 < a_3 < a_4$, satisfying

(2) $[a_i, a_j] = T,$ $1 \leqq i < j \leqq 4.$

To see this assume that (2) holds. Put $T/a_i = b_i$, $1 \leqq i \leqq 4$. Clearly $b_j \mid a_i$ for $j \neq i$ and $(b_i, b_j) = 1$, $1 \leqq i < j \leqq 4$. Finally from $x/2 < a_1 < a_2 < a_3 < a_4 < x$ we obtain $b_2 < b_3 < b_4 < 2b_2$. Thus a_1 would have three divisors $b_2 < b_3 < b_4 < 2b_2$, $(b_i, b_j) = 1, 2 \leqq i < j \leqq 4$, which contradicts our assumptions. Hence $F(4, x) > cx$ as stated.

Thus we only have to prove Theorem 1. First we show the following:

Lemma 1. *Let $1 < u_1 < \cdots$ be any sequence of integers. Denote by d the density, and by $\bar{d}(u_1, \ldots)$ the upper density of the integers having at least one divisor amongst the u's. Assume that for every $\varepsilon > 0$ there is a k satisfying*

(3) $\bar{d}(u_{k+1}, \ldots) < \varepsilon.$

Then $d(u_1, \ldots)$ exists and is less than 1.

A theorem of Behrend [2] states that if $a_1 < \cdots < a_k$ and $b_1 < \cdots < b_l$ are any two sequences of integers then

(4) $1 - d(a_1, \ldots, a_k, b_1, \ldots, b_l) \geqq (1 - d(a_1, \ldots, a_k))(1 - d(b_1, \ldots, b_l)).$

From (3) and (4) we obtain by a simple limiting process that

(5) $$1 - \bar{d}(u_1, \ldots) > (1 - \eta)(1 - d(u_1, \ldots, u_k)).$$

Inequality (5) easily implies Lemma 1. (The term $d(u_1, \ldots, u_k)$ clearly exists for every finite set $u_1, \ldots, u_k$.)

Now let $u_1 < \cdots$ be the sequence of integers which can be written in the form

(6) $$b_1 b_2 b_3, b_1 < b_2 < b_3 < 2b_1, (b_i, b_j) = 1, 1 \leq i < j \leq 3.$$

To prove Theorem 1 it suffices to show that the u's satisfy (3). Denote by $m_1 < \cdots$ the integers which are divisible by at least one u_i, $i > k$; we have to show that for $k > k_0(\varepsilon)$ the upper density of the m's is less than ε. A theorem of mine states [8] that for every ε and δ there is an l such that the density of integers n which for some $t > l$ do not satisfy

$$(1 - \delta)\log \log t < V(n, t) < (1 + \delta)\log \log t$$

is less than $\varepsilon/2$. Thus to prove that the u's satisfy (3) it will suffice to show that for sufficiently small δ and $k > k_0(\varepsilon, \delta, l)$ the upper density of the m's satisfying

(7) $$V(m_i, t) < (1 + \delta)\log \log t$$

for every $t > l$ is less than $\varepsilon/2$.

Showing that this statement is true will be the main difficulty of our proof. First of all observe that for $k > k_0(l)$ (6) implies that every m_j has a divisor of the form

(8) $$b_1 b_2 b_3, l < 2^s < b_1 < b_2 < b_3 < 2^{s+2}, (b_i, b_j) = 1, 1 \leq i < j \leq 3.$$

We now prove

Lemma 2. *The upper density of the integers m_j satisfying (7) and having a divisor of the form (8) is $O(1/s^{1+c})$.*

Since $\sum_{s=1}^{\infty} 1/s^{1+c}$ converges, it immediately follows from Lemma 2 that the upper density of the m's satisfying (7) is less than $\varepsilon/2$. Thus to complete the proof of Theorem 1 we only have to prove Lemma 2. Some of the elementary computations needed in this proof, we will not carry out in full detail.

Clearly every m_j satisfying (7) can be written in the form

(9) $$b_1 b_2 b_3 t_1 t_2, P(t_1) < 2^{s+2}, \quad p(t_2) > 2^{s+2}$$

where the b's satisfy (8) and

(10) $$V(b_1 b_2 b_3 t_1) < (1 + 2\delta)\log s,$$

It easily follows from the sieve of Eratosthenes and the well-known theorems of Mertens that the upper density of the integers of the form (9) and (10) is at most

(11)
$$\sum{}' \frac{1}{b_1 b_2 b_3 t_1} \prod_{p < 2^{s+2}} \left(1 - \frac{1}{p}\right) < \frac{c}{s} \sum{}' \frac{1}{b_1 b_2 b_3 t_1}$$

where the prime indicates that $b_1 b_2 b_3$ satisfies (8) and $b_1, b_2, b_3 t_1$ satisfies (10).

Thus to complete the proof of Lemma 2 we only have to prove that for a sufficiently small c

(12)
$$\sum{}' \frac{1}{b_1 b_2 b_3 t_1} = O\left(\frac{1}{s^c}\right).$$

Now clearly (in $\sum_{r}$, $V(t) = r$, $t < 2^{s+2}$)

(13)
$$\sum_r \frac{1}{t} < \left(\sum_{p^\alpha < 2^{s+2}} \frac{1}{p^\alpha} \right) \Big/ r! < (\log s + c_1)^r / r!.$$

A well-known theorem of Hardy and Ramanujan states [13] that

(14)
$$\Pi_r(x) < c_2 \, x \, \frac{(\log \log x + c_3)^{r-1}}{(r-1)! \log x},$$

where $\Pi_r(x)$ denotes the number of integers $t < x$ satisfying $V(t) = r$.

From (14) we obtain (in $\sum_r{}' V(b) = r$, $2^s < b < 2^{s+2}$)

(15)
$$\sum_r{}' \frac{1}{b} < \Pi_r(2^{s+2})/2^s < c_4 \frac{(\log s + c_3)^{r-1}}{(r-1)! s}.$$

From (13) and (15) we obtain

(16)
$$\sum{}' \frac{1}{b_1 b_2 b_3 t'} \le \sum_1 \left(\sum_{r_1}{}' \frac{1}{b_1} \sum_{r_2}{}' \frac{1}{b_2} \sum_{r_3}{}' \frac{1}{b_3} \sum_{r_4}{}' \frac{1}{t_1} \right)$$

where in $\sum_1$ $r_1 + r_2 + r_3 + r_4 < (1 + 2\delta)\log s$. Using (13) and (15) we obtain by a simple calculation that the terms of the inner sum on the right side of (16) are maximal if

(17)
$$r_i = (1 + o(1))(\tfrac{1}{4} + \delta/2)\log s, \qquad i = 1, 2, 3, 4.$$

From (13), (14), (16), and (17) we easily obtain by a simple computation $(r_i = (1 + o(1))(\tfrac{1}{4} + \delta/2)\log s, \; \eta = \eta(\delta)$ tends to 0 as $\delta \to 0)$

(18)
$$\sum{}' \frac{1}{b_1 b_2 b_3 t_1} < c_5 \frac{(\log s)^{c_6}}{s^3} \prod_{i=1}^{4} \frac{(\log s)^{r_i}}{r_i!}$$

$$< c_5 \frac{(\log s)^{c_6}}{s^3} \prod_{i=1}^{4} \frac{(\log s)^{r_i} e^{r_i}}{r_i^{r_i}} < \frac{s^\eta}{s^2} 4^{(1+o(1))(1+2\delta)\log s} < \frac{1}{s^c}$$

for sufficiently large s if δ and $\eta = \eta(\delta)$ are sufficiently small. Relation (18) proves (12) and thus the proof of Theorem 1 is complete.

It is easy to see that by our method we can construct a sequence A of positive upper density so that there are no four integers $a_i \in A$ which have pairwise the same least common multiple. On the other hand, it is easy to see that if $x > x_0(c, k)$ then

$$(19) \qquad \qquad \sum_{a_i < x} \frac{1}{a_i} > c \log x$$

implies that there are k a's which have pairwise the same least common multiple. In fact (19) implies that for $x > x_0(c, k)$ there is a t such that

$$(20) \qquad \qquad t = a_i p$$

has at least k solutions. To see this observe that if (20) had fewer than k solutions we would have

$$k \sum_{t < x^2} \frac{1}{t} > \sum_{a_i < x} \frac{1}{a_i} \sum_{p < x} \frac{1}{p} > c \log x \log \log x,$$

an evident contradiction.

I do not know how much (19) can be weakened so that there should always be k a's every two of which have the same least common multiple. This question seems connected with the following combinatorial problem: Let $\mathscr{S}$ be a set of n elements, $A_i \subset \mathscr{S}$, $1 \leqq i \leqq m(n, k)$. What is the smallest value of $m(n, k)$ for which we can be sure that there are k A's which have pairwise the same union? An asymptotic formula for $m(n, k)$ would also be of some interest.

Before concluding this section I would like to say a few words about equation (20). Assume first that our sequence is such that (19) has only one solution for every t, in other words the integers a_i/p_j, $p_j | a_i$ are distinct for all i and j. It is not difficult to prove that in this case

$$(21) \qquad \max A(x) = \frac{x}{\exp((c + o(1))(\log x \log \log x)^{1/2})}.$$

The proof of (21) uses methods similar to those in [9] and will not be discussed here.

By the methods used in proving Theorem 1, it is not difficult to prove that there is a sequence A of positive upper density such that (20) has for every t at most two solutions.

It would be of some interest to obtain best possible (or at least good) inequalities on $\sum_{a_i < x} 1/a_i$ which ensure that (20) has at least k solutions for some t.

2.

Let $a_1 < \cdots$ be a sequence of integers no one of which divides any other. I proved [10] that there exists an absolute constant c such that

$$(22) \qquad \sum_i \frac{1}{a_i \log a_i} < c$$

and Behrend [3] proved that

$$(23) \qquad \sum_{a_i < x} \frac{1}{a_i} < \frac{c \log x}{(\log \log x)^{1/2}}.$$

Alexander [1] and later Sárközi, Szemerédi, and I strengthened [22] in the following sense: There is an absolute constant c_1 such that if $a_1 < \cdots$ is any sequence such that

$$(24) \qquad a_i t = a_j, \qquad p(t) > a_i$$

is unsolvable, then

$$(25) \qquad \sum_i \frac{1}{a_i \log a_i} < c_1.$$

Inequality (25) easily implies that if a sequence of integers satisfies (24) then it also satisfies

$$(26) \qquad \sum_{a_i < x} \frac{1}{a_i} = o(\log x).$$

Now we show that (26) is best possible. In other words if $f(x) \to \infty$ as slowly as we wish there always exists an infinite sequence satisfying (24) such that for infinitely many x

$$(27) \qquad \sum_{a_i < x} \frac{1}{a_i} > \frac{\log x}{f(x)}.$$

Equation (27) is indeed very easy to see. Let $x_1 < x_2 < \cdots$ tend to infinity sufficiently fast. Let our sequence A consist of the integers in $(x_i^{1/2}, x_i)$ which have no prime factor less than x_{i-1} but have a prime factor greater than $x_i^{1/2}$. A simple argument shows that our sequence satisfies (24), and if $x_i \to \infty$ sufficiently fast then it also satisfies (27) for $x = x_i$.

We can now ask, if $a_1 < \cdots < a_k < x$ satisfy (24) what is the value of

$$\max_A \frac{1}{\log x} \sum_{a_i < x} \frac{1}{a_i},$$

where the maximum is taken over all such sequences? The maximum is clearly less than 1.

It is well known that the upper density of any sequence of integers no one of which divides any other is less than $\frac{1}{2}$ and any number $\alpha < \frac{1}{2}$ can be the upper density of such a sequence [4]. Similarly one can show that the upper density of any sequence satisfying (24) has upper density less than 1 and any $\beta < 1$ can be the upper density of such a sequence.

It is well known and is easy to see [10] that if $a_1 < \cdots < a_k \leq x$ is such that no a divides any other then

$$\max A(x) = \left[\frac{x+1}{2}\right].$$

Now let $a_1 < \cdots < a_l \leq x$ be a sequence which satisfies (24). We outline the proof that

(28) $$\max A(x) = x - \frac{x}{\exp((\log x)^{1/2+o(1)})}.$$

We can in fact easily write down the sequence $A = \{a_1 < \cdots < a_l \leq x\}$ which maximizes l. $a_i \in A$ if and only if $a_i = p_1, p_2, \ldots, p_j, p_1 \leq \cdots \leq p_j$ and $p_1 \cdots p_j \leq x < p_1 \cdots p_j p_{j+1}$ where p_{j+1} is the least prime greater than p_j. Our sequence clearly satisfies (24). To show that it maximizes l, let $A' = \{a_1' < \cdots < a_{l'} \leq x\}$ be a sequence of integers satisfying (24). It suffices to show that if A' contains r integers not contained in A then A contains at least r integers not in A'. To see this let $u_1 < \cdots < u_s$ be the integers not in A. There clearly is a $p^{(i)} > p(u_i)$ so that $u_i p^{(i)} \in A$. Now these integers must be all distinct. To see this observe that

(29) $$u_i p_1 \neq u_j p_2 \quad \text{where} \quad p_1 > p(u_i), p_2 > p(u_j).$$

To prove (29) observe that we can assume $p_1 \neq p_2$. Thus without loss of generality we have $p_2 > p_1$. But then if (29) did not hold we would have $p_2 | u_i$ which contradicts $p_2 > p_1 > p(u_i)$.

Now it is easy to prove (28). On the one hand consider all the integers n satisfying

(30) $$n < \frac{x}{\exp(2(\log x)^{1/2})}, \qquad p(n) < \exp((\log x)^{1/2}).$$

It is easy to see that none of the integers (30) belong to A and a simple computation gives that their number is greater than $x/\exp((\log x)^{\frac{1}{2}+\varepsilon})$ for every $\varepsilon > 0$ if $x > x_0(\varepsilon)$.

To prove the opposite inequality split the integers not in A into two classes. In the first class are the integers n with $P(n) < \exp((\log x)^{1/2})$. By the results of de Bruijn [5] and others the number of these integers not exceeding x is less than $x/\exp((\log x)^{1/2+o(1)})$. If n is in the second class we have $P(n) \geq \exp((\log x)^{1/2})$. But then since n is not in A we must have $nP(n) < 2x$, or $n < 2x/\exp((\log x)^{1/2})$, which completes the proof of (28).

3.

In this section we investigate some properties of the divisors of n. Let $1 = u_1 < \cdots < u_{d(n)} = n$ be the net of all divisors of n. Denote by A_t the set of those n for which t can be represented as the distinct sum of divisors of n. Clearly if n is in A_t then any multiple of n is also in A_t, and it is easy to see that every integer in A_t is a multiple of an integer in A_t not exceeding $t!$. Thus it easily follows that A_t has a density d_t. It is a little less easy to see that $d_t \to 0$ as $t \to \infty$. To see this we split the integers of A_t into two classes. In the first class are the integers which have a divisor in $(t/(\log t)^2, t)$. I proved [11] that the density of these integers tends to 0 as $t \to \infty$ (in fact the density is $O(1/(\log t)^{c_1})$). The integers of the second class have no divisor in

$$(t/(\log t)^2, t).$$

Thus if t is the sum of divisors of n we must have ($d_t(n)$ denotes the number of divisors of n not exceeding t)

(31) $$d_t(n) > (\log t)^2$$

But clearly

(32) $$\sum_{n=1}^{x} d_t(n) \leq \sum_{u=1}^{t} \frac{x}{u} < 2x \log t.$$

From (32) we obtain that the number of integers $n \leq x$ satisfying (31) is less than $2x/\log t$, or the density of integers of the second class is not greater than $2/\log t$. Hence $d_t \to 0$ (and in fact $d_t < 1/(\log t)^{c_1}$ for $t > t_0$. We can prove that for $t > t_0$, $d_t > 1/(\log t)^{c_2}$. Perhaps

(33) $$d_t = (1 + o(1))c_3/(\log t)^{c_4},$$

but (33) if true may not be quite easy to prove.

An integer n is said to have property P if all the $2^{d(n)}$ distinct sums formed from its $d(n)$ divisors are distinct. One's first guess might be that the integers having property P have density 0. But we prove

Theorem 2. *The density of integers having property P exists and is positive.*

The proof will be similar to [12]. Clearly if m does not have property P then all the multiples of m also do not have property P. Let $m_1 < m_2 < \cdots$ be the sequence of integers which do not have property P but every divisor of them has property P. ($m_1 = 6$.) n has property P if and only if it is not divisible by any of the m's. Thus to prove Theorem 2 we have to show that the density of the integers not divisible by any of the m's exists and is less than 1.

If we could prove that

(34)
$$\sum_{i=1}^{\infty} \frac{1}{m_i} < \infty$$

then as in [12] it would follow that the density of integers having property P exists and is greater than 0. Inequality (34) is quite possibly true but I cannot prove it. Thus we have to argue in a more roundabout way. We split the m's into two classes. In the first class are the $m_i^{(1)}$'s satisfying

(35)
$$V(m_i^{(1)}) > (1 + \varepsilon)\log \log m_i^{(1)}.$$

The $m_i^{(2)}$'s of the second class do not satisfy (35).

Now we prove (see [3])

(36)
$$d(m_1^{(1)}, m_2^{(1)}, \ldots) = \alpha < 1$$

and

(37)
$$d(m_1^{(2)}, m_2^{(2)}, \ldots) = \beta < 1.$$

Using (4) (as in Section 1) we obtain from (36) and (37) that $d(m_1, m_2, \ldots)$ exists and satisfies

(38)
$$1 - d(m_1, m_2, \ldots) \geq (1 - \alpha)(1 - \beta) > 0.$$

In other words the density of integers having property P exists and is positive.

Thus to prove Theorem 2 we only have to prove (36) and (37). Expression (36) indeed follows from my result in [8] as in Section 1. Expression (37) will follow as in [12] from

(39)
$$\sum_i \frac{1}{m_i''} < \infty.$$

To prove (39) it will suffice to show that

(40)
$$\sum_{m_i'' < x} 1 = O\left(\frac{x}{(\log x)^2}\right).$$

To prove (40) we split the $m_i'' < x$ again into two classes. In the first class are the m_i'' satisfying

(41)
$$P(m_i'') < \exp(\log x/(\log \log x)^2).$$

It is well known [12] that the number of integers $m_i'' \leq x$ satisfying (41) is $O(x/(\log x)^2)$.

Thus henceforth it suffices to consider the integers of the second class (not satisfying (41)). Consider the integers $m_i''/P(m_i'')$. They are all less than $x(\exp(\log x/(\log \log x)))^{-1} = x/L$.

Now we prove that for every $t < x/L$ the number of solutions of

(42) $$m_i''/P(m_i'') = t$$

is less than $\exp(\log x/2(\log \log x)^2) = L_1$.

Suppose we already proved that (42) has fewer than L_1 solutions; then we evidently have (in $\sum'$, m_i'' belong to the second class, i.e., they do not satisfy (41))

(43) $$\sum_{m_i'' < x}' 1 < \frac{xL_1}{L} = O(x/(\log x)^2).$$

Expression (43) completes the proof of (40) and hence of Theorem 2.

Let

(44) $$m_{i_r}''/P(m_{i_r}'') = t, \qquad r = 1, \ldots, s$$

be the set of all solutions of (42). Put $P(m_{i_r}'') = p_r$, $r = 1, \ldots, s$. These s primes are clearly all distinct. By our assumptions t has property P but the integers

$$m_{i_r}'' = tp_r, \qquad r = 1, \ldots, s$$

do not have property P. Hence for every r there are divisors $d_j^{(r)}$ of tp_r satisfying

(45) $$\sum_j E_j d_j^{(r)} = 0, E_j = \pm 1,$$

and in the sum (45) at least one $d_j^{(r)}$ must be a multiple of p_r (for otherwise all the $d_j^{(r)}$ would be divisors of t and t would not have property P). Thus for every p_r there is a sum (different from 0) satisfying

(46) $$\sum_u E_u d_u^{(r)} \equiv 0 \pmod{p_r}, \quad d_u^{(r)} | t, \quad E_u = \pm 1.$$

Now since m_i'' does not satisfy (35) we have $V(t) < (1 + \varepsilon)\log \log x$. Hence the number of sums (46) is less than

(47) $$3^{d(t)} < 3^{2^{(1+\varepsilon) \log \log x}} < \exp((\log x)^{1-c}).$$

Each of the sums (46) has fewer than $\log x$ prime divisors, thus from (46) and (44) we have

$$s < \log x \exp((\log x)^{1-c}) < L_1$$

which completes the proof of Theorem 2.

By the same method we can prove that the density of integers n for which n is the sum of distinct proper divisors of n exists and is between 0 and 1. Several other related results can be proved by this method.

REFERENCES

1. R. Alexander, Density and multiplicative structure of sets of integers, *Acta Arith.* **12** (1967), 321–332.
2. F. Behrend, Generalization of an inequality of Heilbronn and Rohrbach, *Bull. Amer. Math. Soc.* **54** (1948), 681–684.
3. ———, On sequences of numbers not divisible one by another, *J. London Math. Soc.* **10** (1935), 42–44.
4. A. S. Besicovitch, On the density of certain sequences, *Math. Ann.* **110** (1934), 336–341.
5. N. G. de Bruijn, On the number of positive integers $\leq x$ and free of prime factors $\geq y$, *Nederl. Akad. Wetensch. Ind. Math.* **13** (1951), 50–60.
6. P. Erdős, On a problem in elementary number theory and a combinatorial problem, *Math. Comp.* **18** (1964), 644–646.
7. ———, On some applications of probability to analysis and number theory, *J. London Math. Soc.* **39** (1964), 692–696.
8. ———, On the distribution function of additive functions, *Ann. of Math.* **47** (1946), 1–20.
9. ———, On primitive abundant numbers, *J. London Math. Soc.* **10** (1935), 49–58.
10. ———, Note on sequences of integers no one of which is divisible by any other, *J. London Math. Soc.* **10** (1935), 126–128.
11. ———, A generalization of a theorem of Besicovitch, *J. London Math. Soc.* **11** (1936), 92–98.
12. ———, On the density of abundant numbers, *J. London Math. Soc.* **9** (1934), 278–282.
13. G. H. Hardy, P. V. Seshu Aiyar, and B. M. Wilson, Eds., *Collected Papers of Srinivasa Ramanujan*, Cambridge University Press, Cambridge, England, 1927, pp. 262–275.

(Received March 20, 1968)

A Theorem Concerning the Real Part of a Power Series

T. M. Flett

Department of Mathematics
The University of Sheffield
Sheffield
England

1.

Let ϕ be a function regular in the unit disk $\Delta = \{z \in \mathbf{C}: |z| < 1\}$, and let

$$\phi(z) = \sum_{n=0}^{\infty} c_n z^n \qquad (z \in \Delta).$$

For $0 < p < +\infty$ we write

$$M_p(\phi; \rho) = \left\{ \frac{1}{2\pi} \int_{-\pi}^{\pi} |\phi(\rho e^{i\theta})|^p \, d\theta \right\}^{1/p}.$$

It is familiar that $M_p(\phi; \rho)$ increases with ρ, and therefore tends to a finite or infinite limit as $\rho \to 1-$. We define

$$(1.1) \qquad \mathfrak{M}_p(\phi) = \lim_{\rho \to 1-} M_p(\phi; \rho),$$

the value $+\infty$ being allowed. The class of ϕ for which the limit in (1.1) is finite is, of course, the class H^p.

As in [1], for any $\beta > 0$ we define the fractional derivative $\vartheta^\beta \phi$ of ϕ of order β by

$$\vartheta^\beta \phi(z) = \sum_{n=1}^{\infty} n^\beta c_n z^n \qquad (z \in \Delta).$$

We write also

$$J_{k,\beta}(\phi) = \left\{ \int_{-\pi}^{\pi} \int_{0}^{1} (\log 1/\rho)^{k\beta-1} |\vartheta^{\beta}\phi(\rho e^{i\theta})|^{k} \rho^{-1} \, d\rho \, d\theta \right\}^{1/k},$$

where $k > 0$ and $\beta > 0$.

The following theorem concerning the fractional derivative $\vartheta^{\beta}\phi$ is known (for a proof and further references see [1]).

Theorem A. *Let ϕ be regular in the unit disk. If $\phi(0) = 0$ and $0 < k \leq 2$, $\beta > 0$, then**

(1.2) $$\mathfrak{M}_{k}(\phi) \leq A(k, \beta)J_{k,\beta}(\phi).$$

Now let u be the real part of ϕ, so that u is harmonic in the unit disk, and for $\beta > 0$ let $\vartheta^{\beta}u$ be the real part of $\vartheta^{\beta}\phi$ (thus if

$$u(\rho, \theta) = \tfrac{1}{2}a_{0} + \sum_{n=1}^{\infty} (a_{n} \cos n\theta + b_{n} \sin n\theta)\rho^{n},$$

then

(1.3) $$\vartheta^{\beta}u(\rho, \theta) = \sum_{n=1}^{\infty} n^{\beta}(a_{n} \cos n\theta + b_{n} \sin n\theta)\rho^{n}.$$

Let also

$$J_{k,\beta}(u) = \left\{ \int_{-\pi}^{\pi} \int_{0}^{1} (\log 1/\rho)^{k\beta-1} |\vartheta^{\beta}u(\rho, \theta)|^{k} \rho^{-1} \, d\rho \, d\theta \right\}^{1/k},$$

where $k > 0$ and $\beta > 0$. It is an immediate consequence of M. Riesz's theorem on conjugate functions (see, for example, [4, p. 253]) that if $k > 1$, then

(1.4) $$J_{k,\beta}(\phi) \leq A(k)J_{k,\beta}(u).$$

Thus when $1 < k \leq 2$ we may replace $J_{k,\beta}(\phi)$ on the right of (1.2) by $J_{k,\beta}(u)$. In this note we show that this result holds also for $0 < k \leq 1$, so that we have

Theorem 1. *Let ϕ be regular in the unit disk, and let u be the real part of ϕ. If $\phi(0) = 0$ and $0 < k \leq 2$, $\beta > 0$, then*

(1.5) $$\mathfrak{M}_{k}(\phi) \leq A(k, \beta)J_{k,\beta}(u).$$

We deduce immediately the following corollary.

---

* The inequality (1.2) and other similar inequalities are to be interpreted as meaning "if the right side is finite, then the left side is finite and satisfies the inequality." We use $A(b, c, \ldots)$ to denote a positive constant depending only on $b, c, \ldots$, not necessarily the same on any two occurrences; A by itself will denote a positive absolute constant.

Corollary. *Let u be harmonic in the unit disk, let*

$$M_k(u; \rho) = \left\{ \frac{1}{2\pi} \int_{-\pi}^{\pi} |u(\rho, \theta)|^k \, d\theta \right\}^{1/k},$$

and let $\vartheta^\beta u$ be defined by (1.3). If $u(0) = 0$ and $0 < k \leq 2$, $\beta > 0$, then

(i) $\displaystyle \sup_{0 \leq \rho < 1} M_k(u; \rho) \leq A(k, \beta) J_{k, \beta}(u);$

(ii) u *has a radial limit* $u(\theta) = \lim_{\rho \to 1} u(\rho, \theta)$ *for almost all θ whenever* $J_{k, \beta}(u)$ *is finite.*

We have also the following result concerning $J_{k, \beta}(\phi)$ and $J_{k, \beta}(u)$.

Theorem 2. *Let ϕ be regular in the unit disk, and let u be the real part of ϕ. If $\phi(0) = 0$ and $k > 0$, $\gamma > \beta > 0$, then*

(1.6) $$A(k, \beta, \gamma) J_{k, \gamma}(\phi) \leq J_{k, \beta}(\phi) \leq A(k, \beta, \gamma) J_{k, \gamma}(\phi),$$

(1.7) $$A(k, \beta) J_{k, \beta}(\phi) \leq J_{k, \beta}(u) \leq J_{k, \beta}(\phi).$$

A result effectively equivalent to the case $k > 1$, of (1.6) has been proved by Hardy and Littlewood [3, Theorems 8, 10].[*] Our proof of the left-hand inequality in (1.6) is essentially the same as that of Hardy and Littlewood, but our proof of the right-hand inequality appears to be new.

The right-hand inequality in (1.7) is immediate, and the case $k > 1$ of the left-hand inequality is simply (1.4). The case of the left-hand inequality in which $0 < k \leq 1$ and $k\beta = 1$ has been stated by Hardy and Littlewood [2, Theorem 5], and our proof of the general case follows the line of argument indicated in their paper.[*]

The inequality (1.5) of Theorem 1 follows immediately from (1.7) and Theorem A. To complete the proofs of Theorems 1 and 2 it is therefore enough to prove (1.6) for $k > 0$ and the left-hand inequality in (1.7) for $0 < k \leq 1$.

2.

We begin with the proof of the left-hand inequality in (1.6). We note first that for any ϕ regular in the unit disk and for $k > 0$, $\delta > 0$ we have $M_k(\vartheta^\delta \phi; \rho^2) \leq A(k, \delta)\rho(1 - \rho)^{-\delta} M_k(\phi; \rho) \leq A(k, \delta)(\log 1/\rho)^{-\delta} M_k(\phi; \rho)$ ([1, Theorem 1, Corollary 2]). Since $\vartheta^\delta(\vartheta^\beta \phi) = \vartheta^{\beta + \delta}\phi$, this gives

$$M_k(\vartheta^{\beta + \delta}\phi; \rho^2) \leq A(k, \delta)(\log 1/\rho)^{-\delta} M_k(\vartheta^\beta \phi; \rho).$$

---

[*] Hardy and Littlewood use a different definition of fractional derivative.

[*] In [2], Hardy and Littlewood prove that if $M_p(u; \rho) \leq (1 - \rho)^{-a}$, where $0 < p \leq 1$ and $a > 0$, then $M_p(\phi; \rho) \leq A(a, p)(1 - \rho)^{-a}$. The case $k\beta = 1$ of (1.7) is stated with only the comment that the reasoning is similar to that employed for the other result.

Taking $\delta = \gamma - \beta$, where $\gamma > \beta > 0$, and noting that

$$J_{k,\gamma}^k(\phi) = 2^{k\gamma} \int_0^1 (\log 1/\rho)^{k\gamma - 1} M_k^k(\vartheta^\gamma \phi; \rho^2)\rho^{-1} \, d\rho,$$

we obtain immediately that

$$J_{k,\gamma}^k(\phi) \leq A(k, \beta, \gamma) \int_0^1 (\log 1/\rho)^{k\beta - 1} M_k^k(\vartheta^\beta \phi; \rho)\rho^{-1} \, d\rho = A(k, \beta, \gamma)J_{k,\beta}^k(\phi),$$

as required.

Consider now the right-hand inequality in (1.6). It has been proved in [1, Theorem 2] that if $k \geq 1$ and $\gamma > \beta > 0$, then

$$\int_0^1 (\log 1/\rho)^{k\beta - 1} |\vartheta^\beta \phi(\rho e^{i\theta})|^k \rho^{-1} \, d\rho$$

$$\leq A(k, \beta, \gamma) \int_0^1 (\log 1/\rho)^{k\gamma - 1} |\vartheta^\gamma \phi(\rho e^{i\theta})|^k \rho^{-1} \, d\rho,$$

and this trivially implies the desired result for $k \geq 1$.

Suppose next that $0 < k \leq 1$, and let $\delta = \gamma - \beta$ as before, where $\gamma > \beta > 0$. Applying Theorem A with β replaced by δ and ϕ replaced by the function sending z to $\vartheta^\beta \phi(\rho z)$, and using the fact that $\vartheta^\delta(\vartheta^\beta \phi) = \vartheta^\gamma \phi$, we obtain

$$M_k^k(\vartheta^\beta \phi; \rho) \leq A(k, \beta, \gamma) \int_{-\pi}^\pi \int_0^1 (\log 1/r)^{k\delta - 1} |\vartheta^\gamma \phi(\rho r e^{i\theta})|^k r^{-1} \, dr \, d\theta$$

$$= A(k, \beta, \gamma) \int_{-\pi}^\pi \int_0^\rho (\log \rho/\sigma)^{k\delta - 1} |\vartheta^\gamma \phi(\sigma e^{i\theta})|^k \sigma^{-1} \, d\sigma \, d\theta.$$

Hence $J_{k,\beta}(\phi)$ does not exceed

$$A(k, \beta, \gamma) \int_{-\pi}^\pi d\theta \int_0^1 (\log 1/\rho)^{k\beta - 1} \rho^{-1} \, d\rho \int_0^\rho (\log \rho/\sigma)^{k\delta - 1} |\vartheta^\gamma \phi(\sigma e^{i\theta})|^k \sigma^{-1} \, d\sigma$$

$$= A(k, \beta, \gamma) \int_{-\pi}^\pi d\theta \int_0^1 |\vartheta^\gamma \phi(\sigma e^{i\theta})|^k \sigma^{-1} \, d\sigma \int_\sigma^1 (\log 1/\rho)^{k\beta - 1} (\log \rho/\sigma)^{k\delta - 1} \rho^{-1} \, d\rho$$

On substituting $x = \log 1/\rho$, $y = \log 1/\sigma$, we see easily that the innermost integral in this last integral is equal to $A(k, \beta, \delta)y^{k(\beta + \delta) - 1}$. The whole integral is therefore equal to $A(k, \beta, \gamma)J_{k,\gamma}^k(\phi)$, and this completes the proof.

3.

We turn now to the proof of the left-hand inequality in (1.7). By virtue of the right-hand inequality in (1.6), it is enough to prove that

$$J_{k,\beta + 1}(\phi) \leq A(k, \beta)J_{k,\beta}(u)$$

for $0 \leq k \leq 1$, $\beta > 0$. Since $\vartheta^{\beta + 1}\phi = \vartheta^1(\vartheta^\beta \phi)$, it is therefore enough to prove

Theorem 3. *Let ϕ be regular in the unit disk, and let $\phi(0) = 0$. Let also u be the real part of ϕ, and let*

$$(3.1) \qquad C_{k,\alpha} = \left\{ \int_{-\pi}^{\pi} \int_{0}^{1} (\log 1/\rho)^{\alpha} |u(\rho, \theta)|^{k} \rho^{-1} d\rho \, d\theta \right\}^{1/k}.$$

If $0 < k \leq 1$ and $\alpha > -1$, then

$$\int_{-\pi}^{\pi} \int_{0}^{1} (\log 1/\rho)^{\alpha+k} |\vartheta\phi(\rho e^{i\theta})|^{k} \rho^{-1} \, d\rho \, d\theta \leq A(k, \alpha) C_{k,\alpha}^{k},$$

where $\vartheta\phi(z) = \vartheta^{1}\phi(z) = z\phi'(z)$.

4.

We divide the proof of Theorem 3 into several lemmas.

Lemma 1. *Let ϕ be regular in the unit disk, let $\phi(0) = 0$, let u be the real part of ϕ, and let $C = C_{k,\alpha}$ be defined by (3.1). If $0 < k \leq 1$ and $\alpha > -1$, then*

$$(4.1) \qquad \sup_{\theta} |\phi(\rho e^{i\theta})| \leq A(k, \alpha) C (1 - \rho)^{A(k, \alpha)}.$$

The argument here is essentially similar to that of [2, Lemma 5], though the details are more complicated. If c_n is the nth Taylor coefficient of ϕ, and

$$M(\phi; \rho) = \sup_{\theta} |\phi(\rho e^{i\theta})|,$$

then

$$\pi |c_n| \rho^n \leq \int_{-\pi}^{\pi} |u(\rho, \theta)| \, d\theta \leq M^{1-k}(\phi; \rho) \int_{-\pi}^{\pi} |u(\rho, \theta)|^{k} \, d\theta.$$

Since $M(\phi; \rho)$ increases with ρ, this gives

$$(4.2) \qquad \pi |c_n| \rho^n \leq M^{1-k}(\phi; r) \int_{-\pi}^{\pi} |u(\rho, \theta)|^{k} \, d\theta$$

for $0 \leq \rho \leq r < 1$. Multiplying both sides of (4.2) by $(\log 1/\rho)^{\alpha} \rho^{-1}$ and integrating with respect to ρ we obtain

$$(4.3) \qquad \pi |c_n| \int_{0}^{r} (\log 1/\rho)^{\alpha} \rho^{n-1} \, d\rho \leq C^{k} M^{1-k}(\phi; r).$$

We observe now that

$$(4.4) \qquad \int_{0}^{r} (\log 1/\rho)^{\alpha} \rho^{n-1} \, d\rho = n^{-\alpha-1} \int_{0}^{r^n} (\log 1/s)^{\alpha} \, ds.$$

Further, if $0 < t < 1$, then

$$(4.5) \qquad \int_0^t (\log 1/s)^\alpha \, ds \geqq A(\alpha) t (1 + \log 1/t)^\alpha.$$

This last inequality is immediate if $\frac{1}{2} \leqq t < 1$. For the remaining case we note that $\log 1/s$ decreases as s increases, and hence if $t^2 \leqq s \leqq t$,

$$(\log 1/s)^\alpha \geqq \min\{(\log 1/t)^\alpha, (\log 1/t^2)^\alpha\} = A(\alpha)(\log 1/t)^\alpha.$$

If now $0 < t \leqq \frac{1}{2}$, then $t - t^2 \geqq \frac{1}{2}t$, and $(\log 1/t)^\alpha \geqq A(\alpha)(1 + \log 1/t)^\alpha$, whence (4.5) follows.

Combining (4.3), (4.4), and (4.5), we obtain, for $0 < r < 1$,

$$|c_n| \leqq A(\alpha) C^k M^{1-k}(\phi; r)(n^{-1} + \log 1/r)^\alpha n r^{-n}$$
$$\leqq A(\alpha) C^k M^{1-k}(\phi; r)(B + \log 1/r)^\alpha n r^{-n},$$

where $B = B(\alpha)$ is 1 or 0 according as $\alpha \geqq 0$ or $-1 < \alpha < 0$. Hence if $0 < \rho \leqq r < 1$ then

$$(4.6)$$

$$M(\phi; \rho) \leqq \sum_{n=1}^{\infty} |c_n| \rho^n \leqq A(\alpha) C^k M^{1-k}(\phi; r)(B + \log 1/r)^\alpha \sum_{n=1}^{\infty} n(\rho/r)^n$$

$$= A(\alpha) C^k M^{1-k}(\phi; r)(B + \log 1/r)^\alpha r\rho(r - \rho)^{-2}.$$

Following [2], we now write $r = e^{-1/t}$, $\rho = e^{-1/\tau}$, where $t > \tau > 0$, $t > \frac{1}{2}$, and we set

$$g(\tau) = \log M(\phi; \rho), \qquad b = 1 - k.$$

It is easily verified that for $x > 0$

$$\log 1/(1 - e^{-1/x}) \leqq A + \log^+ x,$$

where $\log^+ x = \max\{\log x, 0\}$, and therefore

$$\log \frac{r\rho}{(r - \rho)^2} \leqq 2 \log \frac{r}{r - \rho} \leqq A + 2 \log^+ \frac{t\tau}{t - \tau}$$

$$\leqq A + 2 \log^+ t + 2 \log^+ \tau + 2 \log^+ 1/(t - \tau).$$

Further, since $t > \frac{1}{2}$, we have

$$\log(B + \log 1/r)^\alpha \leqq \begin{cases} \alpha \log 3 & \text{if} \quad \alpha \geqq 0, \\ -\alpha \log^+ t & \text{if} \quad \alpha < 0. \end{cases}$$

Hence (4.6) gives

$$g(\tau) - bg(t) \leqq A(\alpha) + k \log C + A(\alpha) \log^+ t + 2 \log^+ 1/(t - \tau).$$

Substituting successively the pairs $(\tau, \tau + 1/2)$, $(\tau + 1/2, \tau + 1/2 + 1/2^2)$, ... for τ and t, we obtain that

$$g(\tau) - b^n g(\tau + 1/2 + \ldots + 1/2^n)$$
$$\leq \{A(\alpha) + k \log C + A(\alpha) \log(\tau + 1)\} \sum_{m=0}^{n-1} b^m + 2 \log 2 \sum_{m=0}^{n-1} (m+1)b^m.$$

Making $n \to \infty$, we thus have

$$g(\tau) \leq \{A(k, \alpha) + k \log C + A(\alpha) \log(\tau + 1)\}/k + A(k),$$
$$\leq A(k, \alpha) + \log C + A(k, \alpha) \log 1/(1 - \rho),$$

and this is equivalent to (4.1).

5.

The remainder of the proof of Theorem 3 depends on

Lemma 2. *Let f be a real-valued function defined for $x > x_0 > 0$ such that*

$$f(x) \leq \lambda + \mu(f(cx))^\gamma \qquad (x > x_0),$$

where $\lambda > 0$, $\mu > 0$, $0 < \gamma < 1$, $c > 1$. Then either

$$f(x) \leq \xi \qquad (x > x_0),$$

where ξ is the positive root of the equation $\xi = \lambda + \mu \xi^\gamma$, or

$$\limsup_{x \to +\infty} x^{-\delta} f(x) = +\infty$$

for all positive δ.

This is Lemma 4 of [2].

Lemma 3. *Let ϕ be regular in the unit disk, and let u be the real part of ϕ. Let also $0 < k \leq 1$, and let $\sigma = \rho^{1/2} - \rho$. Then there exists a positive number K, depending only on k, such that for $0 < \eta < \frac{1}{4}$*

$$(5.1) \quad \sigma^k M_k{}^k(\vartheta\phi; \rho) \leq \eta^{-k}\{M_k{}^k(u; \rho + \eta\sigma)$$
$$+ 3M_k{}^k(u; \rho)\} + K\eta^k \sigma^k M_k{}^k(\vartheta\phi; \rho^{1/4}).$$

This is an easy consequence of equation (2.6.3) of [2].

6.

Now let $C_{k,\alpha}$ be defined by (3.1), let $D = C_{k,\alpha}^k$, let $0 < r < 1$, and let

$$h(r) = \int_0^r (\log 1/\rho)^{\alpha+k} M_k{}^k(\vartheta\phi; \rho)\rho^{-1} \, d\rho.$$

We have to show that $h(r) \leqq A(k, \alpha)D$.

Multiplying both sides of (5.1) by $(\log 1/\rho)^\alpha \rho^{-1}$ and integrating with respect to ρ, we obtain

(6.1) $\displaystyle\int_0^r \sigma^k (\log 1/\rho)^\alpha M_k^{\ k}(\vartheta\phi; \rho)\rho^{-1}\, d\rho$

$$\leqq \eta^{-k}\left\{\int_0^1 (\log 1/\rho)^\alpha M_k^{\ k}(u; \rho + \eta\sigma)\rho^{-1}\, d\rho + 3D\right\}$$

$$+ K\eta^{-k}\int_0^r \sigma^k (\log 1/\rho)^\alpha M_k^{\ k}(\vartheta\phi; \rho^{1/4})\rho^{-1}\, d\rho.$$

It is easily verified that the second integral on the right of (6.1) does not exceed $A(k)h(r^{1/4})$. Further, putting $w = \rho + \eta\sigma$ in the first integral on the right of (6.1) and noting that $r < w < r^{1/2}$ and that

$$\frac{r}{w}\frac{dw}{dr} \geqq \frac{3}{4}$$

we see that this first integral does not exceed $A(\alpha)D$. Finally, the argument of [3, Lemma γ] shows that the integral on the left of (6.1) exceeds $A(\alpha)h(r)$. We thus obtain

(6.2) $h(r) \leqq B_1 \eta^{-k}D + B_2 \eta^k h(r^{1/4}),$

where B_1 and B_2 depend only on k and α.

We now put

$$\eta^{2k} = B_1 D/\{4^{2k}B_1 D + B_2 h(r^{1/4})\},$$

so that $\eta < \frac{1}{4}$. It follows easily from (6.2) that

(6.3) $\displaystyle\frac{h(r)}{D} \leqq 2^{2k+1}B_1 + 2(B_1 B_2)^{1/2}\left(\frac{h(r^{1/4})}{D}\right)^{1/2}.$

Writing $r = e^{-1/x}$, $f(x) = h(r)/D$, we now have

$$f(x) \leqq 2^{2k+1}B_1 + 2(B_1 B_2)^{1/2}(f(4x))^{1/2}.$$

Applying Lemma 2, we deduce that either $f(x) \leqq A(k, \alpha)$, i.e., that $h(r) \leqq A(k, \alpha)D$, or that

$$\limsup_{r\to 1-}\ (\log 1/r)^\eta h(r)/D = +\infty$$

for all positive η. The latter is impossible, by Lemma 1, and this completes the proof of Theorem 3.

REFERENCES

1. T. M. Flett, Mean values of power series, *Pacific J. Math.* **25** (1968), 463–494.
2. G. H. Hardy and J. E. Littlewood, Some properties of conjugate functions, *J. für Math.* **167** (1931), 405–423.
3. ———, Theorems concerning mean values of analytic or harmonic functions, *Quart. J. Math. Oxford* **12** (1941), 221–256.
4. A. Zygmund, *Trigonometric series*, 2nd ed., vol. I, Cambridge University Press, Cambridge, England, 1959.

(*Received February 20, 1968*)

Quasiconformal Mappings which Hold the Real Axis Pointwise Fixed

F. W. Gehring*

Department of Mathematics
University of Michigan
Ann Arbor, Michigan 48104
U.S.A.

1. Introduction

Given a function μ which is measurable and bounded in absolute value by $k < 1$ in the extended complex plane C, we let f^μ denote the unique normalized quasiconformal mapping of C onto itself which has μ as its complex dilatation. That is, f^μ is a homeomorphism of C onto itself which is a generalized solution of the Beltrami equation $w_{\bar{z}} = \mu w_z$ in C and has $0, 1, \infty$ as fixed points.

It is well known that if f^μ maps the real axis onto itself, then so does $f^{t\mu}$ for $0 < t < 1$.

Conjecture. *If f^μ holds the real axis pointwise fixed, then so does $f^{t\mu}$ for $0 < t < 1$.*

This conjecture was raised in a letter to the author by I. Kra. The purpose of this note is to exhibit a dilatation μ^* such that f^{μ^*} holds the real axis pointwise fixed while $f^{t\mu^*}$ does not for each t in $0 < t < 1$.

* This research was supported in part by the National Science Foundation, Contract GP–7234.

2. Main result

Let F denote the class of quasiconformal mappings f of C onto itself which hold the real axis pointwise fixed. Each $f \in F$ maps the upper half-plane H onto itself. For each pair of points $z_1, z_2 \in H$, let $F(z_1, z_2)$ denote the subclass of $f \in F$ for which $f(z_1) = z_2$. Then there exists an extremal mapping $f^* \in F(z_1, z_2)$ such that

$$K(f^*) = \inf K(f), \qquad f \in F(z_1, z_2),$$

where $K(f)$ is the maximal dilatation of f. Moreover, from the work of Teichmüller [2] it follows that f^* is unique and that

$$\log K(f^*) = \phi(h(z_1, z_2)),$$

where h denotes the hyperbolic distance in H and where for $0 \leq t < \infty$

(1) $$\phi(t) = 4 \operatorname{arctanh}(\exp(-v(\tanh t))).$$

Here $v(s)$ denotes for $0 \leq s < 1$ the modulus of the unit disk $|z| < 1$ slit from $z = 0$ to $z = s$.

Theorem. *If $f^* = f_2 \circ f_1$ where $f_1 \in F$ and if neither f_1 nor f_2 is the identity, then*

(2) $$K(f^*) < K(f_2)K(f_1).$$

3. A preliminary result

The function defined in (1) is strictly increasing in $0 \leq t < \infty$. We require the following additional information.

Lemma. *$\phi(t)/t$ is strictly decreasing in $0 < t < \infty$.*

Proof. One can undoubtedly establish this fact by means of direct calculation. We prove it by means of conformal mapping and the Schwarz lemma.

Let D_1 denote the curvilinear triangle in H bounded by the lines $x = 0$, $x = 1$, and the circle $|z - \frac{1}{2}| = \frac{1}{2}$. Next let τ denote the local inverse of the elliptic modular function which maps H conformally onto D_1 so that $0, 1, \infty$ correspond, respectively, to $\infty, 0, 1$. Then

(3) $$v(s) = -i\frac{\pi}{2}\tau(s^2)$$

for $0 < s < 1$. Now $\tanh z$ maps D_2, the strip $0 < x < \infty$, $0 < y < \pi/2$, conformally onto D_3, the quadrant $0 < x < \infty$, $0 < y < \infty$, so that $0, \infty, (\pi/2)i$ correspond, respectively, to $0, 1, \infty$. Hence

(4) $$\omega(z) = \tau((\tanh z)^2)$$

maps D_2 conformally onto D_1 so that 0, ∞, $(\pi/2)i$ correspond, respectively, to ∞, 0, 1. By means of repeated reflections in the lines $y = 2^{n-2}\pi, n = 1, 2, \ldots$, we can extend ω so that it maps D_3 conformally into the strip $0 < x < 1$, $0 < y < \infty$. Then

$$\psi(z) = 4 \operatorname{arctanh}\left(\exp\left(i\frac{\pi}{2}\omega(z)\right)\right)$$

maps D_3 onto a proper subdomain of itself so that the positive half of the real axis L corresponds to itself. If we reflect in L and apply the half-plane form of the Schwarz lemma, we obtain

(5) $$\frac{d}{dx}\psi(x) < \frac{\psi(x)}{x}, \qquad \frac{d}{dx}\left(\frac{\psi(x)}{x}\right) < 0$$

for $0 < x < \infty$, and the desired conclusion follows from (3), (4), and (5).

4. Proof of the Theorem

Suppose that $f^* = f_2 \circ f_1$ where $f_1 \in F$ and that neither f_1 nor f_2 is the identity. Then $f_1 \in F(z_1, z_0)$ and $f_2 \in F(z_0, z_2)$ where $z_0 = f_1(z_1)$. If $h(z_1, z_2) \leqq h(z_1, z_0)$, then

$$\log K(f^*) \leqq \log K(f_1) < \log K(f_1)K(f_2)$$

since f_2 is not conformal, and we obtain (2). A similar argument yields (2) if $h(z_1, z_2) \leqq h(z_0, z_2)$. Excluding these two cases, we obtain

$$\log K(f^*) = \phi(h(z_1, z_2)) < \phi(h(z_1, z_0)) + \phi(h(z_0, z_2)) \leqq \log K(f_1)K(f_2)$$

from the Lemma and the triangle inequality for h.

5. Counterexample

Choose $z_1, z_2 \in H$ so that $z_1 \neq z_2$, let μ^* be the complex dilatation for the corresponding extremal mapping $f^* = f^{\mu^*}$, and for $0 < t < 1$ let $f_1 = f^{t\mu^*}$ and $f_2 = f^* \circ f_1^{-1}$. Then $f^* = f_2 \circ f_1$, $K(f^*) = K(f_2)K(f_1)$, and since μ^* does not vanish identically, neither f_1 nor f_2 is the identity. Thus $f_1 \notin F$ by the Theorem.

E. Reich has pointed out in a letter to the author how one may find simple counterexamples to the Conjecture of Section 1 using a result due to Ahlfors [1, Lemma 1].

REFERENCES

1. L. V. Ahlfors, Some remarks on Teichmüller's space of Riemann surfaces, *Ann. of Math.* **74** (1961), 171–191.
2. O. Teichmüller, Ein Verschiebungssatz der quasikonformen Abbildung, *Deutsche Math.* **7** (1944), 336–343.

(Received February 15, 1968)

Some Examples Related to the cosπρ Theorem

W. K. Hayman

Department of Mathematics
Imperial College
London, S.W.7
England

1. Introduction and statement of results

Let $f(z)$ be an entire function of order ρ and lower order λ, where $0 \leq \lambda \leq \rho \leq 1$. A classical Theorem of Wiman [8] and Valiron [7] asserts that, given $\varepsilon > 0$, the inequality

$$(1.1) \qquad \log \mu(r) > (\cos \pi\rho - \varepsilon)\log M(r)$$

holds for a sequence $r = r_n \to \infty$, where

$$(1.2) \qquad \mu(r) = \inf_{|z|=r} |f(z)|, \qquad M(r) = \sup_{|z|=r} |f(z)|.$$

A very extensive literature has since sprung up about this subject, whose full description would take us too far. However I should like to mention the following result:

Theorem A. *If $\rho < \alpha < 1$, and if*

$$(1.3) \qquad E = \{r \mid \log \mu(r) > \cos \pi\alpha \log M(r)\}$$

then we have

$$(1.4) \qquad \underline{\log \text{ dens }} E \geq 1 - \rho/\alpha,$$

and if f has very regular growth then

(1.5)   strong log dens $E \geq 1 - \rho/\alpha$.

These results are due to Barry (for (1.4) see [1]; for (1.5) see [2]). Besicovitch [3] had earlier proved the weaker result that

(1.6) $\overline{\text{dens}}\ E \geq 1 - \rho/\alpha$

At this stage it seems convenient to define the various densities involved. If E is any set on the positive real axis and $E(r, R)$ is the part of E in the interval $[r, R]$, then

$$\overline{\text{dens}}\ E = \varlimsup_{R \to \infty} \frac{1}{R-1} \int_{E(1, R)} dt,$$

$$\underline{\text{dens}}\ E = \varliminf_{R \to \infty} \frac{1}{R-1} \int_{E(1, R)} dt;$$

$$\overline{\text{log dens}}\ E = \varlimsup_{R \to \infty} \frac{1}{\log R} \int_{E(1, R)} \frac{dt}{t},$$

$$\underline{\text{log dens}}\ E = \varliminf_{R \to \infty} \frac{1}{\log R} \int_{E(1, R)} \frac{dt}{t};$$

$$\overline{\text{strong log dens}}\ E = \varlimsup_{r \to \infty,\, R \to \infty,\, R/r \to \infty} \frac{1}{\log(R/r)} \int_{E(r, R)} \frac{dt}{t},$$

$$\underline{\text{strong log dens}}\ E = \varliminf_{r \to \infty,\, R \to \infty,\, R/r \to \infty} \frac{1}{\log R/r} \int_{E(r, R)} \frac{dt}{t}.$$

The inequalities

$$\underline{\text{dens}}\ E \leq \underline{\text{strong log dens}}\ E \leq \underline{\text{log dens}}\ E$$

$$\leq \overline{\text{log dens}}\ E \leq \overline{\text{strong log dens}}\ E \leq \overline{\text{dens}}\ E$$

are easily proved. For example, to prove the last one we suppose that for any positive ε

$$m(R) = \int_{E(1, R)} dt \leq (d + \varepsilon)(R - 1), \qquad R > R_0(\varepsilon).$$

Then if $R_0 < r < R$, we have

$$\int_{E(r, R)} \frac{dt}{t} = \int_r^R \frac{1}{t}\, dm(t) = \frac{m(R)}{R} - \frac{m(r)}{r} + \int_r^R \frac{m(t)\, dt}{t^2}$$

$$\leq (d + \varepsilon) \log \frac{R}{r} + O(1), \qquad \text{as } r \to \infty \ \text{ and } \ R \to \infty.$$

This yields

$$\text{strong } \overline{\log \text{ dens } E} \leq d + \varepsilon,$$

for every positive ε. Taking $d = \overline{\text{dens}} \ E$, we deduce that

$$\text{strong } \overline{\log \text{ dens } E} \leq \overline{\text{dens}} \ E.$$

By considering the complements of E instead of E we deduce that

$$\underline{\text{dens}} \ E \leq \text{strong} \underline{\log \text{ dens }} E.$$

The other inequalities are proved similarly.

We also note that a function $f(z)$ is said to have very regular growth of order ρ if

$$0 < \varliminf_{r \to \infty} \frac{\log M(r)}{r^\rho} \leq \varlimsup_{r \to \infty} \frac{\log M(r)}{r^\rho} < \infty.$$

If $\lambda = \rho$, then $f(z)$ is said to have regular growth.

We can now state our first result.

Theorem 1. *Given any numbers* ρ, α, *such that* $0 < \rho < \alpha < 1$, *there exists a function* $f(z)$ *of order* ρ *and regular growth such that*

$$(1.7) \qquad \underline{\log \text{ dens }} E = \overline{\log \text{ dens }} E = 1 - \rho/\alpha,$$

where E *is the set defined by* (1.3). *Further given* $\varepsilon > 0$ *there exists a function* $f(z)$ *having very regular growth of order* ρ, *such that*

$$(1.8) \qquad 1 - \frac{\rho}{\alpha} \leq \text{strong} \underline{\log \text{ dens }} E \leq \text{strong } \overline{\log \text{ dens } E} \leq 1 - \frac{\rho}{\alpha} + \varepsilon.$$

It should be said that in the case $\rho = \frac{1}{2}$, (1.8) was previously proved by Kjellberg [5]. The results show that in Barry's Theorem A, (1.4) and (1.5) are both sharp, (1.4) in the sense that the sign $\geq$ cannot be replaced by $>$, and (1.5) at any rate in the weaker sense that $1 - \rho/\alpha$ cannot be replaced by $1 - \rho/\alpha + \varepsilon$ for any positive ε. The question of whether (1.6) is sharp is left open.

1.1

Our second class of examples is concerned with problems of regularity. Suppose that $f(z)$ is an entire function satisfying

$$(1.9) \qquad \log \mu(r) \leq \cos \pi\rho \log M(r), \qquad r > r_0,$$

where $0 < \rho < 1$. Then it follows from the theorem of Wiman and Valiron that $f(z)$ has order at least ρ. More recently Kjellberg [6] has shown that in fact

(1.10) $$\frac{\log M(r)}{r^\rho} \to C, \quad \text{as} \quad r \to \infty,$$

where $0 < C \leqq \infty$. Thus if $f(z)$ has minimal growth subject to (1.9), namely mean type of order ρ, so that C is finite, then $\log M(r)$ has quite strong regularity of growth and in fact so do other quantities connected with $f(z)$, such as the characteristic function $T(r, f)$ and the counting function $n(r, f)$ of the number of zeros in $|z| < r$.

D. Drasin* recently raised the question whether regularity theorems of this type can hold under weaker assumptions than for instance, (1.9) and the finiteness of C in (1.10). In particular he asked whether, if, for instance, (1.9) holds and $f(z)$ has order exactly ρ, can we deduce that $\log M(r)$ shows certain regularity of growth so that for instance

(1.11) $$\lim_{r \to \infty} \frac{\log M(Kr)}{\log M(r)} = K^\rho$$

for every r? (Since $f(z)$ has regular growth of order ρ in this case, the limit must be K^ρ if it exists.)

We shall introduce a class of examples which will show that a result of this type cannot be true in general at least in the case $\rho = \frac{1}{2}$, which is easiest to handle and probably typical. We first introduce a rather general class of functions satisfying (1.9) and then specialize these to show that if $\log M(r)/r^\rho$ is allowed to grow to infinity however slowly then (1.11) need not hold. Our results are

Theorem 2. *Suppose that $\rho(t)$, $\psi(t)$ are positive nondecreasing functions of t for $1 \leqq t < \infty$, that $\psi(t) = 0$, $0 < t < 1$, and*

(1.12) $$v(t) = \psi(t)t^{\rho(t)}, \quad 0 \leqq t < \infty$$

and further

(1.13) $$v(t) = O(t^{1-\delta}), \quad \text{as} \quad t \to \infty$$

for some positive δ.

Then there exists an entire function $f(z)$ of genus zero, whose counting function $n(r, f)$ satisfies

(1.14) $$v(r) - 9 \leqq n(r, f) \leqq v(r), \quad 0 < r < \infty$$

and further, for all sufficiently large r,

(1.15) $$\log \mu(r) < \pi \cot(\pi\rho(r)) \, n(r) < 2\{\cot(\pi\rho(r)) \log M(r)\}$$

We deduce

* Oral communication.

Theorem 3. *There exists an integral function of bounded minimum modulus and with real negative zeros for which $r^{-1/2} \log M(r)$ tends to infinity arbitrarily slowly, while at the same time*

$$(1.16) \qquad \overline{\lim_{r \to \infty}} \, \frac{\log M(Kr)}{\log M(r)} = K, \qquad \text{for every } K > 1.$$

Theorem 4. *Given $\varepsilon(t)$ so that $\varepsilon(t)\log t \to \infty$, there exists $f(z)$ of order $\frac{1}{2}$ and either minimal type, mean type, or maximal type as desired, and such that for sufficiently large r*

$$(1.17) \qquad \log \mu(r) < O\{\varepsilon(r)\log M(r)\},$$

while at the same time (1.16) holds.

2. Proof of Theorem 1: Preliminary results

Let $f(z)$ be an entire function of genus zero, all of whose zeros are real and negative and such that $f(0) = 1$. If $n(t)$ is the number of zeros of $f(z)$ in the interval $[-t, 0]$, then for $|\arg z| < \pi$ we have [4, p. 21]

$$(2.1) \qquad \log f(z) = \int_0^\infty \log\left(1 + \frac{z}{t}\right) dn(t) = z \int_0^\infty \frac{n(t)\, dt}{t(z + t)}.$$

To construct the required function $f(z)$ we construct $n(t)$. We work first with continuous increasing functions $v(t)$ and then define $n(t)$ from $v(t)$. We need the following

Lemma 1. *If $0 < \alpha < 1$, $K_1 \geqq K \geqq 2$, and $K_2 \geqq K \geqq 2$, and*

$$I(z) = \mathscr{R}\left\{z \int_{t_0/K_1}^{K_2 t_0} \frac{t^{\alpha - 1}\, dt}{t + z}\right\}.$$

Then for $z = t_0 e^{i\theta}$, $|\theta| < \pi$, we have

$$\left| I(z) - \frac{\pi}{\sin \pi\alpha} t_0^\alpha \cos \alpha\theta \right| < \varepsilon(\alpha, K)t_0^\alpha,$$

where

$$(2.2) \qquad \varepsilon(\alpha, K) = 2/(\alpha K^\alpha) + 2/\{(1 - \alpha)K^{1-\alpha}\}.$$

We note that

$$z \int_0^\infty \frac{t^{\alpha - 1}\, dt}{t + z} = \frac{\pi}{\sin \pi\alpha} z^\alpha, \qquad |\arg z| < \pi.$$

Also if $|z| = t_0$, then if $K_1 \geq 2$, $t \leq t_0/K_1$, we have $|z/(t + z)| \leq 2$, so that

$$\left| z \int_0^{t_0/K_1} \frac{t^{\alpha-1}\, dt}{t + z} \right| \leq 2 \int_0^{t_0/K_1} t^{\alpha-1}\, dt = \frac{2}{\alpha}\left(\frac{t_0}{K_1}\right)^\alpha \leq \frac{2}{\alpha}\left(\frac{t_0}{K}\right)^\alpha.$$

Similarly for $t \geq K_2 t_0$, we see that $|t + z| \geq \frac{1}{2}t$, so that

$$\left| z \int_{K_2 t_0}^\infty \frac{t^{\alpha-1}\, dt}{t + z} \right| \leq 2t_0 \int_{K_2 t_0}^\infty t^{\alpha-2}\, dt = \frac{2t_0^\alpha}{(1 - \alpha)K_2^{1-\alpha}} \leq \frac{2t_0^\alpha}{(1 - \alpha)K^{1-\alpha}}.$$

Thus for $z = t_0 e^{i\theta}$, where $|\theta| < \pi$, we have

$$\left| I(z) - \frac{\pi}{\sin \pi\alpha} t_0^\alpha \cos \alpha\theta \right| < 2t_0^\alpha\left(\frac{1}{\alpha K^\alpha} + \frac{1}{(1 - \alpha)K^{1-\alpha}}\right) = \varepsilon(\alpha, K)t_0^\alpha.$$

This proves Lemma 1. We have next

Lemma 2. *Suppose that* $n(t)$ *is a real function of* t *such that*

$$|n(t) - ct^\alpha| < 1, \quad 1 \leq t_1 < t < t_2, \quad where \quad c \geq 0, 0 < \alpha < 1.$$

Then if $K \geq 2$, *and* $Kt_1 \leq t_0 \leq t_2/K$, $z = t_0 e^{i\theta}$, *where* $|\theta| < \pi$ *and*

$$I(z) = \mathscr{R}\left\{ z \int_{t_1}^{t_2} \frac{n(t)\, dt}{t(t + z)} \right\},$$

we have

$$\left| I(z) - \frac{\pi c}{\sin \pi\alpha} t_0^\alpha \cos \alpha\theta \right| < c\varepsilon(\alpha, K)t_0^\alpha + 2 \log t_0 + 4 \log\left(\frac{10}{\pi - |\theta|}\right) + 1,$$

where $\varepsilon(\alpha, K)$ *is given by* (2.2).

We write

$$n(t) = ct^\alpha + n_1(t),$$

where $|n_1(t)| < 1$. Then

$$I(z) = \mathscr{R}\left\{ cz \int_{t_1}^{t_2} \frac{t^{\alpha-1}\, dt}{t + z} \right\} + \mathscr{R}\left\{ \int_{t_1}^{t_2} \frac{zn_1(t)\, dt}{t(t + z)} \right\}$$

$$= I_1(z) \qquad\qquad + I_2(z).$$

It follows from Lemma 1, that

$$(2.3) \qquad \left| I_1(z) - \frac{\pi c}{\sin \pi\alpha} t_0^\alpha \cos \alpha\theta \right| < \varepsilon(\alpha, K)ct_0^\alpha.$$

It remains to estimate $I_2(z)$. We divide the integral in three parts. If $t \leq \frac{1}{2}t_0$, then $|z/(t + z)| \leq 2$. Thus

(2.4) $$\left| \int_{t_1}^{t_0/2} \frac{zn_1(t)\, dt}{t(t+z)} \right| \le 2 \int_{t_1}^{t_0/2} \frac{dt}{t} = 2 \log \frac{t_0}{t_1} \le 2 \log t_0.$$

Again if $t \ge 2t_0$, then $|z+t| \ge \frac{1}{2}t$ and so

(2.5) $$\left| \int_{2t_0}^{t_2} \frac{zn_1(t)\, dt}{t(z+t)} \right| \le 2t_0 \int_{2t_0}^{t_2} \frac{dt}{t^2} < 2t_0 \int_{2t_0}^{\infty} \frac{dt}{t^2} = 1.$$

It remains to consider the range $\frac{1}{2}t_0 \le t \le 2t_0$. Suppose first that $|\theta| \le \frac{1}{2}\pi$. Then $|z+t| \ge t$, and so in this case

(2.6) $$\left| \int_{t_0/2}^{2t_0} \frac{zn_1(t)\, dt}{t(z+t)} \right| \le t_0 \int_{t_0/2}^{2t_0} \frac{dt}{t^2} < 2.$$

Finally suppose that $z = t_0 e^{i\theta}$, where $\frac{1}{2}\pi < |\theta| < \pi$, and set $z = x + iy$. Then

$$|z+t| = \{(t+x)^2 + y^2\}^{1/2},$$

(2.7) $$\left| \int_{t_0/2}^{2t_0} \frac{zn_1(t)\, dt}{t(z+t)} \right| < 2 \int_{t_0/2}^{2t_0} \frac{dt}{\{(t+x)^2 + y^2\}^{1/2}}$$

$$= 2 \int_{t_0/2+x}^{2t_0+x} \frac{d\tau}{(\tau^2 + y^2)^{1/2}} < 4 \int_0^{2t_0} \frac{d\tau}{(\tau^2 + y^2)^{1/2}}$$

$$= 4 \log\left\{ \frac{2t_0 + (4t_0^2 + y^2)^{1/2}}{|y|} \right\} < 4 \log \frac{5t_0}{|y|}$$

$$= 4 \log \frac{5}{\sin |\theta|} < 4 \log\left(\frac{10}{\pi - |\theta|} \right).$$

By (2.6) this inequality remains true for $|\theta| \le \frac{1}{2}\pi$. Thus in all uses we obtain from (2.4), (2.5), and (2.7)

$$|I_2(z)| < 2 \log t_0 + 4 \log\left(\frac{10}{\pi - |\theta|} \right) + 1.$$

On combining this with (2.3) we deduce Lemma 2.

2.1

We now construct a function $v(t)$ as follows. We suppose given a sequence r_m, such that

$$r_0 = 1, \qquad r_{m+1}/r_m \text{ increases with } m$$

and

(2.8) $$r_{m+1} \ge 4r_m.$$

Further let ρ be positive and let α_m be a decreasing sequence of positive numbers such that

$$0 < \rho < \alpha_m \leqq \alpha_0 < 1.$$

We define $v(t)$ so that $v(t) = 0$, $t < 1$,

(2.9) $$v(r_m) = r_m^\rho, \quad m = 0 \text{ to } \infty,$$

(2.10) $$\frac{v(r)}{r^{\alpha_m}} = \text{constant}, \quad r_m \leqq r \leqq r_m',$$

(2.11) $$v(r) = \text{constant}, \quad r_m' \leqq r \leqq r_{m+1}.$$

Thus r_m' is given by

$$\frac{v(r_m')}{r_m'^{\alpha_m}} = \frac{v(r_{m+1})}{r_m'^{\alpha_m}} = \frac{v(r_m)}{r_m^{\alpha_m}},$$

that is

(2.12) $$\left(\frac{r_m'}{r_m}\right)^{\alpha_m} = \left(\frac{r_{m+1}}{r_m}\right)^\rho.$$

We then define $n(t) = [v(t)]$, where $[x]$ is the integral part of x. In the applications r_{m+1}/r_m will either be constant in which case α_m and ρ_m will also be constant or r_{m+1}/r_m will tend to infinity slowly with m and α_m, ρ_m will also vary slowly with m. With this definition of $n(t)$, (2.1) defines an entire function of genus zero, which we propose to investigate. We have

Lemma 3. *Suppose that $K \geqq 2$, that $Kr_m \leqq r \leqq r_m'/K$, and that $z_0 = re^{i\theta}$, where $|\theta| < \pi$. Then if $f(z)$ is the entire function defined as above we have*

$$\left| \log |f(z_0)| - \frac{\pi}{\sin \pi \alpha_m} v(r) \cos \alpha_m \theta \right| < 2 \log r + 4 \log \frac{10}{(\pi - |\theta|)} + 1$$

$$+ v(r)\left[\left(\frac{2K^{\alpha_0 - 1}}{1 - \alpha_0} + \varepsilon(\alpha_m, K) + \frac{8}{\rho} K^{-\rho} \log\left(\frac{r_m}{r_{m-1}}\right)\right)\right].$$

We have

$$\log |f(z)| = \mathcal{R}\left\{z \int_1^\infty \frac{n(t)\,dt}{t(t+z)}\right\} = \mathcal{R}\left\{\int_1^{r_m} \frac{zn(t)\,dt}{t(t+z)}\right\} + \mathcal{R}\left\{\int_{r_m}^{r_m'} \frac{zn(t)\,dt}{t(t+z)}\right\}$$

$$+ \mathcal{R}\left\{\int_{r_m'}^\infty \frac{zn(t)\,dt}{t(t+z)}\right\}$$

$$= I_1(z) + I_2(z) + I_3(z),$$

say. By (2.10) we have that

$$v(t) = c_m t^{\alpha_m}, \qquad r_m \leqq t \leqq r'_m,$$

where c_m is constant, and $v(t) - 1 \leqq n(t) \leqq v(t)$. Thus by Lemma 2

(2.13)

$$\left| I_2(z) - \frac{\pi}{\sin \pi \alpha_m} v(r) \cos \alpha_m \theta \right| < \varepsilon(\alpha_m, K) v(r) + 2 \log r + 4 \log \left(\frac{10}{\pi - \theta} \right) + 1.$$

Next for $t \geqq r'_m$, $|t + z| \geqq \frac{1}{2} t$, so that

$$|I_3(z)| \leqq 2r \int_{r'_m}^{\infty} \frac{n(t)\, dt}{t^2} \leqq 2r \int_{r'_m}^{\infty} \frac{v(t)\, dt}{t^2}.$$

It is a consequence of (2.10) and (2.11) that $v(r)/r^{\alpha_0}$ decreases for all r. Thus

(2.14)

$$|I_3(z)| \leqq 2rv(r) \int_{r'_m}^{\infty} \left(\frac{t}{r} \right)^{\alpha_0} \frac{dt}{t^2} = \frac{2rv(r)}{r^{\alpha_0}} \cdot \frac{(r'_m)^{\alpha_0 - 1}}{1 - \alpha_0}$$

$$= \frac{2v(r)}{1 - \alpha_0} \left(\frac{r}{r'_m} \right)^{1 - \alpha_0} \leqq \frac{2K^{\alpha_0 - 1}}{1 - \alpha_0} v(r).$$

Finally for $t \leqq r_m$, we have $|z/(t + z)| \leqq 2$, so that

$$I_1(z) \leqq 2 \int_1^{r_m} \frac{v(t)\, dt}{t} \leqq 2 \sum_{\mu=1}^{m} v(r_\mu) \log \frac{r_\mu}{r_{\mu-1}} \leqq 2 \log \frac{r_m}{r_{m-1}} \sum_{\mu=1}^{m} r_\mu^\rho,$$

since $r_\mu/r_{\mu-1}$ increases. Also in view of (2.8) this yields

$$I_1(z) \leqq 2r_m^\rho \left(\log \frac{r_m}{r_{m-1}} \right) \sum_{k=0}^{\infty} 4^{-k\rho}.$$

Further

$$\sum_{k=0}^{\infty} 4^{-k\rho} = \frac{1}{1 - 4^{-\rho}} = \frac{4^\rho}{4^\rho - 1} < \frac{4}{\rho},$$

and by (2.9) and (2.10) we have

$$r_m^\rho = v(r_m) = \left(\frac{r_m}{r} \right)^{-\alpha_m} v(r) \leqq K^{-\alpha_m} v(r) \leqq K^{-\rho} v(r).$$

Thus we obtain

(2.15)

$$I_1(z) \leqq \frac{8}{\rho} K^{-\rho} \left(\log \frac{r_m}{r_{m-1}} \right) v(r).$$

On combining (2.13), (2.14), and (2.15) we deduce Lemma 3.

3. Proof of Theorem 1, (1.8)

We proceed to prove (1.8), which is a little simpler than (1.7), since here we can take α_m and r_{m+1}/r_m constant in the construction of $v(r)$. We proceed to do this and set $\alpha_m = \alpha_0 = $ constant,

$$(3.1) \qquad\qquad r_m = K_1^m$$

in the construction of Section 2.1, where K_1 is assumed to be a large positive number, K is also assumed to be large, small compared with any power of K_1 but much bigger than $\log K_1$.

We easily see that in view of (2.9) we have for $r_m < r < r_{m+1}$

$$r_m^\rho < v(r) < r_{m+1}^\rho,$$

i.e., $K_1^{-\rho} r^\rho < v(r) < K_1^\rho v(r)$.

Thus $f(z)$ has very regular growth of order ρ, since $v(r) - 1 \leq n(r) \leq v(r)$. Next suppose that

$$(3.2) \qquad\qquad Kr_m \leq r \leq \frac{1}{K} r_m'$$

Then taking $\theta = 0$ in Lemma 3 we see that

$$\log M(r) = \frac{\pi}{\sin \pi\alpha_0} v(r)[1 + O(K^{-\alpha_0}) + O(K^{1-\alpha_0}) + O\{K^{-\rho} \log K_1\} + o(1)].$$

Taking $\theta = \pi - r^{-1}$ in Lemma 3, we see similarly that

$$\log \mu(r) \leq \frac{\pi}{\sin \pi\alpha_0} v(r)[\cos \alpha_0 \pi + O(K^{-\alpha_0}) + O(K^{1-\alpha_0}) + O(K^{-\rho} \log K_1) + o(1)].$$

Thus given $\varepsilon > 0$, we deduce that if K_1 is a sufficiently large positive constant and

$$(3.3) \qquad\qquad K = (\log K_1)^{2/\rho},$$

we have

$$(3.4) \qquad\qquad \log \mu(r) \leq [\cos \pi\alpha_0 + \varepsilon]\log M(r)$$

throughout the range of values given by (3.2) for $m > m_0$. We proceed to estimate the strong logarithmic density of this range.

Suppose that R_1, R_2 and R_2/R_1 are large. Let m_1 be the least integer such that $r_{m_1} > R_1$ and let m_2 be the biggest integer such that $r_{m_2} < R_2$. Let $F(R_1, R_2)$ be the range of all those values of r such that $R_1 < r < R_2$ and

(3.4) holds. Then $F(R_1, R_2)$ includes all the ranges (3.2) with $m_1 \leq m \leq m_2 - 1$, where r'_m is given by (2.12) with $\alpha_m = \alpha_0$, i.e.,

$$\log \frac{r'_m}{r_m} = \frac{\rho}{\alpha_0} \log \frac{r_{m+1}}{r_m} = \frac{\rho}{\alpha_0} \log K_1.$$

Thus

$$\int_{F(R_1, R_2)} \frac{dt}{t} \geq (m_2 - m_1) \int_{Kr_m}^{r'_m/K} \frac{dt}{t} = (m_2 - m_1)\left\{\frac{\rho}{\alpha_0} \log K_1 - 2 \log K\right\}.$$

On the other hand

$$\int_{R_1}^{R_2} \frac{dt}{t} \leq \int_{r_{m_1 - 1}}^{r_{m_2 + 1}} \frac{dt}{t} = \{m_2 - m_1 + 2\} \log K_1.$$

Hence as R_1, R_2, and R_2/R_1 all tend to infinity we deduce that

$$\varliminf \frac{1}{\log(R_2/R_1)} \int_{F(R_1, R_2)} \frac{dt}{t} \geq \varliminf \frac{m_2 - m_1}{m_2 - m_1 + 2}\left\{\frac{\rho}{\alpha_0} - \frac{2 \log K}{\log K_1}\right\}$$

$$= \frac{\rho}{\alpha_0} - \frac{2 \log K}{\log K_1};$$

that is, if F is the set of all those values of r satisfying (3.4) we have

(3.5) $$\underline{\text{strong log dens } F} \geq \frac{\rho}{\alpha_0} - \frac{2 \log K}{\log K_1}.$$

Given ρ, α, such that $0 < \rho < \alpha < 1$ and $\eta > 0$, we first choose $\alpha_0 > \alpha$, but such that

$$\frac{\rho}{\alpha_0} > \frac{\rho}{\alpha} - \tfrac{1}{2}\eta,$$

$$\cos \pi\alpha_0 + \varepsilon < \cos \pi\alpha.$$

We then choose the constants K_1, K related by (3.3) so large that (3.4) holds in the set F described above and in addition so that

$$\frac{2 \log K}{\log K_1} = \frac{4}{\rho} \frac{\log \log K_1}{\log K_1} < \tfrac{1}{2}\eta.$$

Then (3.5) gives

$$\underline{\text{strong log dens } F} \geq \frac{\rho}{\alpha_0} - \tfrac{1}{2}\eta > \frac{\rho}{\alpha} - \eta,$$

and on the set F we have by (3.4)

$$\log \mu(r) < \cos \pi\alpha \log M(r).$$

Hence if E is the complementary set of values on which (1.3) holds, we deduce that

$$\overline{\text{strong log dens }} E \leq 1 - \frac{\rho}{\alpha} + \eta.$$

In view of (1.5) this gives (1.8), since η may be chosen as small as we please.

4. Proof of Theorem 1, (1.7)

In order to prove (1.7) we take α_m and the quantities r_{m+1}/r_m and K variable. We set

(4.1) $$K'_m = r_{m+1}/r_m = 4 + \log m, \qquad m \geq 1$$

and define

(4.2) $$K_m = (\log K'_m)^{2/\rho}.$$

Then if $r_m \leq r \leq r_{m+1}$, we deduce that

$$v(r) \leq r_{m+1}^\rho \leq r^\rho \left(\frac{r_{m+1}}{r_m} \right)^\rho \leq r^\rho (4 + \log m)^\rho$$

$$\leq r^\rho (4 + \log \log r)^\rho,$$

since $4^m \leq r_m \leq r$. Similarly

$$v(r) \geq r_m^\rho \geq r^\rho \left(\frac{r_m}{r_{m+1}} \right)^{-\rho} \geq r^\rho (4 + \log \log r)^{-\rho}.$$

Thus $v(r)$ has lower order and order equal to ρ, i.e., regular growth of order ρ and hence so does $f(z)$.

Suppose next that

(4.3) $$K_m r_m \leq r \leq r'_m / K_m,$$

where r'_m is defined by (2.12). We deduce from Lemma 3 that in this range we have

$$\log M(r, f) \geq v(r) \left\{ \frac{\pi}{\sin \pi \alpha_m} + O(K_m^{-\delta}) + O(K_m^{-\rho/2}) \right\} + O(\log r),$$

where δ is a positive constant depending only on α_0 and ρ. Taking

$$\theta = (\pi - r^{-2\rho}),$$

we deduce also that

$$\log \mu(r, f) \leq v(r) \{ \pi \cot \pi \alpha_m + O(K_m^{-\delta}) + O(K_m^{-\rho/2}) \} + O(\log r).$$

We now choose

(4.4) $\qquad\qquad \alpha_m = \alpha + \frac{1}{2}(1 - \alpha)(1 + \log^+ \log^+ \log^+ m)^{-1}.$

Then the above inequalities yield, with $\delta_0 = \inf(2\delta/\rho, 1)$,

$$\frac{\log \mu(r, f)}{\log M(r, f)} \leq \cos \pi\alpha_m + O(\log \log m)^{-\delta_0} + O\left\{\frac{\log r}{\nu(r)}\right\}$$

$$\leq \cos \pi\alpha_m + O(\log \log m)^{-\delta_0}$$

$$< \cos \pi\alpha_m + o(\alpha_m - \alpha);$$

that is,

(4.5) $\qquad\qquad \log \mu(r, f) < \cos \pi\alpha \log M(r, f), \qquad m \geq m_0.$

Next let F be the union of all the ranges (4.3) for $m \geq m_0$, so that (4.5) holds on F. We proceed to estimate the lower logarithmic density of F.

For this purpose let R be a large positive number and let m_1 be the largest integer such that $r_{m_1} \leq R$. Let $F(R)$ be the part of R lying in the range r_{m_0}, r_{m_1}. Then

$$\int_{F(R)} \frac{dt}{t} \geq \sum_{m=m_0}^{m_1 - 1} \int_{K_m r_m}^{r_m'/K_m} \frac{dt}{t}$$

$$= \sum_{m_0}^{m_1 - 1} \left\{\log \frac{r_m'}{r_m} - 2 \log K_m\right\},$$

where r_m' is given by (2.12). This gives

$$\int_{F(R)} \frac{dt}{t} = \sum_{m_0}^{m_1 - 1} \left\{\frac{\rho}{\alpha_m} \log K_m' - 2 \log K_m\right\}.$$

In view of (4.2)

$$\log K_m = o(\log K_m'), \qquad \text{as} \quad m \to \infty.$$

Also $\alpha_m \to \alpha$ as $m \to \infty$ by (4.4). Hence given $\varepsilon > 0$, we can choose $N = N(\varepsilon)$, so that for $m_1 > N$

$$\int_{F(R)} \frac{dt}{t} > \frac{\rho}{\alpha}(1 - \varepsilon) \sum_N^{m_1 - 1} \log K_m' = \frac{\rho}{\alpha}(1 - \varepsilon) \log \frac{r_{m_1}}{r_N}.$$

It follows from (4.1) that

$$\log r_{m+1} = \log r_m + O(\log \log m) \sim \log r_m, \qquad \text{as} \quad m \to \infty.$$

Thus for all sufficiently large R

$$\int_{F(R)} \frac{dt}{t} > \frac{\rho}{\alpha} (1 - \varepsilon)^2 \log R,$$

$$\overline{\log \text{ dens } F} \geqq \frac{\rho}{\alpha} (1 - \varepsilon)^2.$$

Since ε is an arbitrary positive number independent of F we deduce that

$$\overline{\log \text{ dens } F} \geqq \rho/\alpha.$$

Further the inequality (4.5) holds in F. Hence the set E of r for which (4.5) is false must lie in the complement of F and so this set satisfies

$$\overline{\log \text{ dens } E} \leqq 1 - \frac{\rho}{\alpha}.$$

In view of Theorem A (1.4) we now obtain (1.7). This completes the proof of Theorem 1.

5. Proof of Theorem 2

Let $v(t)$ be the function of Theorem 2, and set

(5.1)
$$\phi(z) = \mathcal{R}\left\{ z \int_0^\infty \frac{v(t) \, dt}{t(z + t)} \right\}.$$

Since $v(t) = 0$, $t < 1$, and (1.13) holds, the integral converges. We set $z = x + iy = re^{i\theta}$, so that

(5.2)
$$\phi(z) = \int_1^\infty \frac{\{x(x + t) + y^2\} v(t) \, dt}{\{(x + t)^2 + y^2\} t}.$$

Our result is based on the following observation

Lemma 4. *Suppose that $x < 0$, and that $t_0 = t_0(z)$ is given by*

$$x(x + t_0) + y^2 = 0,$$

that is,

$$t_0 = \frac{-(x^2 + y^2)}{x} = -r \sec \theta.$$

Then with the hypotheses of Theorem 2, we deduce that

(5.3)
$$\phi(z) \leqq \frac{\pi c r^\rho \cos \rho\theta}{\sin \pi\rho},$$

where $\rho = \rho(t_0)$, $c = \psi(t_0)$.

We note that

$$v(t) \leq ct^\rho, \qquad t \leq t_0$$

$$v(t) \geq ct^\rho, \qquad t \geq t_0.$$

In fact these inequalities are evident if $t_0 \leq 1$, since then $v(t) = 0$, $0 < t \leq t_0$. If $t_0 > 1$, and $1 < t < t_0$, we deduce, since $\psi(t)$ and $\rho(t)$ are increasing functions of t,

$$v(t) = \psi(t)t^{\rho(t)} \leq \psi \, t_0)t^{\rho(t_0)} = ct^\rho.$$

Similarly we have $v(t) \geq ct^\rho$, for $t \geq t_0$. We now note that

$$x(x + t) + y^2 < 0, \; t > t_0, \; x(x + t) + y^2 > 0, \; t < t_0.$$

Thus

$$\phi(z) \leq \int_0^{t_0} \frac{x(x + t) + y^2}{(x + t)^2 + y^2} \frac{ct^\rho \, dt}{t} + \int_{t_0}^\infty \frac{x(x + t) + y^2}{(x + t)^2 + y^2} \frac{ct^\rho \, dt}{t}$$

$$= \mathcal{R}\left\{ cz \int_0^\infty \frac{t^{\rho - 1} \, dt}{t + z} \right\} = \frac{\pi c r^\rho \cos \rho\theta}{\sin \pi\rho},$$

as required. This proves Lemma 4.

We next set $n_1(r) = [v(r)]$, where $[x]$ denotes the integral part of x and define $f_1(z)$ by (2.1) with $n_1(t)$ instead of $n(t)$. Thus

$$\phi(z) - \log |f_1(z)| = \mathcal{R}\left\{ \int_1^\infty \frac{z(v(t) - [v(t)]) \, dt}{t(t + z)} \right\}.$$

We now apply Lemma 2, with $c = 0$, $t_1 = 1$, and let t_2 tend to infinity. Writing $r = |z|$ for t_0 in that Lemma, we deduce that

$$|\phi(z) - \log |f_1(z)|| \leq 2 \log r + 4 \log\left(\frac{10}{\pi - |\theta|}\right) + 1, \qquad r \geq 2.$$

Thus Lemma 4 yields

(5.4)

$$\log |f_1(z)| \leq \frac{\pi c r^\rho \cos \rho\theta}{\sin \pi\rho} + 2 \log r + 4 \log\left(\frac{10}{\pi - |\theta|}\right) + 1, \qquad r > 2.$$

We write $\theta = \pi - \eta$ and define η by the equation

$$t_0 - \frac{1}{t_0} = r = t_0 \cos \eta,$$

that is,

(5.5) $$\cos \eta = (1 - t_0^{-2}),$$

so that

$$t_0^{-2} = 1 - \cos \eta = 2 \sin^2 \tfrac{1}{2}\eta > 2\eta^2/\pi^2.$$

Thus

$$1/t_0 < \eta < 3/t_0.$$

Then

$$|r^\rho \cos \rho\theta - t_0^\rho \cos \rho\pi|$$
$$\leq |r^\rho \cos \rho\theta - t_0^\rho \cos \rho\theta| + |t_0^\rho \cos \rho\theta - t_0^\rho \cos \rho\pi|$$
$$\leq \rho r^{\rho-1}(t_0 - r) + \rho\eta t_0^\rho < (r^{\rho-1} + 3t_0^{\rho-1}) < 5t_0^{\rho-1},$$

for large t_0, since $\rho < 1$. Thus

$$\left| \frac{\pi c r^\rho \cos \rho\theta}{\sin \pi\rho} - \frac{\pi c t_0^\rho \cos \rho\pi}{\sin \pi\rho} \right| < \frac{5c\pi}{\sin \pi\rho} t_0^{\rho-1}$$

$$= \frac{5\pi v(t_0)}{\sin(\pi\rho)t_0} \to 0, \qquad \text{as} \quad t_0 \to \infty,$$

by (1.13). Thus we deduce from (5.4) for all sufficiently large t_0, and $r = t_0 - 1/t_0$,

(5.6) $$\log|f_1(re^{i\theta})| \leq \pi c t_0^\rho \cot \pi\rho + 2 \log t_0 + 4 \log(10/\eta) + 2$$
$$\leq \pi v(t_0)\cot \pi\rho + 7 \log t_0.$$

5.1

We need finally

Lemma 5. *With the hypotheses of Theorem 2 we have as $r \to \infty$, uniformly for $\tfrac{3}{4}\pi \leq |\theta| < \pi$ and $z = re^{i\theta}$,*

$$\left| \frac{f_1'(z)}{f_1(z)} \right| = \frac{O(r^{-\delta})}{|\sin \theta|}.$$

In fact it follows from (2.1) that

$$\frac{f_1'(z)}{f_1(z)} = \int_0^\infty \frac{n_1(t)\, dt}{(z+t)^2} = \int_0^{2r} \frac{n_1(t)\, dt}{(z+t)^2} + \int_{2r}^\infty \frac{n_1(t)\, dt}{(z+t)^2}$$

$$= I_1 + I_2,$$

say. We set $z = x + iy$. Then

$$|I_1| \leq n_1(2r) \int_0^\infty \frac{dt}{(x+t)^2 + y^2} \leq n_1(2r) \int_{-\infty}^{+\infty} \frac{dt}{(x+t)^2 + y^2} = \frac{\pi n_1(2r)}{y}$$

$$= O\left\{\frac{r^{-\delta}}{|\sin \theta|}\right\},$$

in view of (1.13). Also in I_2 we have $|z + t| \geq \frac{1}{2}t$, so that

$$|I_2| \leq 4 \int_{2r}^\infty \frac{n_1(t)\,dt}{t^2} = O\left\{\int_{2r}^\infty t^{-1-\delta}\,dt\right\} = O(r^{-\delta}).$$

This proves Lemma 5.

We now deduce that with $\theta = \pi - \eta$ defined as in (5.5) we have

$$\log |f_1(t_0 e^{i\theta})| \leq \log |f_1(re^{i\theta})| + \int_r^{t_0} \left|\frac{f_1'(te^{i\theta})}{f_1(te^{i\theta})}\right| dt$$

$$\leq \log |f_1(re^{i\theta})| + O\,|t_0 - r|\,r^{-\delta}\,|\text{cosec }\theta|$$

$$\leq \log |f(re^{i\theta})| + O(t_0^{-\delta}).$$

Thus (5.6) yields for all sufficiently large t_0

(5.7) $\qquad \log \mu(t_0, f_1) \leq \log|f_1(t_0 e^{i\theta})| \leq \pi v(t_0)\cot \pi\rho + 7 \log t_0 + 1.$

We now define

$$f(z) = \frac{f_1(z)}{\prod\limits_{s=1}^{8}(z - z_s)},$$

where $z_1, z_2, \ldots, z_8$ are the eight first zeros of $f_1(z)$. Since $v(t) \to \infty$ with t, it is clear that $f_1(z)$ has eight zeros. Then if $n(t)$ denotes the number of zeros of $f(z)$ in $|z| \leq t$, we see that

$$v(t) - 9 \leq n(t) \leq v(t).$$

This is (1.14). Also for large t_0 we deduce from (5.7) that

$$\log \mu(t_0, f) = \log \mu(t_0, f_1) - 8 \log t_0 + O(1)$$

$$\leq \pi v(t_0)\cot \pi\rho - \log t_0 + O(1)$$

$$< \pi n(t_0)\cot \pi\rho - \log t_0 + O(1),$$

since $\cot[\pi\rho(t_0)]$ is bounded as $t_0 \to \infty$. This proves the first inequality in (1.15), when we write r instead of t_0.

To prove the second inequality, we note that by (2.1)

$$\log M(r,f) = r \int_0^\infty \frac{n(t)\,dt}{t(r+t)} \geq r \int_r^\infty \frac{n(t)\,dt}{t(r+t)} \geq rn(r) \int_r^\infty \frac{dt}{t(r+t)}$$

$$= n(r) \log 2.$$

This completes the proof of Theorem 2.

6. Proof of Theorems 3 and 4

Let ρ_k be an increasing sequence of positive numbers, such that

(6.1) $$\rho_1 = 1,$$

(6.2) $$\frac{\rho_{k+1}}{\rho_k} \to \infty, \quad \text{as} \quad k \to \infty.$$

We take for C_k an increasing sequence of positive numbers such that

(6.3) $$C_1 = 1,$$

(6.4) $$\frac{C_{k+1}}{C_k} \to \infty, \quad \text{as} \quad k \to \infty,$$

set

(6.5) $$\psi(t) = 0, \qquad 0 \leq t \leq 1,$$

(6.6) $$\psi(t) = C_k, \qquad \rho_k \leq t < \rho_{k+1},$$

and define $v(t)$ by (1.12) with a function $\rho(r)$, such that $\frac{1}{4} \leq \rho(r) \leq \frac{1}{2}$. We assume in addition that

(6.7) $$\log C_k = o(\log \rho_k), \quad \text{as} \quad k \to \infty.$$

Then (1.13) holds with any δ such that $0 < \delta < \frac{1}{2}$, since then

(6.8) $$\psi(t) = O(t^{1/2+\delta}).$$

We write

$$I(r) = \int_0^\infty \frac{v(t)\,dt}{t(r+t)}$$

and suppose that $\rho_{k-1} < r < \rho_k$. Then

$$I(r) = \int_1^{\rho_k} \frac{v(t)\,dt}{t(r+t)} + \int_{\rho_k}^\infty \frac{v(t)\,dt}{t(r+t)}$$

$$= I_1(r) + I_2(r),$$

say. Since $\rho(r) \geq \frac{1}{4}$, $t^{-1/4}v(t) = t^{\rho(r)-1/4}\psi(t)$ is an increasing function of t. Thus

$$I_1(r) \leq \frac{v(\rho_k - 0)}{\rho_k^{1/4}} \int_0^{\rho_k} \frac{t^{-3/4}\, dt}{r+t} \leq \frac{4v(\rho_k - 0)}{r}.$$

Set

(6.9) $$J_k = \int_{\rho_k}^{\infty} \frac{v(t)\, dt}{t^2} > v(\rho_k) \int_{\rho_k}^{\infty} \frac{dt}{t^2} = \frac{v(\rho_k)}{\rho_k}.$$

Then if $r \leq \lambda\rho_k$, where $0 < \lambda < 1$, we deduce that $t \leq r + t \leq (1 + \lambda)t$ in $I_2(r)$, so that

$$\frac{1}{1+\lambda} J_k \leq I_2(r) \leq J_k.$$

Thus

$$\frac{J_k}{1+\lambda} < I(r) < J_k + \frac{4v(\rho_k - 0)}{r}.$$

Suppose now that λ_1, λ_2 are fixed positive numbers, such that $\lambda_2 = K\lambda_1$, where $K > 1$. Then for $\lambda_1 \rho_k \leq r \leq \lambda_2 \rho_k$, we have

$$\frac{J_k}{1+\lambda_2} < I(r) < J_k + \frac{4v(\rho_k - 0)}{\lambda_1 \rho_k}.$$

By (6.4) and (6.6)

$$\frac{v(\rho_k)}{v(\rho_k - 0)} \geq \frac{C_k}{C_{k-1}} \to \infty, \qquad \text{as} \quad k \to \infty.$$

Hence from this and (6.9) we deduce that in our range we have as $k \to \infty$

$$\frac{J_k}{1+\lambda_2} < I(r) < [1 + o(1)]J_k.$$

Also if $f(z)$ is the function of Theorem 2, then $\log f(z)$ is given by (2.1), so that

(6.10) $$\log M(r, f) = r \int_0^{\infty} \frac{n(t)\, dt}{t(r+t)} = rI(r) + r \int_1^{\infty} \frac{[n(t) - v(t)]\, dt}{t(r+t)}$$

$$= rI(r) + O\left\{ r \int_1^{\infty} \frac{dt}{t(r+t)} \right\} = rI(r) + O(\log r).$$

Thus

$$\frac{rJ_k}{1 + \lambda_2} + O(\log r) \leqq \log M(r, f) \leqq [1 + o(1)]rJ_k + O(\log r).$$

Taking $r = \lambda_1 \rho_k$, $\lambda_2 \rho_k$ in turn, we deduce that

$$\varlimsup_{r \to \infty} \frac{\log M(Kr, f)}{\log M(r, f)} \geqq \lim_{r \to \infty} \frac{KrJ_k}{(1 + \lambda_2)rJ_k} = \frac{K}{1 + \lambda_2}.$$

Here λ_2 can be chosen as small as we please, so that (1.16) holds.

To prove Theorem 3, we now take $\rho(r) = \frac{1}{2}$, fix $C_k = k!$, so that (6.4) holds and allow ρ_k to tend to infinity so rapidly that (6.2) and (6.7) hold and further that given a function $\phi(t)$ which tends to infinity with t, we have

$$\phi(t) > C_k, \quad t > \rho_k, \quad k = 2, 3, \ldots.$$

Thus

$$\psi(t) < \phi(t), \qquad t > \rho_2,$$

so that $\psi(t)$ and hence $t^{-1/2}n(t) \leqq \psi(t)$ can be chosen to tend to infinity as slowly as we please. In view of (6.10) we deduce that $r^{-1/2} \log M(r, f)$ can be chosen to tend to infinity as slowly as we please. Also in this case (1.15) implies

$$\mu(r) < 0, \qquad r > r_0,$$

which is stronger than (1.17). Thus we have proved Theorem 3 and have also obtained a function of order $\frac{1}{2}$ maximal type satisfying (1.16) and (1.17).

It remains to complete the proof of Theorem 4, by constructing a function of order 1 mean type or minimal type and satisfying (1.16) and (1.17). For this purpose let $\varepsilon(r)$ be the function of Theorem 4. We may suppose without loss in generality that $\varepsilon(r)$ is continuous to the left and decreases to zero as $r \to \infty$. For otherwise we can first replace $\varepsilon(r)$ by the smaller function

$$\varepsilon_1(r) = \inf\{\varepsilon(r), \log(A + r)^{-1/2}\}$$

which tends to zero as $r \to \infty$ and then by the still smaller function

(6.11) $$\varepsilon_2(r) = \inf_{t < r+1} \varepsilon_1(r),$$

which is decreasing and continuous to the left. By choosing $A = e^{16}$, we may also suppose that $\varepsilon_2(r) \leqq \frac{1}{4}$.

We now set $\rho(r) = \frac{1}{2} - \varepsilon(r)$, so that $\rho(r)$ increases with r, and construct $\psi(t)$ as before. Then (1.16) continues to hold and (1.15) yields

$$\log \mu(r) < O\{\tan\{\pi\varepsilon(r)\}\log M(r)\},$$

which yields (1.17). To make sure that $f(z)$ has order $\frac{1}{2}$ minimal type, we require that

$$v(t) = o(t^{1/2})$$

that is,

$$\psi(t) = o\{t^{\varepsilon(t)}\} = o\{\exp[\varepsilon(t)\log t]\},$$

which is possible, since we may let $\psi(t)$ tend to infinity arbitrarily slowly. To construct $f(z)$ to have mean type we choose C_k, ρ_k to satisfy

$$C_k = \rho_k{}^{\varepsilon(\rho_k)}.$$

Then (6.7) still holds, and given C_k, we choose C_{k+1} so large that (6.2) holds and in addition (6.4). This is possible, since $\varepsilon(t)\log t$ tends to infinity with t. Thus for $t = \rho_k$, we have

$$v(t) = \rho_k^{1/2-\varepsilon(\rho_k)}\rho_k^{\varepsilon(\rho_k)} = t^{1/2},$$

and for $\rho_k < t < \rho_{k+1}$, we have

(6.12)
$$v(t) = v(\rho_k)t^{1/2-\varepsilon(t)}/\rho_k^{1/2-\varepsilon(\rho_k)} = \rho_k^{\varepsilon(\rho_k)}t^{1/2-\varepsilon(t)} \leqq t^{1/2},$$

provided that

$$t^{\varepsilon(t)} \geqq \rho_k^{\varepsilon(\rho_k)}, \qquad t > \rho_k.$$

Since $\varepsilon(t)\log t$ tends to infinity with t and is continuous to the left by the construction (6.11), (6.12) is satisfied for some arbitrarily large values of $\rho = \rho_k$, and so given ρ_{k-1} we can always choose ρ_k to satisfy all the above conditions. Thus $v(r)$ has order $\frac{1}{2}$ mean type and hence so does $n(r)$ and $f(z)$ (see, e.g., [4, p. 27]). This completes the proof of Theorem 4.

We note that for functions of genus zero with real negative zeros the upper limit in (1.16) cannot be greater than K. For by (2.1)

$$\frac{1}{r}\log M(r) = \int_0^\infty \frac{n(t)\,dt}{t(r+t)}$$

is clearly a decreasing function of r.

REFERENCES

1. P. D. Barry, On a theorem of Besicovitch, *Quart. J. Math. Oxford Ser.* 2 **14** (1963), 293–302.
2. ———, Some theorems related to the $\cos\pi\rho$-theorem, to appear in *Proc. London Math. Soc.*

3. A. S. Besicovitch, On integral functions of order < 1, *Math. Ann.* **97** (1927), 677–695.
4. W. K. Hayman, *Meromorphic functions*, Oxford, Clarendon Press, 1964.
5. B. Kjellberg, *On certain integral and harmonic functions*, Dissertation, Uppsala, 1948.
6. ——, A theorem on the minimum modulus of entire functions, *Math. Scand.* **12** (1963), 5–11.
7. G. Valiron, Sur les fonctions entières d'ordre nul et d'ordre fini et en particulier les fonctions à correspondance régulière, *Ann. Fac. Sci. Univ., Toulouse* **5** (1913), 117–257.
8. A. Wiman, Über eine Eigenschaft der ganzen Funktionen von der Höhe Null, *Math. Ann.* **76** (1915), 197–211.

(*Received December 28, 1967*)

On a Theorem of Study Concerning Conformal Maps with Convex Images

Maurice Heins
University of Illinois
Urbana, Illinois 61801
U.S.A.

1.

It will be convenient to make in advance the following notational and terminological agreements. Given a complex and $r > 0$, we denote the disk $\{z : |z - a| < r\}$ by $\Delta(a; r)$. We denote the unit disk $\Delta(0; 1)$ more simply by Δ. Given complex numbers a and b, by the *segment* $[a, b]$ we understand

$$\{(1 - t)a + tb : 0 \leq t \leq 1\}.$$

Unqualified topological terms are to be construed in the sense of the standard topology of the complex plane. A topological term prefixed by "Δ-" is to be taken as the term in question in the sense of the relative topology of Δ.

2.

The theorem of Study with which we are concerned is the following [3, p. 109]:

Let f be analytic on Δ, be univalent, and have a convex image. Then $f[\Delta(0; r)]$ is convex, $0 < r < 1$.

A proof of this theorem based on the lemma of Schwarz was given by T. Radó (see [2, p. 224]). The following brief proof is closely related to Radó's. We introduce a, b, t, where $|a|$, $|b| < 1$ and $0 \leq t \leq 1$, and thereupon

$$g(z) = f^{-1}[(1 - t)f(az) + tf(bz)],$$

$|z| < 1$. Noting that $g(0) = 0$ and that with a and b fixed we have $|g(z)| \leq m$, $|z| < 1$, where m satisfies $0 < m < 1$ and is independent of t, we conclude, applying the lemma of Schwarz to $m^{-1}g$ and setting $z = r$, that the segment $[f(ar), f(br)]$ lies in $f[\Delta(0; r)]$. Given the arbitrariness of a and b we conclude that $f[\Delta(0; r)]$ is convex.

The theorem of Study discerns a family of parts of Δ each member of which has the property that its image with respect to each f of the theorem of Study is convex, namely the family of disks $\Delta(0; r)$, $0 < r < 1$. (Of course, we may adjoin Δ for which the property holds trivially.) In this paper *we shall characterize the parts X of Δ which share with the disks $\Delta(0; r)$, $0 < r \leq 1$, the property* (S): $f(X)$ *is convex for each f admitted in the theorem of Study.*

Given (η, r) satisfying $|\eta| = 1$, $0 < r < +\infty$, we let $A(\eta, r)$ denote the set of z satisfying

$$\mathrm{Re}\left(\frac{\eta + z}{\eta - z}\right) \geq r.$$

Given a, b ($\neq a$) $\in \Delta$, there exist precisely two distinct sets of the form $A(\eta, r)$ such that a and b belong to the frontier of each, say $A(\eta_1, r_1)$ and $A(\eta_2, r_2)$. Necessarily $\eta_1 \neq \eta_2$. We let $D(a, b)$ denote the intersection of the $A(\eta_k, r_k)$. The set $D(a, b)$ is the convex hull of the union of the subarcs γ_k with endpoints a and b of the sets

$$\left\{ \mathrm{Re}\left(\frac{\eta_k + z}{\eta_k - z}\right) = r_k \right\},$$

$k = 1, 2$. The assertions stated here have an elementary geometric character and are easily proved. Given $X \subset \Delta$, we say that X is *D-convex* provided that $D(a, b) \subset X$ when a, b ($\neq a$) $\in X$.

We now state our principal result.

Theorem. *A part of Δ has the property* (S) *if and only if it is D-convex.*

The necessity of the stated condition is readily proved. Indeed, X having the property (S) is convex since the identity map on Δ is an admitted f. Further given a, b ($\neq a$) $\in X$, each of the maps

$$f_k : z \to \frac{\eta_k + z}{\eta_k - z}, \qquad |z| < 1,$$

$k = 1, 2$, where the η_k are the first components of the (η_k, r_k) associated above with a and b, is admitted by the theorem of Study. Hence

$$f_k(\gamma_k) = [f_k(a), f_k(b)] \subset f_k(X),$$

$k = 1, 2$. We conclude that $\gamma_k \subset X$, $k = 1, 2$, and consequently, since X is convex, $D(a, b) \subset X$. The D-convexity of X is thereby established.

As it might be expected, the proof of the sufficiency will involve more serious considerations. We shall treat the question by showing that a Δ-closed D-convex set admits a representation as the intersection of a family of sets $A(\eta, r)$. This will be achieved with the aid of standard results concerning the representation of closed convex sets in the complex plane as the intersection of closed half-planes (see [1, p. 6]). The Study theorem combined with some elementary observations yields the result that intersections of sets $A(\eta, r)$ have the property (S). After the case of Δ-closed D-convex sets will have been cared for in this fashion, the proof of the sufficiency will be completed by referring the study of an unrestricted D-convex set to that of its Δ-closure.

3.

A lemma. We suppose that X is a nonempty Δ-closed, D-convex, proper subset of Δ and show

Lemma. *X admits a representation as an intersection of a family of sets $A(\eta, r)$ where the family has the property that each Δ-frontier point of X is a frontier point of a member of the family and the frontier of each member of the family contains a Δ-frontier point of X.*

Proof. On referring to the classical theory of convex sets [1, p. 6] we see that X admits a representation of the form

(3.1)
$$X = \Delta \cap \bigcap_{H \in \mathfrak{H}} H,$$

where $\mathfrak{H}$ is a family of closed half-planes which has the property that the frontier of each member half-plane contains a Δ-frontier point of X and also each Δ-frontier point of X is a frontier point of some member half-plane. It is essential to observe that if a point ζ of the unit circumference belongs to $\overline{X}$, then $A(\zeta, r) \subset X$ for some r, $0 < r < +\infty$, as we see using the D-convexity of X and a limiting argument. It suffices to consider $D(a, b)$ with a fixed in X and b in X tending to ζ. A closed half-plane H containing $\overline{X}$ such that $\zeta \in \text{fr } H$ satisfies $\Delta \subset H$, as we see with the aid of the fact that $A(\zeta, r) \subset X$. It will emerge from the developments below that the frontier of each half-plane belonging to $\mathfrak{H}$ contains exactly one frontier point of X.

Given $H \in \mathfrak{H}$ and ξ a Δ-frontier point of X belonging to the frontier of H, let A denote the unique $A(\eta, r)$ satisfying the condition that it lie in H and its frontier be tangent to fr H at ξ. We show that

$$(3.2) \qquad\qquad X \subset A,$$

whence it follows that there is a unique point, say $\xi(H)$, common to the frontier of H and the (Δ-) frontier of X. The image of $H \rightarrow \xi(H)$ is the Δ-frontier of X. We shall denote by $A(H)$ the A associated above with H. It is, of course, unique. From

$$X \subset A(H) \subset \Delta \cap H,$$

we conclude from (3.1) that

$$(3.3) \qquad\qquad X = \bigcap_{H \in \mathfrak{H}} A(H).$$

In order to demonstrate (3.2) we refer the problem to the open right half-plane, $\{\operatorname{Re} z > 0\}$, by the Möbius transformation

$$L : z \rightarrow \frac{\eta + z}{\eta - z},$$

where η is the first component of the parameter pair (η, r) associated with A. We show that

$$(3.4) \qquad\qquad L(X) \subset \{\operatorname{Re} z \geq \operatorname{Re} L(\xi) = r\}.$$

The conclusion (3.2) follows when the inverse of L is applied. If (3.4) did not hold, there would exist $\zeta \in X$ such that $\operatorname{Re} L(\zeta) < \operatorname{Re} L(\xi)$. We note that

$$[L(\xi), L(\zeta)] \subset L[D(\xi, \zeta)]$$

by virtue of the convexity of the set on the right. Using the D-convexity of X and the fact that $X \subset H$ we conclude that

$$[L(\xi), L(\zeta)] \subset L(H).$$

However the points of the segment $[L(\xi), L(\zeta)]$ which are near but different from $L(\xi)$ do not lie in $L(H)$, as we see on noting that $L(\text{fr } H)$ is a circular circumference passing through -1 and $L(\xi)$ and lying in $\{\operatorname{Re} z \leq \operatorname{Re} L(\xi)\}$ less the point -1. Contradiction. The lemma is established.

4.

We note some rules which permit us to conclude that a part of Δ has the property (S).

(a) If X has the property (S) and α is a conformal automorphism of Δ, then $\alpha(X)$ also has the property (S).

It suffices to observe that

$$[\alpha(X)] = f \circ \alpha(X)$$

and that when f satisfies the conditions imposed in Study's theorem, so does $f \circ \alpha$.

(b) If X_λ has the property (S) for each $\lambda \in \Lambda$ $(\neq \varnothing)$, then so does $\bigcap_{\lambda \in \Lambda} X_\lambda$.

This follows from the standard formula of set theory for the image with respect to a suitably restricted univalent map of the intersection of a family of sets and the fact that convexity is closed with respect to intersection.

(c) If X_n has the property (S) and $X_n \subset X_{n+1}$, $n = 0, 1, \ldots$, then $\bigcup_0^\infty X_n$ has the property (S).

This is easily concluded with the use of the formula for the image of a union and the definition of convexity.

On applying these rules taken together with Study's theorem, we find that the family of parts of Δ having the property (S) contains "much more" than the disks $\Delta(0; r)$, $0 < r \leq 1$. Thus (i) $\Delta(a; r)$, where $|a| < 1$ and $0 < r < 1 - |a|$ has the property (S). Indeed, $\Delta(a; r)$ is the image of a disk $\Delta(0; \rho)$, $0 < \rho < 1$, with respect to a conformal automorphism of Δ and (i) follows with the aid of (a). In addition, (ii) the $A(\eta, r)$ and their interiors have the property (S). Here we note that int $A(\eta, r)$ admits a representation as a union of the form

$$\bigcup_0^\infty \Delta(a; r_n)$$

where $|a| < 1$ and

$$r_n = \frac{n+1}{n+2}(1 - |a|).$$

Hence using (i) and (c) we conclude that int $A(\eta, r)$ has the property (S). Noting that

$$A(\eta, r) = \bigcap_0^\infty \text{int } A\left(\eta, \frac{n+1}{n+2} r\right)$$

and using (b) we see that $A(\eta, r)$ has the property (S).

At this stage we conclude with the aid of the lemma of Section 3 that a Δ-closed part of Δ has the property (S) if and only if it is D-convex. Actually, we may obtain somewhat more. Thus using the fact that the $A(\eta, r)$ have the property (S), (b), and the lemma of Section 3, we conclude that the Δ-closed parts of Δ having the property (S) are precisely the intersections of families of $A(\eta, r)$. (Here the intersection with empty index set is understood to be Δ.) Consequently, a proper Δ-closed part of Δ having the property (S) is either an $A(\eta, r)$ or else a compact subset of Δ.

5.

To complete our study it remains to be shown that an unrestricted D-convex set has the property (S). We consider such a set, say X, supposing that X is a proper part of Δ containing more than one point—the omitted cases are trivial. Using the fact that X is convex in the classical sense, we infer that the Δ-closure of X, which we shall denote by Y, is a proper part of Δ. Since Y is D-convex, it has the property (S).

To continue we make use of the representation (3.3) applied to Y. Our goal is to show that given $a, b \, (\neq a) \in X$, and an admitted f, then

$$(5.1) \qquad\qquad [f(a), f(b)] \subset f(X).$$

In any case (5.1) holds with Y replacing X, and, in addition,

$$\text{int } f(Y) = f(\text{int } Y) = f(\text{int } X) = \text{int } f(X).$$

Hence we are led to the following alternatives: either every point of $[f(a), f(b)]$ with the possible exceptions $f(a)$, $f(b)$ lies in int $f(Y)$, or else

$$[f(a), f(b)] \subset f(Y) \cap \text{fr } f(Y) = f(\Delta \cap \text{fr } Y).$$

In the former case (5.1) holds. In the latter case let $f(c)$ be a point of $[f(a), f(b)]$ distinct from $f(a)$ and $f(b)$ and let A be an $A(H)$ in a representation of Y of the form (3.3) which is such that $c \in \text{fr } A$. We introduce

$$\gamma = f^{-1}([f(a), f(b)])$$

and note that

$$c \in \gamma \subset A.$$

In addition, $a, b \in \text{fr } A$ for otherwise $f(c) \in f(\text{int } A)$ and $c \notin \text{fr } A$. Suppose that

$$d \in \gamma \cap \text{int } A.$$

Then every point of $[f(a), f(d)]$ save $f(a)$ and every point of $[f(d), f(b)]$ save $f(b)$ would lie in int $f(A) = f(\text{int } A)$. This would imply that $f(c) \in f(\text{int } A)$. This is impossible. Hence $\gamma \subset \text{fr } A$. Since X is D-convex, $\gamma \subset X$ and consequently (5.1) holds. It follows that a D-convex set has the property (S).

REFERENCES

1. T. Bonnesen and W. Fenchel, *Theorie der konvexen Körper*, Ergebnisse der Mathematik und Ihrer Grenzgebiete, Bd. III, no.1, Springer-Verlag, Berlin, 1934.
2. S. Saks and A. Zygmund, *Analytic Functions* (trans. by E. J. Scott), Warsaw-Wrocław, 1952.
3. E. Study, *Vorlesungen über ausgewählte Gegenstände der Geometrie*, Heft 2, Konforme Abbildung einfach zusammenhängender Bereiche, B. G. Teubner, Leipzig and Berlin, 1913.

(Received December 23, 1967)

On Groups of Automorphisms of Certain Sequence Spaces

V. Ganapathy Iyer

Department of Mathematics
Annamalai University
Annamalai Nagar
India

1.

Let P denote the class of all power series of the form $f(u) = 1 + \sum_1^\infty a_n u^n$. The power series $1/f(u) = 1 + \sum_1^\infty A_n u^n$ is uniquely defined in the sense that the numbers A_n are uniquely determined from the relations

(1) $$A_n + a_1 A_{n-1} + \cdots + a_n = 0, \qquad n = 1, 2, 3, \ldots$$

and, for Cauchy multiplication, $[f(u)]\,[1/f(u)] \equiv 1$. Since Cauchy multiplication in P is associative, the set P forms a group for this multiplication. The subset P_1 of P consisting of power series with positive radii of convergence form a subgroup. The set P_2 of all elements of P for which both $1 + \sum |a_n|$ and $1 + \sum |A_n|$ converge form a subgroup of P as well as of P_1.

1.1

With each power series $f \in P$, we associate the upper triangular matrix $M_f = (a_{nk})$ where $a_{nk} = a_{k-n}$, $k, n = 0, 1, 2, \ldots$ and $a_0 = 1$, $a_n = 0$ for $n < 0$. Since $M_f M_g = M_{fg}$ and multiplication among upper triangular matrices is associative, the set of matrices M_f form a group for multiplication isomorphic to P. We shall use the same symbol P to denote this group and similarly for the groups of matrices corresponding to the elements in P_1 and P_2. Note that all these groups are commutative.

1.2

We shall denote by B the space of all bounded sequences $\xi = (x_0, x_1, \ldots)$ of complex numbers with norm $\|\xi\| = \sup_{(n)} |x_n|$. With this norm, B is a Banach space. We shall denote by c and c_0 the space of convergent sequences and sequences converging to zero, respectively. With the same norm, these are closed in B and c_0 is closed in c.

1.3

We denote by Γ the space of all entire functions specified by their co-efficients in the Taylor expansion around the origin, that is, sequences $\xi = (x_n)$ with $|x_n|^{1/n} \to 0$ as $n \to \infty$. It is known that with the metric

$$D(\xi, \eta) = \sup(|x_0 - y_0|, |x_n - y_n|^{1/n}, n \geq 1),$$

Γ is a complete linear metric space—more precisely an FK-space in the terminology of Zeller ([2], 4).

2.

The object of this paper is to show that each matrix in P_2 defines an automorphism (a one-to-one bicontinuous onto linear mapping) of the space B and that each of the spaces c and c_0 are left invariant by the elements of P_2. Further we shall show that P_2 can be mapped in a one-to-one manner onto an open subset of the space l of absolutely convergent series with the usual norm. Similarly, the elements of P_1 define automorphisms of the space Γ and these lead to a large class of bases in Γ.

2.1

The above results are related to the solution of the infinite system of equations

$$(2) \qquad x_n + \sum_{k=1}^{\infty} a_k x_{n+k} = y_n, \qquad n = 0, 1, 2, \ldots,$$

where the sequences (a_n) and (y_n) are given and the x_n's are to be found. If $f \in P_2$, for each $\eta = (y_n) \in B$, there is a unique $\xi = (x_n) \in B$ satisfying the above equation. Similarly when $f \in P_1$, the system (2), has for each $\eta \in \Gamma$ a unique $\xi \in \Gamma$ satisfying (2).

2.2

To prove the above results, we require the following known theorems [1, pp. 41, 54]:

(A) If T is a continuous linear transformation of a Banach space X into another Banach space Y, then for a suitable constant K, $\|Tx\| \leq K\|x\|$ for $x \in X$, $Tx \in Y$, and conversely.

(B) A one-to-one continuous linear transformation of a Banach space onto another is bicontinuous. More generally this is true if X and Y are FK-spaces.

2.3

The following results on matrices and FK-spaces are true [4, p. 41]:

(C) If X and Y are FK-spaces and for a matrix M, the transform $\eta = M(\xi)$ exists for every $\xi \in X$ and $\eta \in Y$, then M defines a continuous linear transformation from X into Y.

(D) A matrix $M = (a_{nk})$ is said to be row bounded if for a suitable constant H independent of n, $\sum_{k=0}^{\infty} |a_{nk}| \leq H$, $n = 0, 1, 2, \ldots$. Multiplication among row-bounded matrices is associative. A bounded sequence will be regarded as a row-bounded matrix whose first column consists of the terms of the sequence and whose other entries are zero.

(E) An element of P will transform c into c if and only if it is row bounded, that is $\sum_{1}^{\infty} |a_n|$ is convergent. The same is necessary and sufficient for c_0 being transformed into c_0. Obviously, each such element of P transforms B into B and the converse is also true. Hence each matrix $M_f \in P$ which is row bounded defines a continuous linear transformation of B into B leaving invariant the subspaces c and c_0.

3.

Theorem 1. *Each matrix of the group P_2 defines an automorphism of the space B and leaves invariant the subspaces c and c_0.*

Proof. Let $M_f \in P_2$ so that both $\sum |a_n|$ and $\sum |A_n|$ are convergent. Let $\xi \in B$ and let $\eta = M_f(\xi)$. The matrices M_f and $M_{1/f}$ are row bounded and ξ can be regarded as a row-bounded matrix and so by (D), $M_{1/f}(\eta) = \xi$. Hence M_f defines a one-to-one linear transformation of B onto B. Moreover $\|M_f(\xi)\| \leq H\|\xi\|$, where $H = 1 + \sum |a_n|$. Hence by (A), M_f defines a continuous transformation. So by (B), the transformation is bicontinuous and hence an automorphism.

3.1

Theorem 2. *If an element $M_f \in P$ defines a one-to-one transformation of B onto B, then $M_f \in P_2$.*

Proof. Since $M_f(\xi)$ exists for each $\xi \in B$, M_f is row bounded by (E). Since $\|M_f(\xi)\| \leq (1 + \sum |a_n|)\|\xi\|$, M_f defines a continuous linear transformation of B onto B and since by hypothesis the transformation is one-to-one, the inverse transformation is continuous by (A). Let T be the inverse transformation. Then there is a constant K such that $\|T(\eta)\| \leq K\|\eta\|$. Let, for a complex number u, sgn(u) be defined by $u\,\text{sgn}(u) = |u|$ if $u \neq 0$ and sgn(u) = 0 when $u = 0$. Using the notation in Section 1, define $\eta = (y_n)$ as follows: $y_0 = 1$, $y_k = \text{sgn}(A_k)$, $k = 1, 2, \ldots, p$, and $y_k = 0$ for $k > p$. Define $\xi = (x_n)$ by $x_0 = y_0 + A_1 y_1 + \cdots + A_p y_p$, $x_1 = y_1 + A_1 y_2 + \cdots + A_{p-1} y_p$, $\ldots$, $x_p = y_p$, and $x_k = 0$ for $k > p$. Then by using relations (1), we see that $M_f(\xi) = \eta$ so that $T(\eta) = \xi$. Also, by definition $\|T(\eta)\| = 1 + \sum_1^p |A_k|$ and $\|\eta\| = 1$. So we get $1 + \sum_1^p |A_k| \leq K$. Hence $\sum |A_n|$ converges and $M_f \in P_2$.

3.2

We now define a map ϕ of P_2 into the space l of absolutely convergent series by the relation, if $f = 1 + \sum a_n u^n$, $\phi(M_f) = (a_1, a_2, \ldots)$, which belongs to l. This is a one-to-one map of P_2 onto $\phi(P_2)$. We shall now show that $\phi(P_2)$ is an open set in l.

Theorem 3. *The set $\phi(P_2)$ is open in the space l.*

Proof. Since $M_f \in P_2$, both $\sum |a_n|$ and $\sum |A_n|$ converge where $f = 1 + \sum_1^\infty a_n u^n$ and $1/f = 1 + \sum_1^\infty A_n u^n$. To prove the theorem, it is enough to show that if $g(u) = \sum_1^\infty b_n u^n$, then in the expansion $[f(u) + g(u)]^{-1} = 1 + \sum_1^\infty c_n u^n$, the series $\sum_1^\infty |c_n|$ converges if $\sum_1^\infty |b_n|$ is sufficiently small. Formally,

$$[f(u) + g(u)]^{-1} = \sum_{k=0}^\infty [1 + \sum_1^\infty A_n u^n]^{k+1} [-\sum_1^\infty b_n u^n]^k.$$

This shows that if $(\sum |b_n|)(1 + \sum |A_n|) < 1$, the series $\sum_1^\infty |c_n|$ will be convergent. Hence the result.

Remark. If $f(u) = 1 + \sum_1^\infty a_n u^n$ has radius of convergence greater than one and does not vanish in $|u| \leq 1$, then $M_f \in P_2$. Also if f has positive radius of convergence and $g(u) = f(xu)$, then $M_g \in P_2$ if $|x|$ is sufficiently small.

4.

We shall next prove that each $M_f \in P_1$ defines an automorphism of the space Γ. First we note that if $\xi = (x_n) \in \Gamma$ and $M_f \in P_1$, then $M_f(\xi) = \eta = (y_n)$ exists. To prove that $\eta \in \Gamma$, let d be a positive number less than the radius of convergence of f and let $0 < \delta < d$. Then $|y_n| \leq \delta^n D(f)$, where $D(f) = 1 + \sum |a_n| d^n$. This proves that $|y_n|^{1/n} \to 0$ as $n \to \infty$. So $\eta \in \Gamma$.

Theorem 4. *Each $M_f \in P_1$ defines an automorphism of Γ.*

Proof. As proved above, M_f defines a linear transformation of Γ into itself and by (C) the transformation is continuous. So by (B), M_f will define an automorphism, if we can show that the transformation is one-to-one and onto Γ. Let $\eta = (y_n) \in \Gamma$. Then $\xi = (x_n) = M_{1/f}(\eta) \in \Gamma$ by what we have proved above since $1/f$ has a positive radius of convergence. Let d be a positive number less than the radius of convergence and f and $1/f$. We can find H so that $|y_n| \leq Hd^n$, $n = 0, 1, 2, \ldots$. The double series

$$\sum_{k=0}^{\infty} |a_k| \sum_{n=0}^{\infty} |A_n| |y_{p+k+n}|$$

converges for any given integer $p = 0, 1, 2, \ldots$ because

$$\sum |A_n| |y_{p+n+k}| < Hd^{p+k} D(1/f)$$

and $Hd^p \sum |a_k| d^k$ converges. So we have the relation

$$\sum_0^{\infty} a_k x_{p+k} = \sum_0^{\infty} a_k \sum_0^{\infty} A_n y_p y_{p+k+n} = \sum_0^{\infty} y_{p+n}(a_0 A_n + \cdots + a_n A_0) = y_p$$

by using Equations (1). This proves that M_f is one-to-one and onto Γ. Hence M_f defines an automorphism of Γ and $M_{1/f}$ is its inverse.

4.1

A sequence (α_n) of elements of Γ is said to form a base if every $\alpha \in \Gamma$ can be uniquely represented in the form $\alpha = \sum t_n(\alpha)\alpha_n$ where $t_n(\alpha)$ are complex numbers. If $\alpha_n \equiv u^n$, the sequence forms a base. Also a base is transformed into a base by any automorphism of Γ. So if we transform the base (u^n) by any $M_f \in P_1$ we get a base. The transform is the sequence (α_n) where $\alpha_n = u^n + a_1 u^{n-1} + \cdots + a_n$. A base (α_n) is called proper if $|t_n(\alpha)|^{1/n} \to 0$ as $n \to \infty$. It can be directly verified that each of the bases obtained from (u^n) by $M_f \in P_1$ is a proper base [3].

5.

We conclude with some remarks on the system of equations (2). As already remarked, for each $\eta \in B$, there is a unique $\xi \in B$ satisfying (2) when $M_f \in P_2$. The question arises whether (2) will have unbounded solutions. Taking $f = 1 - u/2$, we see that $M_f \in P_2$. The sequence $(1, 2, 2^2, \ldots)$ is a solution of (2) for $\eta = (0, 0, \ldots)$, and so for any $\eta \in B$, (2) will have unbounded solutions. The question is open whether for a given $M_f \in P_2$, (2) will always have unbounded solutions. The same example shows that for $M_f \in P_1$, (2) can have solutions not belonging to Γ and here also the general case remains open.

REFERENCES

1. S. Banach *Théorie des Operations Linéaires*, Warsaw, 1932; New York, Stechert, 1932.
2. V. Ganapathy Iyer, Space of Integral Functions—I, *J. Indian Math. Soc.* (2), **12** (1948), 13–30.
3. ———, The space of integral functions—II, *Quart. J. Math. Oxford Ser.* 2 **1** (1950), 86–96.
4. K. Zeller *Limiteirungs Verfahren*, Springer-Verlag, Berlin, 1958.

(*Received January 20, 1968*)

On the Phragmén-Lindelöf Theorem, the Denjoy Conjecture, and Related Results*

James A. Jenkins
Department of Mathematics
Washington University
St. Louis, Missouri 63130
U.S.A.

1.

The classical Phragmén-Lindelöf theorem has been extended in many ways and the method of harmonic majoration has repeatedly been found a very effective method in this connection. Heins, in particular, obtained a very penetrating extension of this result [4]. Moreover he subsequently manifested a very close methodological connection between this result and the Denjoy conjecture [5, 6]. Of course Ahlfors' original proof [1] of the latter used harmonic majoration in an essential manner. Later Kennedy [10, 11], using a development of Ahlfors' technique, sharpened Heins' results to a form which, in a certain sense, he showed best possible.

Some years ago the author gave a proof of the Denjoy conjecture, indeed an extension of this to quasiconformal mappings [9], in which the principal feature was that the use of harmonic majoration was confined to the proof of a single lemma concerning the module of a triad. This lemma then, combined with arguments using the method of the extremal metric, was the basis for the remainder of the proof. It has now been observed that this

* Research supported in part by the National Science Foundation.

same lemma can be used in treating the other results mentioned above. In all cases it leads to very simple proofs and in some instances to results more general than those previously obtained.

2.

The lemma in question deals with a geometrical configuration called a triad. This consists of a simply connected domain D of hyperbolic type, an open boundary arc γ of D (in the natural sense of boundary correspondence) with complementary closed boundary arc γ^* and a distinguished interior point P of D. We denote by $m(P, \gamma, D)$ the module of the class of (locally rectifiable) open arcs lying in $D - \{P\}$, running from γ back to γ and separating P from γ^*. For real numbers a and b, $a < b$, we denote by $S(a, b)$ the strip in the (u, v)-plane defined by $a < u < b$, by $g(a)$ its boundary arc $u = a$ and by $g(b)$ its boundary arc $u = b$.

The results mentioned above have all been proved so as to relate to subharmonic functions generalizing the form for regular functions in which they were originally proved. Our lemma can be stated in these terms also.

Lemma 1. *Let* (D, γ, P) *be a triad where* D *is a domain in the z-plane and let there exist a subharmonic function* $u(z)$, $z \in D$, *such that*

$$\varlimsup_{z \to \zeta} u(z) \leqq a, \qquad \zeta \in \gamma$$

$$\varlimsup_{z \to \zeta} u(z) \leqq b, \qquad \zeta \in \gamma^*$$

with $a < b$ *and* $u(P) > a$. *Then*

$$m(P, \gamma, D) \leqq m(u(P), g(a), S(a, b)).$$

The proof follows at once from the two-constants theorem and the monotonic correspondence between $m(P, \gamma, D)$ and the harmonic measure of γ taken at P with respect to D. A detailed proof is found in [9]. There the result is stated for quasiconformal mappings, and if we wished we could give it a form applicable to a class of functions simultaneously generalizing subharmonic functions and the logarithm of the modulus of quasiconformal mappings. For these the lemma would become a definition with the form

$$m(P, \gamma, D) \leqq Km(u(P), g(a), S(a, b))$$

with $K \geqq 1$. We will not follow this up here but confine ourselves to subharmonic functions. This remark does, however, manifest the questionable validity of Pfluger's criticism of the paper [9] in his *Zentralblatt* review [vol. 100 (1963), p. 76].

Lemma 2. *Let* Δ *denote a hyperbolic simply connected domain in the right-hand half z-plane, part of whose boundary consists of a segment* σ:

$$\{iy, \, -y_1 < y < y_2, \, y_1, y_2 > 0\}.$$

Let the open complementary boundary arc to σ *be denoted by* τ. *Let* $\tilde{\Delta}$ *be the reflection of* Δ *in the imaginary axis and let* $\Xi = \Delta \cup \sigma \cup \tilde{\Delta}$. *Let* Ξ *be mapped conformally on* $|w| < 1$ *by* $w = \phi(z)$ *where* $\phi(z)$ *has Taylor expansion at the origin*

$$\phi(z) = \alpha^{-1}z + \textit{higher powers of } z,$$

$\alpha > 0$. *Then for fixed* $P = x + iy$, $x > 0$,

$$m(P, \sigma, \Delta) = -\frac{1}{\pi} \log x + \frac{1}{\pi} \log 2\alpha + o(1)$$

as $\alpha \to \infty$.

Under the mapping ϕ, P is mapped into the point $Q = x/\alpha + iy/\alpha + O(\alpha^{-2})$. Let R denote the point $w = r$ on the real axis concyclic with i, Q, $-i$. Let E' denote the open right-half unit circle in the w-plane and s the segment of the imaginary axis in $|w| < 1$. Then

$$m(P, \sigma, \Delta) = m(R, s, E').$$

If $Q = u + iv$, $r = u + O(u^2 + v^2)$ therefore

$$r = \frac{x}{\alpha} + O(\alpha^{-2}).$$

Let the unit circle be slit along the segment $[-r, r]$ to produce a doubly connected domain **P** and let the latter have (the usual) module m. Then $m(R, s, E') = 2m$. Replace **P** by the doubly connected domain bounded by the slit $[-2, 2]$ and the circle $|W| = 2/r$ which likewise has module m. Apply the inverse of the mapping $W = Z + Z^{-1}$ to obtain a domain bounded by $|Z| = 1$ and a curve C which will lie between circles of radii $2/r - O(r)$ and $2/r + O(r)$. From this we see at once that

$$m = \frac{1}{2\pi} \log \frac{2}{r} + o(1).$$

Thus finally

$$m(P, \sigma, \Delta) = \frac{1}{\pi} \log \frac{2\alpha}{x} + o(1).$$

Corollary 1. *If $l < \lambda < L$ are real numbers then for l, λ fixed and L tending to infinity*

$$m(\lambda, g(l), S(l, L)) = \frac{1}{\pi} \log \frac{L - l}{\lambda - l} + \frac{1}{\pi} \log \frac{8}{\pi} + o(1).$$

This result improves that of [9, Lemma 2] while providing a simpler proof. Indeed we apply Lemma 2 with Δ the strip $0 < \mathscr{R}z < L - l$, P the point $\lambda - l$, σ the imaginary axis. In this case the value of α is $4\pi^{-1}(L - l)$.

Lemma 3. *Let v_j, $j = 0, \ldots, n$, be arcs lying in the strip $S: 0 < y < 1$ apart from the end points A_j, B_j respectively on $y = 0$, $y = 1$. Let $\xi_j' = \min_{z \in v_j} \mathscr{R}z$, $\xi_j'' = \max_{z \in v_j} \mathscr{R}z$. Let $\xi_j'' \leq \xi_{j+1}'$, $j = 0, \ldots, n$. Let $\mathcal{2}_j$ be the quadrangle with one pair of opposite sides the segments $A_{j-1}A_j$, $B_{j-1}B_j$, the other pair the arcs v_{j-1}, v_j. Let m_j be the module of $\mathcal{2}_j$ for the class of curves joining the first pair of sides. Then*

$$\sum_{j=1}^{n} m_j \leq \xi_n' - \xi_0'' + 2 - \sum_{j=1}^{n-1} f(\xi_j'' - \xi_j'),$$

where f is a monotone increasing function independent of any geometrical properties of the configuration.

We define

$$\rho(z) = 0, \qquad \mathscr{R}z < \xi_0'' - 1, \mathscr{R}z > \xi_n' + 1,$$

$$\rho(z) = (1 + (\tfrac{1}{3}(\xi_j'' - \xi_j'))^2)^{-1/2}, \tfrac{1}{3}(2\xi_j' + \xi_j'') < \mathscr{R}z < \tfrac{1}{3}(\xi_j' + 2\xi_j'')$$
$$j = 1, \ldots, n - 1,$$

$$\rho(z) = 1, \qquad \text{elsewhere in } S.$$

It is readily verified that for ρ restricted to $\mathcal{2}_j$, $\rho(z) |dz|$ is an admissible metric for the module problem defining m_j, $j = 1, \ldots, n$. Thus

$$\sum_{j=1}^{n} m_j \leq \iint_S \rho^2(z) \, dA_z = \xi_n' - \xi_0'' + 2 - \sum_{j=1}^{n-1} \tfrac{1}{9}(\xi_j'' - \xi_j')^3(1 + \tfrac{1}{3}(\xi_j'' - \xi_j')^2)^{-1}.$$

This proves the result of Lemma 3 with $f(t) = \tfrac{1}{9}t^3(1 + \tfrac{1}{3}t^2)^{-1}$ which is verified to be strictly monotone increasing for $t > 0$.

It is interesting to observe that the present technique provides a completely trivial proof of the Ahlfors distortion theorem, even simpler than that devised some years ago by the author and later presented in Ohtsuka's notes [13, pp. 246–248]. The latter exposition perhaps fails to clarify the important point that this proof applies independently of any auxiliary symmetrization of the domain.

3.

We give a generalized form of the Phragmén-Lindelöf theorem in a form parallel to that due to F. and R. Nevanlinna [12].

Theorem 1. *Let $u(z)$ be a function subharmonic in $\Re z > 0$ such that*

$$\overline{\lim_{z \to i\eta}} u(z) \leqq 0$$

for each real η. Let Δ_n be a sequence of domains in $\Re z > 0$ having the properties outlined in Lemma 2 with associated entities σ_n, τ_n, α_n such that $\lim_{n \to \infty} \alpha_n = \infty$. Let $U_n = \text{l.u.b.}_{z \in \bar{\tau}_n} \overline{\lim}_{z \to \zeta} u(z)$. Let $l = \underline{\lim}_{n \to \infty} U_n/\alpha_n$. Then if $l \leqq 0$, $u(z) \leqq 0$ in $\Re z > 0$. If $0 < l < \infty$

$$u(x + iy) \leqq \frac{4l}{\pi} x.$$

Applying Lemma 1, Lemma 2, and Corollary 1 we have for any z such that $u(z) > 0$

$$\frac{1}{\pi} \log \frac{2\alpha_n}{x} + o(1) \leqq \frac{1}{\pi} \log \frac{U_n}{u(z)} + \frac{1}{\pi} \log \frac{8}{\pi} + o(1).$$

The result is immediate.

It should be pointed out that Hersch [7] has presented a generalized version of the Phragmén-Lindelöf theorem. It is much less explicit than that just given and the formulas depend on the particular function relating harmonic measure and module for a triad rather than elementary estimates as here. It should be observed that no greater generality is obtained by considering any other hyperbolic plane domain rather than a half-plane. Hersch's exposition is essentially reproduced in the notes of Ohtsuka [13, pp. 239–243].

4.

Heins [4] obtained an interesting extension of the Phragmén-Lindelöf theorem. We give now a generalized version of it.

Theorem 2. *Let $u(z)$ be a function subharmonic in $\Re z > 0$ such that*

$$\overline{\lim_{z \to i\eta}} u(z) \leqq 0$$

for each real η. Let Δ be a domain with the properties given in Lemma 2 and such that $\text{l.u.b.}_{z \in \Delta} \Re z = 1$. Let $t\Delta$, $t > 0$, denote the domain obtained from Δ

by magnification by the factor t centered at the origin with a similar notation for the other associated entities. Let

$$U(t) = \text{l.u.b.} \overline{\lim_{\substack{\zeta \in \overline{t\tau} \\ z \to \zeta}}} u(z)$$

and let

$$\underline{\lim_{t \to \infty}} U(t)/t = \lambda, \qquad \overline{\lim_{t \to \infty}} U(t)/t = \mu.$$

Then, if $\lambda \geqq 0$, $\lambda = \mu$. If $0 < \lambda < \infty$ then either

$$U(t) < \lambda t$$

for all t or $u(z) = \lambda x$. If Δ does not contain any strip $0 < x < v$, $v > 0$, then $\lambda \geqq 0$.

First let us suppose that $\lambda \geqq 0$. Then there exists a sequence of values t_n tending to infinity with $\lim_{n \to \infty} U(t_n)/t_n = \lambda$. Thus since $t_n \Delta$ has the associated quantity $t_n \alpha$ and the corresponding value of l in Theorem 1 is λ/α

$$u(x + iy) \leqq \frac{4\lambda}{\pi\alpha} x.$$

Let $k = \text{l.u.b.}_{x>0} u(x + iy)/x$. Evidently $\lambda \leqq k \leqq 4\lambda/\pi\alpha$. If $k > \lambda$ choose $0 < \varepsilon < k - \lambda$. Evidently

$$u(x + iy) \leqq (\lambda + \varepsilon)t_n$$

in the strip $t_n \mathscr{S}$ where $\mathscr{S}$ denotes the strip $0 < x < (\lambda + \varepsilon)/k$. The value corresponding to α for a suitable simply connected subdomain of $\Delta \cup \mathscr{S}$ is $\alpha^* \geqq (4/\pi)(\lambda + \varepsilon)/k + \delta$, where $\delta > 0$ can be found independent of ε for ε sufficiently small. Then by Theorem 1 we have

$$u(x + iy) \leqq \frac{4(\lambda + \varepsilon)}{\pi} \left(\frac{4}{\pi} \frac{\lambda + \varepsilon}{k} + \delta \right)^{-1} x$$

where

$$\frac{4(\lambda + \varepsilon)}{\pi} \left(\frac{4}{\pi} \frac{\lambda + \varepsilon}{k} + \delta \right)^{-1} < k$$

contradicting the definition of the latter. Thus $k = \lambda$ and $U(t) \leqq \lambda t$. Hence $\mu = \lambda$. The statement concerning sharpness of the inequality for $0 < \lambda < \infty$ is immediate.

Next suppose $\lambda < 0$. Since $u(z) \not\equiv -\infty$ let $P = x + iy$ be a point where $u(x + iy) \neq -\infty$. There exists a sequence t_n tending to infinity with $U(t_n) < u(x + iy) < 0$. Thus we can apply Lemma 1 to obtain

$$m(P, t_n\tau, t_n\Delta) \leqq m(u(x + iy), g(U(t_n)), S(U(t_n), 0)).$$

Now

$$m(P, t_n\tau, t_n\Delta) = \tfrac{1}{4}(m(P, t_n\sigma, t_n\Delta))^{-1}$$
$$m(u(x + iy), g(U(t_n)), S(U(t_n), 0)) = \tfrac{1}{4}(m(-u(x + iy), g(0), S(0, -U(t_n))))^{-1}.$$

Thus

$$\frac{1}{\pi}\log\frac{8U(t_n)}{\pi u(x + iy)} + o(1) \leqq \frac{1}{\pi}\log\frac{2\alpha t_n}{x} + o(1)$$

or

$$u(x + iy) \leqq \frac{4\lambda}{\pi\alpha}x.$$

Let κ denote the largest positive number such that $u(x + iy) \leqq \kappa\lambda x$ for all $x > 0$. Evidently $\kappa \geqq 4/\pi\alpha$. Given $\varepsilon > 0$ with $\lambda + \varepsilon < 0$ let the t_n above be chosen so that

$$U(t_n) < (\lambda + \varepsilon)t_n.$$

Let $\mathfrak{r}$ denote the strip $0 < x < \kappa^{-1}$. Then

$$u(x + iy) \leqq \lambda t_n$$

on the line $x = \kappa^{-1}t_n$. Let Λ denote the component of $\Delta \cap \mathfrak{r}$ with $z = 0$ on its boundary. Subject to the assumption that Δ does not contain any strip $0 < x < v$, $v > 0$, we have that the value α^* for Λ is strictly less than $4/\pi\kappa$. Thus repeating the previous argument we would obtain a contradiction.

This final condition cannot be eliminated if we are to conclude $\lambda \geqq 0$ and $\lambda = \mu$ as is seen by considering the function $-x + \log|w|$ where w is a suitable root of the equation $w(1 - w)^{-2} = z^{-2}$.

5.

The present techniques were used to prove the Denjoy conjecture in [9] and the proof to follow is essentially the one given there. However it is now observed that it applies in a much more general context.

Theorem 3. *Let $u(z)$ be a function subharmonic for all finite z. Let D_n be a sequence of simply connected plane domains with boundary ∂D_n (relative to the sphere) and with inner conformal radius with respect to the origin R_n where $\lim_{n\to\infty} R_n = \infty$. Let*

$$U_n = \text{l.u.b.} \varlimsup_{z\to\zeta} u(z), \qquad \zeta \in \partial D_n.$$

Let $u(z)$ have k distinct asymptotic values. Then

$$\varlimsup_{n \to \infty} U_n R_n^{-k/2} > 0.$$

The theorem is phrased to allow the possibility that D_n is unbounded.

The standard reduction of the problem enables us to consider the following situation. There exist a circle $|z| = r$, $r > 0$, and k nonintersecting open arcs A_j, $j = 1, \ldots, k$, running in $|z| > r$ from a point T_j on this circle to the point at infinity and dividing $|z| > r$ into k domains D_j, $j = 1, \ldots, k$. Further there is a real constant a such that on $|z| = r$ and on the A_j we have $u(z) \leqq a$. Finally there is in each respective domain D_j, $j = 1, \ldots, k$, a point P_j, $j = 1, \ldots, k$, of modulus β_j for which $u(P_j) > a$. Let $\beta = \max \beta_j$, $j = 1, \ldots, k$.

As soon as n is large enough the circle $|z| = \beta$ will lie in D_n and bound with ∂D_n a doubly connected domain Δ_n. It is readily seen that the module of Δ_n is $(1/2\pi)\log R_n + O(1)$. Let the first point of intersection of A_j with ∂D_n be denoted by $Q_j^{(n)}$. Let $C_j^{(n)}$ denote the open arc which is the union of the open arc $Q_j^{(n)}T_j$ on A_j, the arc $T_j T_{j+1}$ on $|z| = r$, and the open arc $T_{j+1}Q_{j+1}^{(n)}$ on A_{j+1} (here $T_{k+1} = T_1$, etc.). Now $C_j^{(n)}$ bounds with an arc (in the sense of natural boundary correspondence) B_j on ∂D_n a simply connected domain $E_j^{(n)}$. For n large enough P_j will lie in $E_j^{(n)}$ and we will have $u(P_j) < U_n$. Let $\rho_j^{(n)}(z)|dz|$ be the extremal metric in the module problem defining $m(P_j, C_j^{(n)}, E_j^{(n)})$, $j = 1, \ldots, k$.

In Δ_n we define the metric $\rho^{(n)}(z)|dz|$ by

$$\rho^{(n)}(z) = \rho_j^{(n)}(z) \qquad z \in \Delta_n \cap E_j^{(n)}, j = 1, \ldots, k,$$
$$\rho^{(n)}(z) = 0 \qquad \text{elsewhere in } \Delta_n.$$

If the rectifiable Jordan curve γ separates the boundary components of Δ_n we see at once

$$\int_\gamma \rho^{(n)}(z)\,|dz| \geqq k;$$

thus

(1) $$\frac{k^2}{2\pi} \log R_n + O(1) \leqq \sum_{j=1}^{k} m(P_j, C_j^{(n)}, E_j^{(n)}).$$

By Lemma 1, since

$$\varlimsup_{z \to \zeta} u(z) \leqq a \qquad \zeta \in C_j^{(n)}$$

$$\varlimsup_{z \to \zeta} u(z) \leqq U_n \qquad \zeta \in B_j^{(n)}$$

we have

(2) $$\sum_{j=1}^{k} m(P_j, C_j^{(n)}, E_j^{(n)}) \leq \sum_{j=1}^{k} m(u(P_j), g(a), S(a, U_n)).$$

By Lemma 2

(3) $$\sum_{j=1}^{k} m(u(P_j), g(a), S(a, U_n)) \leq \frac{k}{\pi} \log U_n + O(1).$$

Combining (1), (2), (3) we find

$$\frac{k^2}{2\pi} \log R_n + O(1) \leq \frac{k}{\pi} \log U_n.$$

Thus there exists a suitable positive constant K so that

$$U_n \geq K R_n^{k/2}$$

for n large enough.

6.

We come now to the results of Kennedy [10] which extended earlier results of Heins [5, 6]. We work here in essentially the same framework as did Kennedy but the use of the method of the extremal metric produces substantial simplification at many points.

Definitions. *Let A_j, $j = 1, \ldots, n$, be n arcs in the z-sphere joining the origin and the point at infinity and intersecting at these points only. These divide the plane into domains D_j, $j = 1, \ldots, n$, where D_j is bounded by A_j and A_{j+1}, $j = 1, \ldots, n$, (A_{n+1} is taken to be A_1). The intersection of D_j with the circle $|z| = r$, $r > 0$, consists of a possibly countable set of open arcs which divide D_j into a possibly countable set of subdomains. Let D_j' be the unique subdomain with the origin as a boundary point. The set $D_j - \bar{D}_j'$ consists of a possibly countable set of subdomains of D_j of which a unique one D_j'' has the point at infinity on its boundary. It is well known that D_j', D_j'', have a unique common open boundary arc (on $|z| = r$) which we denote by $\alpha_1^{(j)}(r)$. Interchanging the roles of the origin and the point at infinity we obtain an open arc $\alpha_2^{(j)}(r)$, possibly coincident with $\alpha_1^{(j)}(r)$. For $0 < r < R$ let $Q_{kl}^{(j)}(R, r)$, $k, l = 1, 2$ denote the quadrangle bounded by $\alpha_k^{(j)}(R)$, $\alpha_l^{(j)}(r)$ and a pair of arcs on A_j, A_{j+1}. Let $M_{kl}^{(j)}(R, r)$ denote the module of $Q_{kl}^{(j)}(R, r)$ for the class of curves joining the latter pair of opposite sides.*

Theorem 4. *Let $u_j(z)$, $j = 1, \ldots, n$, be a function subharmonic in D_j and such that*

$$\overline{\lim_{z \to \zeta}} u_j(z) \leq 0, \quad z \in D_j, \zeta \text{ finite on } A_j \text{ or } A_{j+1},$$

while

$$u_j(z) > 0 \quad \text{for some} \quad z \in D_j.$$

Let

$$\sigma_j(r) = \text{l.u.b.} \; u_j(z), \qquad z \in D_j \cap \{|z| = r\}.$$

Let

$$\sigma(r) = \max \sigma_j(r), \qquad j = 1, \ldots, n.$$

We assume that

(4)
$$\varlimsup_{r \to \infty} \frac{\sigma(r)}{r^{n/2}} < +\infty.$$

Then as r tends to infinity, $j = 1, \ldots, n$,

(5)
$$\log \sigma_j(r) = \lambda_j + \pi M_{kl}^{(j)}(r, r_0) + o(1)$$

for $k, l = 1, 2$, fixed $r_0 > 0$, and an appropriate constant λ_j,

(6)
$$\log \sigma_j(r) - \tfrac{1}{2} n \log r = o\{(\log r)^{1/2}\}$$

(7)
$$\prod_{j=1}^{n} \sigma_j(r) \sim \gamma r^{n^2/2}$$

for an appropriate positive constant γ.

Let

$$B_k^{(j)}(r) = \text{l.u.b.} \; u_j(z), \qquad z \in \alpha_k^{(j)}(r).$$

Evidently $B_k^{(j)}(r) \leqq \sigma_j(r)$. Let D_j be mapped conformally on the strip $-\tfrac{1}{2}\pi < \eta < \tfrac{1}{2}\pi$ in the $\zeta = \xi + i\eta$-plane by $\zeta = f_j(z)$ so that the origin and point at infinity correspond to the left- and right-hand ends of the strip. Let $\beta_k^{(j)}(r)$ denote the image of $\alpha_k^{(j)}(r)$. As soon as r is sufficiently large we have $B_k^{(j)}(r) > 0$. Then for $R > r$ and $r_1, \ldots, r_m$ chosen arbitrarily with

$$r = r_0 < r_1 < \ldots < r_m < r_{m+1} = R$$

we shall observe

(8)
$$\frac{n^2}{2\pi} \log \frac{R}{r} = \sum_{s=0}^{m} \frac{n^2}{2\pi} \log \frac{r_{s+1}}{r_s} \leqq \sum_{s=0}^{m} \sum_{j=1}^{n} M_{12}^{(j)}(r_{s+1}, r_s)$$

$$\leqq \sum_{j=1}^{n} M_{kl}^{(j)}(R, r) \leqq \sum_{j=1}^{n} \left(\frac{1}{\pi} \log \frac{B_k^{(j)}(R)}{B_l^{(j)}(r)} + \frac{1}{\pi} \log \frac{8}{\pi} + o(1) \right)$$

$$\leqq \frac{n^2}{2\pi} \log R + O(1),$$

the final inequality holding for R tending to infinity through an appropriate sequence of values. The result holds for any choice of k, l among 1, 2. The first inequality is a standard result for modules similar to that used in (1) of Section 5; the second is the quadrangle version of Grötzsch's Lemma (see [8, pp. 22, 23]); the third follows by Lemma 1, Lemma 2, and an analogue of the argument used in (1) of Section 5; and the fourth from our assumption (4).

In particular the difference between any two terms in (8) is bounded. Applying this statement to the third and fourth terms and using Lemma 3 it follows that

$$\max_{\zeta \in \beta_2^{(j)}(R)} \mathscr{R}\zeta - \min_{\zeta \in \beta_1^{(j)}(R)} \mathscr{R}\zeta$$

tends to zero as R tends to infinity (in any manner). Thus if ξ denotes $\mathscr{R}\zeta$ for any point $\beta_k^{(j)}(R)$, $k = 1, 2$,

$$(9) \qquad M_{kl}^{(j)}(R, r) = \frac{1}{\pi}\xi + \kappa_j + o(1)$$

for a constant κ_j which may depend on r, l but not on R and k. On the other hand, by Heins' version of the Phragmén-Lindelöf theorem (contained in Theorem 2) if $v(\zeta) = u_j(f_j^{-1}(\zeta))$

$$B(\xi) = \text{l.u.b. } v(\zeta), \quad \zeta = \xi + i\eta, \quad -\tfrac{1}{2}\pi < \eta < \tfrac{1}{2}\pi,$$

and

$$\beta = \lim_{\xi \to \infty}(\log B(\xi) - \xi)$$

then $\beta > -\infty$ and if $\beta < \infty$

$$\log B(\xi) = \beta + \xi + o(1).$$

That $\beta < \infty$ follows from (9) and the fact that the difference between the fourth and fifth terms in (8) is bounded. Thus

$$\log B_k^{(j)}(R) = \xi + \mu_j + o(1)$$

for a constant μ_j. Consequently using (9) again

$$(10) \qquad \log B_k^{(j)}(r) = \lambda_j + \pi M_{kl}^{(j)}(r, r_0) + o(1)$$

for fixed $r_0 > 0$ and an appropriate constant λ_j.

Moreover it follows from (8) that

$$\sum_{j=1}^{n}\left(M_{kl}^{(j)}(R, r) - \frac{n}{2\pi}\log\frac{R}{r}\right)$$

is a bounded increasing function of R for fixed r which thus has a limit $l(r)$ as R tends to infinity. Combining this with (10) we obtain

$$(11) \qquad \prod_{j=1}^{n} B_k^{(j)}(r) \sim \gamma r^{n^2/2}$$

for an appropriate constant γ.

Let us denote the circular ring $r < |z| < R$ by $C(R, r)$ and let $C^{(j)}(R, r)$ denote the set of $z \in \alpha_2^{(j)}(t)$, $r < t < R$. Let $\rho_j(z)|dz|$ denote the extremal metric for the module problem defining $M_{22}^{(j)}(R, r)$ and let

$$\rho(z) = \frac{1}{n} \rho_j(z), \qquad z \in C^{(j)}(R, r)$$

$$= 0, \qquad\qquad \text{elsewhere.}$$

Let us denote

$$\mu(z) = \frac{1}{2\pi |z|}, \qquad z \in C(R, r)$$

$$= 0, \qquad\qquad \text{elsewhere.}$$

Given $\varepsilon > 0$, for all r sufficiently large and $R > r$

$$(12) \qquad \frac{n^2}{2\pi} \log \frac{R}{r} \leq \sum_{j=1}^{n} M_{22}^{(j)}(R, r) < \frac{n^2}{2\pi} \log \frac{R}{r} + \varepsilon.$$

Moreover

$$\iint_{C(R,r)} \mu^2(z)\, dA_z = \frac{1}{2\pi} \log \frac{R}{r} \leq \iint_{C(R,r)} \rho^2(z)\, dA_z$$

while

$$\iint_{C(R,r)} \rho^2(z)\, dA_z \leq \frac{1}{n^2} \sum_{j=1}^{n} M_{22}^{(j)}(R, r) < \iint_{C(R,r)} \mu^2(z)\, dA_z + \frac{\varepsilon}{n^2}.$$

On the other hand

$$\frac{1}{2} \iint_{C(R,r)} \mu^2(z)\, dA_z + \frac{1}{2} \iint_{C(R,r)} \rho^2(z)\, dA_z = \iint_{C(R,r)} \left(\frac{\rho(z) + \mu(z)}{2}\right)^2 dA_z$$

$$+ \iint_{C(R,r)} \left(\frac{\rho(z) - \mu(z)}{2}\right)^2 dA_z,$$

where

$$\iint\limits_{C(R,r)} \left(\frac{\rho(z)+\mu(z)}{2}\right)^2 dA_z \geqq \iint\limits_{C(R,r)} \mu^2(z)\, dA_z.$$

Thus

(13)
$$\iint\limits_{C(R,r)} \left(\frac{\rho(z)-\mu(z)}{2}\right)^2 dA_z < \frac{\varepsilon}{n^2}.$$

Consequently

(14)
$$\left| \left(\iint\limits_{C^{(j)}(R,r)} \rho^2(z)\, dA_z \right)^{1/2} - \left(\iint\limits_{C^{(j)}(R,r)} \mu^2(z)\, dA_z \right)^{1/2} \right| < \frac{2}{n}\sqrt{\varepsilon}.$$

If $\alpha_2^{(j)}(t)$ has angular measure $\theta_j(t)$ we have

$$\iint\limits_{C^{(j)}(R,r)} \mu^2(z)\, dA_z = \frac{1}{4\pi^2} \int_r^R \frac{\theta_j(t)}{t}\, dt.$$

From (14) follows

(15)
$$M_{22}^{(j)}(R, r) = \frac{n^2}{4\pi^2} \int_r^R \frac{\theta_j(t)}{t}\, dt + o((\log R)^{1/2}).$$

Further

$$\frac{n}{\pi} \log \frac{R}{r} \leq \int_r^R \frac{dt}{t\theta_j(t)} + \frac{n^2}{4\pi^2} \int_r^R \frac{\theta_j(t)}{t}\, dt,$$

and by a familiar argument

$$\int_r^R \frac{dt}{t\theta_j(t)} \leqq M_{22}^{(j)}(R, r).$$

Combining this with (12) and (15) we obtain

$$M_{22}^{(j)}(R, r) = \frac{n}{2\pi} \log \frac{R}{r} + o((\log R)^{1/2}).$$

Finally from (10)

(16)
$$\log B_k^{(j)}(r) = \frac{n}{2\pi} \log r + o((\log r)^{1/2}).$$

Now to complete the proof of Theorem 4 we need only show

$$\log \sigma_j(r) = \log B_2^{(j)}(r) + o(1).$$

Evidently $\log \sigma_j(r)$ differs from $\log B_2^{(j)}(r)$ only when there is a cross cut of D_j on $|z| = r$ distinct from $\alpha_2^{(j)}(r)$ and not separated in D_j from the point at infinity by $\alpha_2^{(j)}(r)$. However such a cross cut will always be separated in D_j from the point at infinity by $\alpha_2^{(j)}(t)$ for all sufficiently large t. Let r_1 be the greatest lower bound of such t. In cases where $r_1 = r$ it is immediate that

$$\log \sigma_j(r) = \log B_2^{(j)}(r) + o(1).$$

Hence we may suppose that $r_1 > r$.

We readily see that there is an absolute constant A such that if $0 < \lambda < L$

$$m(\lambda, g(0), S(0, L)) \leqq \frac{1}{\pi} \log \frac{L}{\lambda} + A.$$

Indeed the module depends only on the ratio L/λ and

$$m(1, g(0), S(0, L)) - \frac{1}{\pi} \log L$$

varies continuously with L, tends to zero as L approaches one, and is bounded as L approaches infinity.

Let K be a positive number such that

(17) $$\frac{n^2}{2} X + \frac{n^2}{2} K\varepsilon^{1/2} + 3\varepsilon + \pi A - \pi\varepsilon^{-1} X^2 < 0$$

for $X \geqq K\varepsilon^{1/2}$ and for all $0 < \varepsilon \leqq 1$. Let $r \leqq t < r_1$ and let γ_t be an arc on $|z| = t$ in D_j separated in D_j from the point at infinity by every $\alpha_2^{(j)}(r_2)$ with $r_2 > r_1$. Then given $\varepsilon > 0$ for r sufficiently large

$$\text{l.u.b.}_{z \in \gamma_t} u(z) \leqq \log B_2^{(j)}(r_2)$$

$$< \log B_2^{(j)}(t) + \pi M_{22}^{(j)}(r_2, r_0) - \pi M_{22}^{(j)}(t, r_0) + \varepsilon$$

$$< \log B_2^{(j)}(t) + \frac{n^2}{2} \log \frac{r_2}{t} + 2\varepsilon.$$

Evidently

$$\text{l.u.b.}_{z \in \gamma_t} u(z) \leqq \log B_2^{(j)}(t) + \frac{n^2}{2} \log \frac{r_1}{t} + 2\varepsilon.$$

Thus if $\log(r_1/t) \leqq K\varepsilon^{1/2}$

$$\text{l.u.b.}_{z \in \gamma_t} u(z) \leqq \log B_2^{(j)}(t) + \frac{n^2}{2} K\varepsilon^{1/2} + 2\varepsilon.$$

In particular if $\log(r_1/r) \leqq K\varepsilon^{1/2}$

$$\log \sigma_j(r) \leqq \log B_2^{(j)}(r) + \frac{n^2}{2} K\varepsilon^{1/2} + 2\varepsilon.$$

Now suppose there is a cross cut γ of D_j on $|z| = r$ distinct from $\alpha_2^{(j)}(r)$ and not separated in D_j from the point at infinity by $\alpha_2^{(j)}(r)$. Given t, $r < t < r_1$, with $\log(r_1/t) \leqq K\varepsilon^{1/2}$, there exists a cross cut γ_t of D_j on $|z| = t$ separating γ from the point at infinity on D_j. Clearly γ_t cannot join A_j, A_{j+1} thus bounds with an arc of A_j or A_{j+1} a domain Λ containing γ but containing no point of any $\alpha_2^{(j)}$. For s with $r < s < t$ there exists a cross cut of D_j on $|z| = s$ in Λ separating γ and γ_t of angular measure $\theta(s)$ and these can be chosen so that their totality forms a measurable set H. By (13) given $\varepsilon > 0$ and r sufficiently large

$$\iint\limits_H \mu^2(z)\, dA_z = \int_r^t \frac{\theta(s)}{s}\, ds < \varepsilon$$

while

$$\left(\log \frac{t}{r}\right)^2 \leqq \int_r^t \frac{\theta(s)}{s}\, ds \int_r^t \frac{1}{s\theta(s)}\, ds;$$

thus

$$\int_r^t \frac{1}{s\theta(s)}\, ds \geqq \frac{1}{\varepsilon} \left(\log \frac{t}{r}\right)^2.$$

On the other hand by Lemma 1 and Lemma 2 if $U(t) = \text{l.u.b.}_{z \in \gamma_t}\, u(z)$, $U(r) = \text{l.u.b.}_{z \in \gamma}\, u(z)$ and the latter is positive

$$\int_r^t \frac{1}{s\theta(s)}\, ds \leqq \frac{1}{\pi} \log(U(t)/U(r)) + A$$

or

$$\log U(r) \leqq \log U(t) + \pi A - \pi\varepsilon^{-1}\left(\log \frac{t}{r}\right)^2$$

$$\leqq \log B_2^{(j)}(t) + \frac{n^2}{2} K\varepsilon^{1/2} + 2\varepsilon + \pi A - \pi\varepsilon^{-1}\left(\log \frac{t}{r}\right)^2$$

$$\leqq \log B_2^{(j)}(r) + \frac{n^2}{2} \log \frac{t}{r} + \frac{n^2}{2} K\varepsilon^{1/2} + 3\varepsilon + \pi A - \pi\varepsilon^{-1}\left(\log \frac{t}{r}\right)^2.$$

If $\log(r_1/r) > K\varepsilon^{1/2}$ we can choose t so that $\log(t/r) > K\varepsilon^{1/2}$; consequently using (17)

$$\log U(r) \leqq \log B_2^{(j)}(r).$$

Thus in any case

$$\log \sigma_j(r) = \log B_2^{(j)}(r) + o(1).$$

7.

The preceding results have some interesting consequences concerning the geometrical properties of the arcs A_j.

Corollary 2. *With the definitions and conditions of Theorem 4, $\alpha_1^{(j)}(r)$ and $\alpha_2^{(j)}(r)$, $j = 1, \ldots, n$, coincide for r sufficiently large, i.e., there is a unique arc on $|z| = r$ connecting A_j and A_{j+1}.*

If $\alpha_1^{(j)}(r)$, $\alpha_2^{(j)}(r)$ are distinct they will bound with arcs on A_j, A_{j+1} a quadrangle $\mathcal{Q}$. Let m denote the module of $\mathcal{Q}$ for the class of curves joining the latter pair of sides. We will show below that $m \geqq 1$, then the result of the corollary follows from inequality (8) using the fact that the difference between the third and fourth terms is bounded.

A single-valued branch of $w = i \log z$ taken in D_j maps the latter conformally on a domain in the w-plane and in particular maps $\alpha_1^{(j)}(r)$, $\alpha_2^{(j)}(r)$ onto segments α_1, α_2 on a horizontal line $\mathcal{L}$: $\mathcal{I}w = \log r$, and $\mathcal{Q}$ onto a quadrangle Q. In Q adjacent to α_1, $\mathcal{I}w > \log r$ and adjacent to α_2, $\mathcal{I}w < \log r$. Thus reflecting Q in $\mathcal{L}$ and joining the copies along the segments α_1, α_2 we obtain a doubly connected Riemann surface $\mathcal{D}$. If w_1, w_2 are the closest end points of α_1, α_2 then an appropriate branch of $\zeta = \xi((w - w_1)/(w - w_2))^{1/2}$ for suitable ξ maps $\mathcal{D}$ onto a doubly connected Riemann domain Δ symmetric with respect to the real axis, the latter meeting Δ in a segment lying in $-1 < \mathcal{R}\zeta < 0$ and one lying in $1 < \mathcal{R}\zeta < \infty$. While Δ may not be schlicht it covers any pair of values ζ, $-\zeta$ at most twice. From this we see at once that the module of Δ is less than or equal to that of the doubly connected domain consisting of the plane slit from the point at infinity to -1 along the negative real axis and from 0 to 1 along the positive real axis. The latter module has the value $\frac{1}{2}$. Thus the module of $\mathcal{D}$ for the class of curves joining its boundary components is greater than or equal to 2. The module m is just half of this, thus $m \geqq 1$.

Kennedy [10, Theorem IV] discussed the asymptotic behavior of arg z on A_j but his proof was not sufficient to justify the original assertation, and he had later [11, Theorem B] to add a subsidiary condition to obtain the desired conclusion. Clunie [13] indicated how this condition could be dropped. We give now a proof of the result in this form.

Corollary 3. *Let* arg z *denote a single-valued branch of the argument chosen on* A_j. *Then as* z *tends to the point at infinity on* A_j

$$\arg z = o((\log |z|)^{1/2}).$$

A single-valued branch of $w = i \log z$, $w = u + iv$, taken in the z-sphere slit along A_j will map this domain conformally on a domain Ξ bounded by two images l_1, l_2 of A_j, one obtained from the other by a horizontal translation of 2π. Let l_1, the one on the left, correspond to the given branch of arg z. By Corollary 2 for t sufficiently large there is a single open arc $\alpha(t)$ on $v = t$ in Ξ joining l_1 and l_2. Let $T > t$, the corresponding open arc being $\alpha(T)$. Then $\alpha(t)$, $\alpha(T)$ together with arcs on l_1, l_2 bound a quadrangle $Q(T, t)$. Let $M(T, t)$ be the module of $Q(T, t)$ for the class of curves joining the latter arcs. Given $\varepsilon > 0$ for t sufficiently large by (12)

(18)
$$M(T, t) \leqq \frac{1}{2\pi} (T - t) + \varepsilon.$$

Let l_1 be sensed by passing from $v = -\infty$ to $v = +\infty$ and let (u_t, t) be the last point of intersection of l_1 with $v = t$. Let

$$U_T^{+} = \max u, \qquad w \in \{v = T\} \cap l_1$$
$$U_T^{-} = \min u, \qquad w \in \{v = T\} \cap l_1$$

let

$$s = \max(U_T^{+} - u_t, u_t - U_T^{-}).$$

Clearly s is nonnegative and we may suppose it large compared to 2π. Let

$$\rho(w) = 1, \qquad w \in Q(T, t), t < \mathscr{I}w < T$$
$$= 0, \qquad \text{elsewhere in } Q(T, t).$$

Then in the ρ-metric, $Q(T, t)$ has area at most $2\pi(T - t)$ while any curve in $Q(T, t)$ joining $\alpha(t)$, $\alpha(T)$ has length at least $((T - t)^2 + (s - 6\pi)^2)^{1/2}$. Thus

$$(M(T, t))^{-1} \leqq 2\pi(T - t)((T - t)^2 + (s - 6\pi)^2)^{-1}$$

so by (18) for t sufficiently large and any $T > t$

$$\frac{1}{2\pi} (T - t)^{-1}((T - t)^2 + (s - 6\pi)^2) \leqq \frac{1}{2\pi} (T - t) + \varepsilon.$$

Therefore with t fixed and letting T tend to infinity

$$s = o(T^{1/2})$$

This implies the result of Corollary 3.

It should be remarked that our techniques can be applied also to the case of asymptotic tracts corresponding to Al-Katifi's extension of Kennedy's results [2].

REFERENCES

1. L. V. Ahlfors, Untersuchungen zur Theorie der konformen Abbildung und der ganzen Funktionen, *Acta Soc. Sci. Fenn.* (Nova Series A) **1** (1930), No. 9, 1–40.
2. W. Al-Katifi, On the asymptotic values and paths of certain integral and meromorphic functions, *Proc. London Math. Soc.* **16** (1966), 599–634.
3. J. Clunie, On a paper of Kennedy, *J. London Math. Soc.* **33** (1958), 118–120.
4. M. Heins, On the Phragmén-Lindelöf principle, *Trans. Amer. Math. Soc.* **60** (1946), 238–244.
5. ———, Entire functions with bounded minimum modulus; subharmonic function analogues, *Ann. of Math.* **49** (1948), 200–213.
6. ———, On the Denjoy-Carleman-Ahlfors theorem, *Ann. of Math.* **49** (1948), 533–537.
7. J. Hersch, Longueurs extrémales et théorie des fonctions, *Comment. Math. Helv.* **29** (1955), 301–337.
8. James A. Jenkins, *Univalent functions and conformal mapping*, Springer-Verlag, Berlin-Göttingen-Heidelberg, 1958; reprinted, 1965.
9. ———, On the Denjoy conjecture, *Canad. J. Math.* **10** (1958), 627–631.
10. P. B. Kennedy, On a conjecture of Heins, *Proc. London Math. Soc.* **5** (1955), 22–47.
11. ———, A class of integral functions bounded on certain curves, *Proc. London Math. Soc.* **6** (1956), 518–547.
12. F. and R. Nevanlinna, Über die Eigenschaften analytischer Funktionen in der Umgebung einer singulären Stelle oder Linie, *Acta Soc. Sci. Fenn.* **50** (1922), No. 5, 1–46.
13. M. Ohtsuka, Dirichlet problem, extremal length, and prime ends, Notes, Washington University, 1962–63.

(*Received February 22, 1968*)

A Strengthened Form of the $\frac{1}{4}$ Theorem for Starlike Univalent Functions

F. R. Keogh*

Department of Mathematics
University of Kentucky
Lexington, Kentucky. 40506
U.S.A.

1.

Suppose† that $f(z) = \sum_1^\infty a_n z^n$ is regular and univalent for $|z| < 1$ and that $g(z)$ is subordinate to $f(z)$, that is, $g(z) = f(\omega(z))$, where $\omega(z)$ is regular for $|z| < 1$ and satisfies the conditions $\omega(0) = 0$, $|\omega(z)| < 1$ for $|z| < 1$. If the transformation $w = f(z)$ maps $|z| < 1$ onto a domain D in the w-plane then every value taken by $g(z)$ lies in D. Conversely, if $g(z)$ is regular for $|z| < 1$, if $g(0) = f(0)$, and if every value taken by $g(z)$ lies in D, then $g(z)$ is subordinate to $f(z)$. We shall express the fact that $g(z)$ is subordinate to $f(z)$ by writing

$$g(z) \prec f(z).$$

Suppose now that $f(z)$ is normalized so that $a_1 = 1$. The well-known Koebe " $\frac{1}{4}$ theorem " is equivalent to

(1) $$\tfrac{1}{4}z \prec f(z),$$

* Research supported by National Science Foundation Grant GP–7377.
† Definitions and facts relevant to this section are to be found, for example, in [4].

and it holds, in particular, if $f(z)$ is starlike. The starlike Koebe function $z(1 - z)^{-2}$, which we shall denote from now on by $k(z)$, shows that the constant $\frac{1}{4}$ is best possible and that, for a complex number λ,

$$\lambda z \prec f(z)$$

for all starlike $f(z)$ if and only if $|\lambda| \leq \frac{1}{4}$.

The main object of this note is to consider the problem of determining necessary and sufficient conditions on complex numbers λ, μ under which, for all starlike $f(z) = z + \sum_2^\infty a_n z^n$, $\lambda z + \mu a_2 z^2$ is starlike and

(2) $$\tfrac{1}{4} z \prec \lambda z + \mu a_2 z^2 \prec f(z),$$

an extended form of (1).

A necessary condition for (2) is that $|\lambda| \geq \frac{1}{4}$.* It is also clear that, for any real θ, (2) holds if and only if it also holds with λ, μ replaced by $\lambda e^{i\theta}$, $\mu^2 e^{i\theta}$, so we may suppose from now on, with no effective loss of generality, that λ is real and positive. If $\lambda z + \mu a_2 z^2$ is to be starlike for all starlike $f(z)$ then, with $f(z) = k(z)$, $\lambda z + 2\mu z^2$ must be starlike, and this implies $\lambda \geq 4|\mu|$ (see Lemma 1). The minimum value of $|\lambda z + 2\mu z^2|$ on $|z| = 1$ is then $\lambda - 2|\mu|$, and so $\tfrac{1}{4} z \prec \lambda z + 2\mu z^2$ implies that

(3) $$\lambda - 2|\mu| \geq \tfrac{1}{4}.$$

An examination of the proof of Theorem 5 shows that, if $\lambda z + \mu a_2 z^2 \prec f(z)$ for all starlike $f(z)$, then

(4) $$\lambda \leq 2R\mu + \tfrac{1}{4}.$$

Relations (3) and (4) show that μ is real and nonnegative and

(5) $$\lambda = 2\mu + \tfrac{1}{4}.$$

Thus we may suppose that both λ and μ are real, $\lambda > 0$, $\mu \geq 0$, and (5) is then a necessary condition for (2) to hold for all starlike $f(z)$. In Section 4 we prove that a further necessary condition is

(6) $$\mu \leq \tfrac{1}{16}$$

(Theorem 6), and that the conditions (5) and (6) together are also sufficient. These facts are restated in the form of Theorem 7. Theorem 4 is given for completeness and because of its relevance to a further problem to be discussed in a subsequent paper.

2.

In this section we introduce a number of lemmas and prove the key theorem.

* [4, p. 228].

Lemma 1.* *The function* $z + cz^2$ *is univalent for* $|z| < 1$ *if and only if* $|c| \leq \frac{1}{2}$, *and it is then starlike.*

Lemma 2.† *If* $\omega(z) = \omega_1 z + \omega_2 z^2 + \cdots$ *is regular and* $|\omega(z)| < 1$ *for* $|z| < 1$, *then*

$$|\omega_1| \leq 1,$$
$$|\omega_2| \leq 1 - |\omega_1|^2.$$

Lemma 3 *Suppose that* b_0, b_1, b_2 *are complex numbers,* $b_2 \neq 0$, *and let* $P(z) = b_0 + b_1 z + b_2 z^2$.
  (i) *If* $|b_0| < |b_2|$ *and*

(7) $$|b_0 \bar{b}_1 - \bar{b}_2 b_1| \leq |b_2|^2 - |b_0|^2,$$

   then the zeros of $P(z)$ *lie on* $|z| \leq 1$.
  (ii) *If the zeros of* $P(z)$ *lie on* $|z| \leq 1$ *then* $|b_0| \leq |b_2|$ *and* (7) *holds.*

This is a slightly modified form of a special case of a more general result of Schur (see, e.g., Marden [3], p. 151). The example $P(z) = 1 + z^2$ shows that in part (ii) the possibility $|b_0| = |b_2|$ can occur.

Proof. (i) If $|b_0| < |b_2|$ then at least one zero of $P(z)$ is in $|z| < 1$. We may suppose that the other zero does not lie on $|z| = 1$, for otherwise there is nothing to prove. Let

$$P^*(z) = \bar{b}_2 + \bar{b}_1 z + \bar{b}_0 z^2.$$

Then

(8) $$P^*(z) = z^2 \overline{P(1/\bar{z})},$$

and for real θ,

(9) $$|P^*(e^{i\theta})| = |P(e^{i\theta})|.$$

Now let

(10) $$P_1(z) = \bar{b}_0 P(z) - b_2 P^*(z).$$

Since $|b_0| < |b_2|$, by (9) we have $|\bar{b}_0 P(e^{i\theta})| < |b_2 P^*(e^{i\theta})|$, so by Rouché's theorem $b_2 P^*(z)$ and $P_1(z)$ have the same number of zeros in $|z| < 1$. But

(11) $$P_1(z) = z(\bar{b}_0 b_1 - b_2 \bar{b}_1) + |b_0|^2 - |b_2|^2$$

and so, by (7), $P_1(z)$ has no zero in $|z| < 1$. It follows that $P^*(z)$ has no zero in $|z| < 1$, and from (8) we conclude that $P(z)$ has no zero in $|z| > 1$.

---

* See, for example, [1].
† See, for example, [2].

(ii) If the zeros of $P(z)$ lie on $|z| \leq 1$ then certainly $|b_0| \leq |b_2|$. Suppose that $|b_0| < |b_2|$ but $|b_0 \bar{b}_1 - \bar{b}_2 b_1| > |b_2|^2 - |b_0|^2$. Then by (11), $P_1(z)$ has its only zero in $|z| < 1$, and it follows from (9) and (10) that $P(z)$, $P^*(z)$ do not vanish on $|z| = 1$. The argument used above can now be repeated and we conclude that $P(z)$ has a zero in $|z| > 1$. This is a contradiction and establishes (7). Suppose, finally, that $|b_0| = |b_2|$. Then both zeros of $P(z)$ are on $|z| = 1$, $P(z)$ is of the form $a(z - e^{i\alpha})(z - e^{i\beta})$, and we easily verify that the left-hand side of (7) vanishes.

We now prove

Theorem 1. *Suppose that $k(\omega(z))$ is any function subordinate to $k(z)$, and $\omega(z) = \omega_1 z + \omega_2 z^2 + \cdots$. Then*

$$(12) \qquad \tfrac{3}{8}\omega_1 z + \tfrac{1}{16}(\omega_2 + 2\omega_1{}^2)z^2 \prec k(z).$$

Proof. Relation (12) is equivalent to the condition that, for all $\alpha \geq \frac{1}{4}$, the zeros of the polynomial $\alpha + \tfrac{3}{8}\omega_1 z + \tfrac{1}{16}(\omega_2 + 2\omega_1{}^2)z^2$ lie on $|z| \geq 1$, for this expresses the fact that no value of the polynomial taken for $|z| < 1$ lies on the slit along the negative real axis from $-\frac{1}{4}$ to ∞. Writing $\beta = 4\alpha$, we see that this in turn is equivalent to the condition that the zeros of $\omega_2 + 2\omega_1{}^2 + 6\omega_1 z + 4\beta z^2$ lie on $|z| \leq 1$ for all $\beta \geq 1$. By Lemma 3, it is sufficient for the proof of this to show that

$$(13) \qquad |\omega_2 + 2\omega_1{}^2| < 4\beta, \qquad \beta \geq 1,$$

and

$$(14) \qquad 6\,|2(2\beta - |\omega_1|^2)\omega_1 - \bar{\omega}_1\omega_2| + |\omega_2 + 2\omega_1{}^2|^2 \leq 16\beta^2, \qquad \beta \geq 1.$$

By Lemma 2,

$$|\omega_2 + 2\omega_1{}^2| \leq |\omega_2| + 2\,|\omega_1|^2 \leq 1 + |\omega_1|^2 \leq 2 < 4\beta,$$

and this establishes (13). If we write $\omega_1 = r_1 e^{i\theta_1}$, $\omega_2 = r_2 e^{i\theta_2}$, $\theta_2 - 2\theta_1 = \phi$, (14) is equivalent to

$$(15) \quad g(\phi) \equiv 6r_1\,|2(2\beta - r_1{}^2) - r_2 e^{i\phi}| + |r_2 + 2r_1{}^2 e^{-i\phi}|^2 \leq 16\beta^2, \qquad \beta \geq 1,$$

and by Lemma 2 it is enough to prove (15) for all ϕ and for all r_1, r_2 subject to $0 \leq r_1 \leq 1$, $0 \leq r_2 \leq 1 - r_1{}^2$.

If $r_1 = 0$ then (15) obviously holds, so we may suppose $r_1 \neq 0$. If $r_2 = 0$ then (15) is equivalent to

$$3r_1(2\beta - r_1{}^2) + r_1{}^4 \leq 4\beta^2, \qquad \beta \geq 1.$$

It is easily shown that the worst case is when $\beta = 1$, and we have then only to note that

$$(r_1 - 1) + r_1(r_1 - 1)^3 - 3(r_1 - 1)^2 \leqq 0.$$

Thus we may further suppose $r_2 \neq 0$. We now have

$$|2(2\beta - r_1{}^2) - r_2 e^{i\phi}| \geqq 4\beta - 2r_1{}^2 - r_2 \geqq 3 - r_1{}^2 > 0,$$

so that $g(\phi)$ is a differentiable function of ϕ, and

(16) $\dfrac{1}{4r_1 r_2} g'(\phi) = 3(2\beta - r_1{}^2)|2(2\beta - r_1{}^2) - r_2 e^{i\phi}|^{-1} \sin \phi - r_1 \sin \phi.$

Since $g(\phi)$ is a periodic function of ϕ, its maximum value for fixed r_1, r_2, β is attained when ϕ is a certain solution of the equation $g'(\phi) = 0$. From (16) we find that $g'(\phi) = 0$ when $\sin \phi = 0$ and when

(17) $r_1{}^2[4(2\beta - r_1{}^2)^2 - 4r_2(2\beta - r_1{}^2)\cos \phi + r_2{}^2] = 9(2 - r_1{}^2)^2.$

We now prove, however, that

(18) $E_1 \equiv r_1{}^2[4(2\beta - r_1{}^2)^2 + 4r_2(2\beta - r_1{}^2) + r_2{}^2] - 9(2\beta - r_1{}^2)^2 < 0,$

and this shows that (17) has no real solution. Using the inequalities $0 \leqq r_1 \leqq 1$, $0 \leqq r_2 \leqq 1 - r_1{}^2$, and writing $r_1{}^2 = x$, we have

$$\frac{1}{4}\frac{\partial E_1}{\partial \beta} = 2(4x - 9)\beta - 4x^2 + 9x + 2xr_2$$

$$\leqq -10\beta - 4x^2 + 9x + 2x(1 - x)$$

$$\leqq -6x^2 + 11x - 10.$$

The latter expression has a negative maximum when $x = \frac{11}{12}$, so E_1 is a decreasing function of β for any fixed r_1, r_2 and it is enough to prove (18) for $\beta = 1$, that is, to prove that

$$E_2 \equiv x[4(2 - x)^2 + 4r_2(2 - x) + r_2{}^2] - 9(2 - x)^2 < 0.$$

We have

$$\frac{1}{2}\frac{\partial E_2}{\partial r_2} = 2x(2 - x) + r_2 x \geqq 0$$

so, for fixed x, E_2 is an increasing function of r_2, and we have now only to show that $E_2 < 0$ when $r_2 = 1 - x$, that is, that

$$E_3(x) \equiv 9x^3 - 39x^2 + 61x - 36 < 0, \qquad 0 \leqq x \leqq 1.$$

This follows since $E_3'(x) = 27x^2 - 78x + 61 > 0$ and $E_3(1) < 0$.

We have thus proved (18) and so shown that $g'(\phi) = 0$ only when $\sin \phi = 0$. The maximum value of $g(\phi)$ therefore occurs either when $\phi = 0$ or when $\phi = \pi$. The simplest procedure now seems to be to prove (15) for both of these cases. With $\phi = 0$, since $2(2\beta - r_1{}^2) - r_2 \geq 3 - r_1{}^2 > 0$, we have to show that

$$6r_1[2(2\beta - r_1{}^2) - r_2] + r_2{}^2 + 4r_1{}^2 r_2 + 4r_1{}^4 - 16\beta^2 \leq 0, \qquad \beta \geq 1.$$

Differentiation with respect to β shows that the worst case is when $\beta = 1$, so we must prove that

$$
\begin{aligned}
(19) \qquad E_4 &\equiv 6r_1[2(2 - r_1{}^2) - r_2] + r_2{}^2 + 4r_1{}^2 r_2 + 4r_1{}^4 - 16 \\
&\equiv 4(r_1 - 1) + 4r_1(r_1 - 1)^3 - 12(r_1 - 1)^2 \\
&\quad + 4r_1 r_2(r_1 - 1) + r_2(r_2 - 2r_1) \\
&\leq 0.
\end{aligned}
$$

The first four terms in the second expression for E_4 are nonpositive, and $r_2 - 2r_1 \leq 1 - r_1{}^2 - 2r_1 \leq 0$ provided that $r_1 \geq \sqrt{2} - 1$, so (19) is valid in this case. Also, dropping the term $-6r_1 r_2$ in the first expression for E_4, and again using the inequality $r_2 \leq 1 - r_1{}^2$, we have

$$E_4 \leq (r_1 - 1)(r_1{}^3 - 11r_1{}^2 - 9r_1 + 15) = (r_1 - 1)E_5(r_1),$$

say. But

$$E_5'(r_1) = 3r_1{}^2 - 22r_1 - 9 < 0,$$

and $E_5(\frac{1}{2}) > 0$, so (19) is also valid for $r_1 \leq \frac{1}{2}$, and this establishes (19) completely.

Finally, with $\phi = \pi$ in (15), we can show as before that the worst case is when $\beta = 1$, and we then have to prove that

$$(20) \qquad E_6 \equiv 12(2 - r_1{}^2)r_1 + 6r_1 r_2 + r_2{}^2 - 4r_1{}^2 r_2 + 4r_1{}^4 - 16 \leq 0.$$

We have

$$\frac{\partial E_6}{\partial r_2} = 6r_1(1 - r_1) + 2r_2 + 2r_1{}^2 \geq 0$$

so, for fixed r_1, E_6 increases with r_2 and we must show that (20) holds when $r_2 = 1 - r_1{}^2$. But (20) then reduces to the valid inequality

$$(r_1 - 1)^2(3r_1{}^2 - 5) \leq 0,$$

and this completes the proof of Theorem 1.*

---

* If the constant $\frac{1}{16}$ in (12) is replaced by $\frac{1}{32}$ then the proof becomes considerably simpler!

3.

Suppose that $g(z) = b_1 z + b_2 z^2 + \cdots$ is any function subordinate to $k(z)$, and that $g(z) = k(\omega(z))$, where $\omega(z) = \omega_1 z + \omega_2 z^2 + \cdots$. Then $b_1 = \omega_1$ and $b_2 = \omega_2 + 2\omega_1^2$, so Theorem 1 is equivalent to the implication

$$(21) \qquad g(z) \prec k(z) \Rightarrow \tfrac{3}{8}b_1 z + \tfrac{1}{16}b_2 z^2 \prec k(z).$$

Next, let $D(\alpha, \psi)$, $\alpha > 0$, denote the domain consisting of the w-plane slit along the ray $\arg w = \psi$ from the point $\alpha e^{i\psi}$ to ∞, and let $k_{\alpha, \psi}(z) = -4\alpha e^{i\psi}k(z)$. Then the transformation $w = k_{\alpha, \psi}(z)$ maps $|z| < 1$ onto $D(\alpha, \psi)$, and if $g(z) = b_1 z + b_2 z^2 + \cdots$ is now any function subordinate to $k_{\alpha, \psi}(z)$, then $-(4\alpha)^{-1}e^{-i\psi}g(z) \prec k(z)$, and by (21) this implies that

$$-(4\alpha)^{-1}e^{-i\psi}[\tfrac{3}{8}b_1 z + \tfrac{1}{16}b_2 z^2] \prec k(z),$$

or

$$\tfrac{3}{8}b_1 z + \tfrac{1}{16}b_2 z^2 \prec k_{\alpha, \psi}(z).$$

Thus we have the following modified form of Theorem 1.

Theorem 1'. *If* $g(z) = b_1 z + b_2 z^2 + \cdots$ *is subordinate to* $k_{\alpha, \psi}(z)$ *then*

$$\tfrac{3}{8}b_1 z + \tfrac{1}{16}b_2 z^2 \prec k_{\alpha, \psi}(z).$$

From this we now deduce

Theorem 2. *If* $f(z) = z + \sum_2^\infty a_n z^n$ *is starlike, then so is* $\tfrac{3}{8}z + \tfrac{1}{16}a_2 z^2$, *and*

$$\tfrac{1}{4}z \prec \tfrac{3}{8}z + \tfrac{1}{16}a_2 z^2 \prec f(z).$$

Proof. Suppose that the transformation $w = f(z)$ maps $|z| < 1$ onto the domain D in the w-plane. Let E be the (nonempty) set of ψ in $0 \le \psi < 2\pi$ such that the ray $\arg w = \psi$ contains a finite point of the boundary of D and, for each ψ in E, let $\alpha(\psi)e^{i\psi}$ be that point of the boundary of D on the ray $\arg w = \psi$ which is closest to the origin. Then for each ψ in E we have $D \subset D(\alpha(\psi), \psi)$, so that

$$f(z) \prec k_{\alpha(\psi), \psi}(z).$$

It follows from Theorem 1' that

$$\tfrac{3}{8}z + \tfrac{1}{16}a_2 z^2 \prec k_{\alpha(\psi), \psi}(z), \qquad \psi \in E.$$

But $D = \bigcap_{\psi \in E} D(\alpha(\psi), \psi)$, so we conclude that

$$\tfrac{3}{8}z + \tfrac{1}{16}a_2 z^2 \prec f(z).$$

By Lemma 1, since $|a_2| \leq 2$, the function $\frac{3}{8}z + \frac{1}{16}a_2 z^2$ is starlike and, on $|z| = 1$,

$$|\tfrac{3}{8}z + \tfrac{1}{16}a_2 z^2| \geq \tfrac{3}{8} - \tfrac{1}{16}|a_2| \geq \tfrac{1}{4},$$

so

$$\tfrac{1}{4}z \prec \tfrac{3}{8}z + \tfrac{1}{16}a_2 z^2.$$

In what follows we suppose always that $\lambda > 0$, $\mu \geq 0$.

Theorem 3. *For all starlike $f(z) = z + \sum_2^\infty a_n z^n$ and all $\mu \leq \frac{1}{16}$,*

$$(2\mu + \tfrac{1}{4})z + \mu a_2 z^2$$

is starlike and

$$\tfrac{1}{4}z \prec (2\mu + \tfrac{1}{4})z + \mu a_2 z^2 \prec \tfrac{3}{8}z + \tfrac{1}{16}a_2 z^2.$$

Proof. By Lemma 1, since $|a_2| \leq 2$ and $\mu \leq \frac{1}{16}$, $(2\mu + \frac{1}{4})z + \mu a_2 z^2$ is starlike and, on $|z| = 1$,

$$|(2\mu + \tfrac{1}{4})z + \mu a_2 z^2| \geq 2\mu + \tfrac{1}{4} - \mu|a_2| \geq \tfrac{1}{4},$$

so

$$\tfrac{1}{4}z \prec (2\mu + \tfrac{1}{4})z + \mu a_2 z^2.$$

It is now sufficient to show that, for each real α and each $\mu < \frac{1}{16}$, the polynomial

(22) $$\tfrac{3}{8}z + \tfrac{1}{16}a_2 z^2 - (2\mu + \tfrac{1}{4})e^{i\alpha} - \mu a_2 e^{2i\alpha}$$

has a zero on $|z| \leq 1$.

We shall show, in fact, that except when $|a_2| = 2$ and α takes a certain value, it has a zero in $|z| < 1$. Suppose that for some α it has no zero in $|z| < 1$. Then the polynomial

$$[(2\mu + \tfrac{1}{4})e^{i\alpha} + \mu a_2 e^{2i\alpha}]z^2 - \tfrac{3}{8}z - \tfrac{1}{16}a_2$$

has both zeros on $|z| \leq 1$, so by Lemma 3,

(23) $$|3a_2/128 + \tfrac{3}{8}[(2\mu + \tfrac{1}{4})e^{-i\alpha} + \mu\bar{a}_2 e^{-2i\alpha}]| \leq |(2\mu + \tfrac{1}{4}) + \mu a_2 e^{i\alpha}|^2$$
$$- |a_2|^2/256.$$

If we write $a_2 = \rho e^{i\phi}$, $\alpha + \phi = \psi$, (23) is equivalent to

(24) $$16|(8\mu + 1) + 4\mu\rho e^{i\psi}|^2 - 6|\rho e^{i\psi} + 4(8\mu + 1) + 16\mu\rho e^{-i\psi}| \geq \rho^2.$$

But, since $\mu < \frac{1}{16}$, we have

(25) $$|\rho e^{i\psi} + 4(8\mu + 1) + 16\mu\rho e^{-i\psi}|$$
$$= |16\mu\rho e^{i\psi} + 4(8\mu + 1) + 16\mu\rho e^{-i\psi} + \rho(1 - 16\mu)e^{i\psi}|$$
$$\geq 4(8\mu + 1) + 32\mu\rho \cos\psi - \rho(1 - 16\mu),$$

and (24), (25) give

$$-(1 - 256\mu^2)\rho^2 - 64\mu\rho(1 - 16\mu)\cos\psi$$
$$- 8(1 + 8\mu)(1 - 16\mu) + 6\rho(1 - 16\mu) \geq 0.$$

Again since $\mu < \frac{1}{16}$, we may divide this inequality by $1 - 16\mu$, and we obtain

$$2\rho(3 - 32\mu \cos\psi) \geq (1 + 16\mu)\rho^2 + 8(1 + 8\mu).$$

This implies that

$$2\rho(3 + 32\mu) \geq (1 + 16\mu)\rho^2 + 8(1 + 8\mu),$$

or

$$(\rho - 2)\left(\rho - 2 - \frac{2}{16\mu + 1}\right) \leq 0.$$

Since $\rho \leq 2$, this is a contradiction and shows that (22) has a zero in $|z| < 1$, unless $\rho = 2$, $\psi = \pi$. But (22) then has a zero $-e^{i\phi}$, and this completes the proof.

Theorem 4. *For all starlike $f(z) = z + \sum_2^\infty a_n z^n$, all $\mu \leq \frac{1}{8}$, and all λ satisfying $4\mu \leq \lambda \leq 2\mu + \frac{1}{4}$, $\lambda z^2 + \mu a_2 z^2$ is starlike and*

$$\lambda z + \mu a_2 z^2 \prec (2\mu + \tfrac{1}{4})z + \mu a_2 z^2.$$

Proof. By Lemma 1 and the fact that $|a_2| \leq 2$, the condition $4\mu \leq \lambda$ ensures that $\lambda z + \mu a_2 z^2$ is starlike. By Theorem 3, to prove the rest it is enough to assume that $\lambda < 2\mu + \frac{1}{4}$ and prove that for each real α and each μ satisfying $0 < \mu \leq \frac{1}{8}$,

(26) $$(2\mu + \tfrac{1}{4})z + \mu a_2 z^2 - \lambda e^{i\alpha} - \mu a_2 e^{2i\alpha}$$

has a zero in $|z| \leq 1$. As in the proof of Theorem 3 we show that, except when $|a_2| = 2$ and λ, α take certain values, there is a zero in $|z| < 1$. Suppose that for some α, λ there is no zero in $|z| < 1$. Then arguing as before, and with the same notation, we have

$$|\lambda + \mu\rho e^{i\psi}|^2 - (2\mu + \tfrac{1}{4})|\mu\rho e^{i\psi} + \lambda + \mu\rho e^{-i\psi}| \geq \mu^2\rho^2,$$

or

$$-(2\mu + \tfrac{1}{4} - \lambda)(\lambda + 2\mu\rho \cos\psi) \geq 0,$$

whence

$$-1 \leq \cos\psi \leq \frac{-\lambda}{2\mu\rho} \leq \frac{-\lambda}{4\mu} \leq -1.$$

This is a contradiction and proves that (26) has a solution in $|z| < 1$ unless $\rho = 2$, $\psi = \pi$, $\lambda = 4\mu$. But (26) then has a zero

$$\tfrac{1}{4}e^{-i\phi}[(\tfrac{1}{16} + \mu - 12\mu^2)^{1/2} - 2\mu - \tfrac{1}{4}]/\mu,$$

and it is easily verified that the condition $\mu \leq \frac{1}{8}$ implies that this lies on $|z| \leq 1$.

Theorem 5. *If, for all starlike $f(z) = z + \sum_{2}^{\infty} a_n z^n$,*

$$\lambda z + \mu a_2 z^2 \prec f(z),$$

then $\lambda \leq 2\mu + \frac{1}{4}$.

Proof. Let $l(z) = z(1 + z^2)^{-1}$ and, for $-1 < \alpha < 1$, let

$$f(z) = \left(\frac{1 + \alpha^2}{1 - \alpha^2}\right)^2 \left[-l\left(\frac{\alpha - z}{1 - \alpha z}\right) + \frac{\alpha}{1 + \alpha^2}\right] = z + \frac{4\alpha}{1 + \alpha^2}z^2 + \cdots.$$

Then $w = f(z)$ is a starlike function and maps $|z| < 1$ onto the w-plane with two slits, one along the negative real axis from $-\frac{1}{2}(1 + \alpha^2)(1 + \alpha)^{-2}$ to ∞, the other along the positive real axis from $\frac{1}{2}(1 + \alpha^2)(1 + \alpha)^{-2}$ to ∞. If $\lambda z + 4\mu\alpha z^2/(1 + \alpha^2) \prec f(z)$ then for real x, $-1 < x < 1$, we must have

$$-\frac{1}{2}\frac{1 + \alpha^2}{(1 + \alpha)^2} < \lambda x + \frac{4\mu\alpha}{1 + \alpha^2}x^2.$$

Allowing $x \to -1$, and then allowing $\alpha \to 1$, we obtain $\lambda \leq 2\mu + \frac{1}{4}$.

Theorem 6. *If, for all starlike $f(z) = z + \sum_{2}^{\infty} a_n z^n$,*

$$(2\mu + \tfrac{1}{4})z + \mu a_2 z^2 \prec f(z),$$

then $\mu \leq \frac{1}{16}$.

Proof. Taking $f(z)$ as before, if $(2\mu + \frac{1}{4})z + 4\mu\alpha z^2/(1 + \alpha^2) \prec f(z)$, then for real x, $-1 < x < 1$, we must have

$$-\frac{1}{2}\frac{1 + \alpha^2}{(1 + \alpha)^2} < (2\mu + \tfrac{1}{4})x + \frac{4\mu\alpha}{1 + \alpha^2}x^2.$$

Allowing $x \to -1$ and dividing through by $(1 - \alpha)^2$, we obtain

$$\mu \leq \frac{1}{8}\frac{1 + \alpha^2}{(1 + \alpha)^2},$$

and then allowing $\alpha \to 1$ we have $\mu \leq \frac{1}{16}$.

Finally, by the argument used in Section 1, combining Theorems 2, 3, and 6, we obtain

Theorem 7. (i) *If, for all starlike* $f(z) = z + \sum_2^{\infty} a_n z^n$, $\lambda z + \mu a_2 z^2$ *is starlike and*

$$\tfrac{1}{4}z \prec \lambda z + \mu a_2 z^2 \prec f(z),$$

then $\lambda = 2\mu + \tfrac{1}{4}$, $\mu \leqq \tfrac{1}{16}$.

(ii) *Conversely, for all starlike* $f(z) = z + \sum_2^{\infty} a_n z^n$ *and all* $\mu \leqq \tfrac{1}{16}$, $(2\mu + \tfrac{1}{4})z + \mu a_2 z^2$ *is starlike and*

$$\tfrac{1}{4}z \prec (2\mu + \tfrac{1}{4})z + \mu a_2 z^2 \prec f(z).$$

REFERENCE

1. J. Clunie and F. R. Keogh, On starlike and convex schlicht functions, *J. London Math. Soc.* **35** (1960), 229–233.
2. P. Dienes, *The Taylor series*, Dover, New York, 1957.
3. M. Marden, *The geometry of the zeros of a polynomial in a complex variable*, American Mathematical Society Mathematical Surveys, Providence, R.I., No. III, 1949.
4. Z. Nehari, *Conformal mapping*, McGraw-Hill, New York, 1952.

(Received February 10, 1968)

An Alternative to Boundedness

Bo Kjellberg

Division of Mathematics
Royal Institute of Technology
Stockholm 70, Sweden

1.

In the theory of analytic and subharmonic functions a usual condition is that the function in question be bounded on some set; say, for subharmonic functions, $u(z) \leqq C$. If $u(z)$ is subharmonic in the plane we define

$$m(r) = \inf_{|z|=r} u(z), \qquad M(r) = \max_{|z|=r} u(z).$$

A condition $u(z) \leqq C$ on some set implies that $m(r) \leqq C$ for some set of values of r. For simplicity, let us take $C = 0$ in the following.

In many cases the classical condition

(A) $\qquad\qquad\qquad m(r) \leqq 0$

can be replaced by the more general condition

(B) $\qquad\qquad\qquad m(r) \leqq (\cos \pi\lambda)M(r),$

where $0 < \lambda < 1$. Here we give some examples illustrating this possibility.

2.

First we consider a nonconstant function $u(z)$, subharmonic in the finite plane. According to a theorem of M. Heins [6] it is true that if (A) is valid for $0 < r < \infty$, then the limit

$$\lim_{r \to \infty} r^{-1/2} M(r)$$

exists and is positive (if this word may be used also for the case $+\infty$). We note that condition (A) is the special case of (B) obtained by setting $\lambda = \frac{1}{2}$.

If we now replace (A) by (B), the result is that the limit

$$\lim_{r \to \infty} r^{-\lambda} M(r)$$

exists and is positive (see Kjellberg [9] for the case of entire functions, Anderson [1] for the subharmonic case).

The proof is based on the following integral inequality:

$$(1) \qquad\qquad r^{-\lambda} M(r) \leqq \int_0^\infty K(r, t) t^{-\lambda} M(t) \, dt,$$

where $K(r, t)$ is a certain positive kernel. A tauberian theorem then gives the result (see Essén [4]).

3.

An essential part of the proof of (1) is to show that, with a given value of $M(R)/R^\lambda$, the quotient $M(r)/r^\lambda$ must be bounded for $0 < r < R$. This leads us to compare the respective effects of (A) and of (B) on a function $u(z)$, subharmonic for $|z| < R$. The only case of interest is when

$$M(R) = \sup_{r < R} M(r) > 0.$$

The classical result is the Milloux-Schmidt inequality: if condition (A) is fulfilled for $0 < r < R$, then

$$(2) \qquad\qquad M(r) \leqq U_0(r) = \frac{4M(R)}{\pi} \arctan\left(\frac{r}{R}\right)^{1/2}$$

(see, for example, [7], p. 108–109).

One consequence of (2) is that

$$(3) \qquad\qquad r^{-1/2} M(r) \leqq (4/\pi) R^{-1/2} M(R).$$

If (A) is replaced by (B) the result is

$$(4) \qquad M(r) \leqq U(r) = \frac{2M(R)}{\pi} \tan\left(\frac{1}{2} \pi\lambda\right) \int_0^{r/R} \frac{t^{\lambda-1} - t^{1-\lambda}}{1 - t^2} \, dt.$$

The inequality corresponding to (3) is

$$(5) \qquad\qquad \frac{M(r)}{r^\lambda} \leqq \frac{\tan(\frac{1}{2}\pi\lambda)}{\frac{1}{2}\pi\lambda} \frac{M(R)}{R^\lambda}.$$

These results will be published by Hellsten, Kjellberg, and Norstad [8].

4.

We now consider a subharmonic function $u(z)$ defined on an angular region. To normalize the situation, let the region, called D, be the complex plane cut along the negative real axis. We make appropriate changes in the previous conditions (A) and (B) as follows:

(A′) $$u(re^{i\pi\pm}) \leq 0,$$

(B′) $$u(re^{i\pi\pm}) \leq (\cos \pi\lambda)u(r),$$

or

(B″) $$u(re^{i\pi\pm}) \leq (\cos \pi\lambda)M(r).$$

Here, of course, $u(re^{i\pi\pm})$ signify lim sup taken from above and from below, respectively. The classical result is that if (A′) holds for $0 < r < \infty$ and $\liminf r^{-1/2}M(r)$ is finite when $r \to \infty$, then there exists a number α such that

(6) $$\lim_{r \to \infty} r^{-1/2}u(re^{i\theta}) = \alpha \cos \tfrac{1}{2}\theta, \qquad |\theta| < \pi,$$

except when θ belongs to a set of logarithmic capacity zero. This theorem is due to Ahlfors and Heins [3].

In a recent publication by Essén [5], a generalization obtained by replacing (A′) by (B′) is given:

If (B′) holds, then either $\lim_{r \to \infty} r^{-\lambda}M(r) = \infty$ or

(7) $$\lim_{r \to \infty} r^{-\lambda}u(re^{i\theta}) = \alpha \cos \lambda\theta, \qquad |\theta| < \pi,$$

except when θ belongs to a set of logarithmic capacity zero. The proof is based on a study of convolution inequalities.

5.

Considering again the angular region D, we have another theorem of Ahlfors [2]. If we set

(8) $$\phi(r) = \int_{-\pi}^{+\pi} u(re^{i\theta}) \cos\left(\frac{1}{2}\theta\right) d\theta,$$

then, if (A′) holds for $0 < r < \infty$, the function $\phi(r)$ has the property of being convex with respect to the family of curves $ar^{1/2} + br^{-1/2}$. A corresponding result with (B″) implying convexity with respect to $ar^{\lambda} + br^{-\lambda}$ will be published by Norstad.

6.

Still more such examples can be expected. For example, David Drasin, John Lewis, and Daniel F. Shea have already obtained results in this field.

REFERENCES

1. J. M. Anderson, Growth properties of integral and subharmonic functions, *J. Analyse Math.* **13** (1964), 355–389.
2. L. Ahlfors, On Phragmén-Lindelöf's principle, *Trans. Amer. Math. Soc.* **41** (1937), 1–8.
3. L. Ahlfors and M. Heins, Questions of regularity connected with the Phragmén-Lindelöf principle, *Ann. of Math.* (2) **50** (1949), 341–346.
4. M. Essén, Note on "A theorem on the minimum modulus of entire functions" by Kjellberg, *Math. Scand.* **12** (1963), 12–14.
5. ——— A generalization of the Ahlfors-Heins theorem, *Trans. Amer. Math. Soc.* **142** (1969), 331–344.
6. M. Heins, Entire functions with bounded minimum modulus; subharmonic analogues, *Ann. of Math.* (2) **49** (1948), 200–213.
7. ——— Selected Topics in the Classical Theory of Functions of a Complex Variable, Holt, Rinehart and Winston, New York, 1962.
8. U. Hellsten, B. Kjellberg and F. Norstad, Subharmonic functions in a circle, to appear in *Arkiv för matematik*.
9. B. Kjellberg, A theorem on the minimum modulus of entire functions, *Math. Scand.* **12** (1963), 5–11.

(*Received June 10, 1969*)

On a Result of A. J. Macintyre

Thomas Kövari

Department of Mathematics
Imperial College
London, S.W.7
England

In 1952 A. J. Macintyre proved the following result [1]:

If $\{\lambda_n\}$ is a strictly increasing sequence of positive integers satisfying the condition

$$(1) \qquad \sum_{n=1}^{\infty} \frac{1}{\lambda_n} = +\infty$$

then there exists an entire function of the form

$$(2) \qquad f(z) = b_0 + \sum_{n=1}^{\infty} b_n z^{\lambda_n}$$

which tends to zero along the positive real axis. In this paper I will give a new and completely elementary proof of this theorem. I use the word elementary in the sense that it is customarily used in the theory of numbers, meaning that the proof does not use complex analysis (nor for that matter does the proof use any nontrivial real analysis either). It seems to me that such a proof may have some interest.

We set out to determine the unique polynomial

$$(3) \qquad p_n(z) = 1 + \sum_{k=1}^{n} a_k z^{\lambda_k}$$

which has a zero of order n at the point $z = 1$. The polynomial

$$q_0(z) = z^{\lambda_n} p_n(1/z)$$
$$= z^{\lambda_n} + a_1 z^{\lambda_n - \lambda_1} + a_2 z^{\lambda_n - \lambda_2} + \cdots + a_{n-1} z^{\lambda_n - \lambda_{n-1}} + a_n$$

also has a zero of order n at the point $z = 1$. The polynomials $q_0'(z)$ and

$$q_1(z) = z^{-(\lambda_n - \lambda_{n-1} - 1)} q_0'(z)$$
$$= \lambda_n z^{\lambda_n - 1} + (\lambda_n - \lambda_1) a_1 z^{\lambda_n - 1 - \lambda_1} + \cdots + (\lambda_n - \lambda_{n-1}) a_{n-1}$$

have (both) a zero of order $n - 1$ at $z = 1$. The polynomial $q_k(z)$ defined by the recursion

$$q_k(z) = z^{-(\lambda_{n-k+1} - \lambda_{n-k} - 1)} q_{k-1}'(z) \qquad (k = 1, \ldots, n - 1)$$

and given explicitly by

$$q_k(z) = \lambda_n \lambda_{n-1} \cdots \lambda_{n-k+1} z^{\lambda_{n-k}} + (\lambda_n - \lambda_1) \cdots (\lambda_{n-k+1} - \lambda_1) a_1 z^{\lambda_{n-k} - \lambda_1}$$
$$+ \cdots + (\lambda_n - \lambda_{n-1}) \cdots (\lambda_{n-k+1} - \lambda_{n-k}) a_{n-k}$$

has a zero of order $n - k$ at $z = 1$. Thus

$$q_k(1) = 0 \quad \text{for} \quad k = 0, 1, 2, \ldots, n - 1$$

which yields the following system of n linear equations for the undetermined coefficients a_k :

$$-1 = \sum_{k=1}^{n} a_k$$

$$-\lambda_n = \sum_{k=1}^{n-1} (\lambda_n - \lambda_k) a_k$$

$$\cdots$$

(4)

$$-\lambda_n \lambda_{n-1} \cdots \lambda_{n-j+1} = \sum_{k=1}^{n-j} (\lambda_n - \lambda_k) \cdots (\lambda_{n-j+1} - \lambda_k) a_k$$

$$\cdots$$

$$-\lambda_n \lambda_{n-1} \cdots \lambda_2 = (\lambda_n - \lambda_1)(\lambda_{n-1} - \lambda_1) \cdots (\lambda_2 - \lambda_1) a_1.$$

The unique solution* of (4) is given by

(5)
$$a_k = - \prod_{\substack{1 \leq i \leq n \\ i \neq k}} \frac{\lambda_i}{\lambda_i - \lambda_k}, \qquad k = 1, 2, \ldots, n.$$

---

* I am greatly indebted to Mr. R. London for suggesting a proof of (5).

We shall now renormalize the polynomials p_n and write

(6) $$P_n(z) = p_n(z(-a_1)^{-1/\lambda_1}) = 1 - z^{\lambda_1} + \sum_{k=2}^{n} b_k^{(n)} z^{\lambda_k},$$

where

(7) $$(-1)^k |b_k^{(n)}| = b_k^{(n)} = a_k(-a_1)^{-\lambda_k/\lambda_1} \qquad (2 \le k \le n).$$

Now

$$\log |b_k^{(n)}| = \log |a_k| - \frac{\lambda_k}{\lambda_1} \log |a_1|$$

$$= \log \prod_{i=1}^{k-1}{}^{(1)} \frac{\lambda_i}{\lambda_k - \lambda_i} - \frac{\lambda_k}{\lambda_1} \log \prod_{i=2}^{k}{}^{(2)} \frac{\lambda_i}{\lambda_i - \lambda_1}$$

$$+ \log \prod_{i=k+1}^{n} \left\{ \frac{\lambda_i}{\lambda_i - \lambda_k} \left(\frac{\lambda_i}{\lambda_i - \lambda_1} \right)^{-\lambda_k/\lambda_1} \right\}$$

$$= \log \prod_{i=1}^{k-1}{}^{(1)} - \frac{\lambda_k}{\lambda_1} \log \prod_{i=2}^{k}{}^{(2)}$$

$$+ \log \prod_{i=k+1}^{\infty}{}^{(3)} \left\{ \frac{\lambda_i}{\lambda_i - \lambda_k} \left(\frac{\lambda_i - \lambda_1}{\lambda_i} \right)^{\lambda_k/\lambda_1} \right\}$$

$$- \log \prod_{i=n+1}^{\infty}{}^{(4)} \left\{ \frac{\lambda_i}{\lambda_i - \lambda_k} \left(\frac{\lambda_i - \lambda_1}{\lambda_i} \right)^{\lambda_k/\lambda_1} \right\}.$$

For large i (for example, if $\lambda_i \ge 2\lambda_k$),

$$\log\{ \quad \} = \frac{\lambda_k}{\lambda_1} \log\left(1 - \frac{\lambda_1}{\lambda_i}\right) - \log\left(1 - \frac{\lambda_k}{\lambda_i}\right)$$

(8) $$= \lambda_k(\lambda_k + \lambda_1) O\left(\frac{1}{\lambda_i^2}\right) = \frac{\lambda_k^2}{\lambda_i^2} O(1).$$

Thus the infinite product $\prod^{(3)}$ converges, and $\log \prod^{(4)}$ tends to zero as $n \to \infty$. We can therefore write

$$\log |b_k^{(n)}| = \log |b_k| - \rho_k^{(n)},$$

where

$$\log |b_k| = \log \prod{}^{(1)} - \frac{\lambda_k}{\lambda_1} \log \prod{}^{(2)} + \log \prod{}^{(3)}$$

is *independent* of n and

(9) $$\rho_k^{(n)} = \log \prod{}^{(4)} \to 0 \quad \text{as} \quad n \to \infty.$$

Further, since $x \log(1 - y) - \log(1 - xy) \geq 0$ for $x > 1$, $0 < xy < 1$, all the factors of $\prod^{(4)}$ are greater than 1, and hence

$$\text{(10)} \qquad \rho_k^{(n)} \geq 0.$$

We first estimate $\prod^{(1)}$:

$$\prod_{i=1}^{k-1}{}^{(1)} \frac{\lambda_i}{\lambda_k - \lambda_i} = \frac{\displaystyle\prod_{i=1}^{k-1} \lambda_i}{\displaystyle\prod_{i=1}^{k-1} (\lambda_k - \lambda_i)} \leq \frac{\lambda_{k-1}^{k-1}}{(k-1)!} \leq \frac{\lambda_k^k}{k!}.$$

Let $\lambda_k/k = q$. Clearly $q \geq 1$. Then

$$\text{(11)} \qquad \log \prod{}^{(1)} \leq k \log(kq) - k(\log k - 1) = k(\log q + 1) \leq kq = \lambda_k.$$

We now estimate $\prod^{(2)}$:

$$
\begin{aligned}
-\frac{\lambda_k}{\lambda_1} \log \prod_{i=2}^{k}{}^{(2)} \frac{\lambda_i}{\lambda_i - \lambda_1} &= \frac{\lambda_k}{\lambda_1} \sum_{i=2}^{k} \log\left(1 - \frac{\lambda_1}{\lambda_i}\right) \\
&= \frac{\lambda_k}{\lambda_1} \sum_{i=2}^{k} \left\{-\frac{\lambda_1}{\lambda_i} + O\left(\frac{1}{\lambda_i^2}\right)\right\} \\
&= -\lambda_k \sum_{i=2}^{k} \frac{1}{\lambda_i} + O(\lambda_k).
\end{aligned}
$$

(12)

Finally, to estimate $\prod^{(3)}$, we split it up as follows:

$$
\begin{aligned}
\log \prod_{i=k+1}^{\infty}{}^{(3)} = \log \prod_{\lambda_k < \lambda_i < 2\lambda_k}{}^{(4)} \frac{\lambda_i}{\lambda_i - \lambda_k} &+ \frac{\lambda_k}{\lambda_1} \sum_{\lambda_k < \lambda_i < 2\lambda_k}^{(5)} \log\left(1 - \frac{\lambda_1}{\lambda_i}\right) \\
&+ \sum_{\lambda_i > 2\lambda_k}^{(6)} \left\{\frac{\lambda_k}{\lambda_1} \log\left(1 - \frac{\lambda_1}{\lambda_i}\right) - \log\left(1 - \frac{\lambda_k}{\lambda_i}\right)\right\}.
\end{aligned}
$$

(13)

Let s denote the number of λ_i's in the interval $\lambda_k < \lambda_i < 2\lambda_k$. Clearly $s < \lambda_k$. Then

$$\prod_{\lambda_k < \lambda_i < 2\lambda_k}{}^{(4)} \frac{\lambda_i}{\lambda_i - \lambda_k} = \frac{\prod \lambda_i}{\prod (\lambda_i - \lambda_k)} \leq \frac{(2\lambda_k)^s}{s!}.$$

Let $\lambda_k/s = q$. We have that $q > 1$, and hence

$$
\begin{aligned}
\text{(14)} \qquad \log \prod{}^{(4)} &\leq s \log 2 + s(\log q + \log s) - s(\log s - 1) \\
&= s(\log 2 + \log q + 1) \\
&\leq s(\log 2 + q) = s \log 2 + \lambda_k < (1 + \log 2)\lambda_k.
\end{aligned}
$$

Next,

$$\frac{\lambda_k}{\lambda_1} \sum_{\lambda_k < \lambda_i < 2\lambda_k}^{(5)} \log\left(1 - \frac{\lambda_1}{\lambda_i}\right) = \frac{\lambda_k}{\lambda_1} \sum_{\lambda_k < \lambda_i < 2\lambda_k} \left\{-\frac{\lambda_1}{\lambda_i} + O\left(\frac{1}{\lambda_i^2}\right)\right\}$$

(15)
$$= -\lambda_k \sum_{i=k+1}^{k+s} \frac{1}{\lambda_i} + O(\lambda_k) = O(\lambda_k)$$

since $s < \lambda_k$.

Finally, using (8) we find

$$\sum_{\lambda_i \geq 2\lambda_k}^{(6)} \left\{\frac{\lambda_k}{\lambda_1} \log\left(1 - \frac{\lambda_1}{\lambda_i}\right) - \log\left(1 - \frac{\lambda_k}{\lambda_i}\right)\right\}$$

(16)
$$= O(\lambda_k^2) \sum_{2\lambda_k \leq \lambda_i} \frac{1}{\lambda_i^2} = O(\lambda_k^2) \sum_{v \geq 2\lambda_k} \frac{1}{v^2} = O(\lambda_k).$$

Combining (14), (15), and (16), we find that

(17)
$$\log \prod^{(3)} = O(\lambda_k).$$

Combining (11), (12), and (17), we obtain that

(18)
$$\log |b_k| = -\lambda_k \sum_{i=2}^{k} \frac{1}{\lambda_i} + O(\lambda_k).$$

Let us write $b_k = (-1)^k |b_k|$ and

$$f(z) = 1 - z^{\lambda_1} + \sum_{k=2}^{\infty} b_k z^{\lambda_k}.$$

In view of (1) and (18), this power series converges absolutely for every z, that is, $f(z)$ is an *entire function*. Further, for $0 \leq x$

$$|f(x) - P_n(x)| \leq \sum_{k=2}^{n} |b_k| (1 - e^{-\rho_k(n)}) x^{\lambda_k} + \sum_{k=n+1}^{\infty} |b_k| x^{\lambda_k}.$$

Hence, for $x \leq y$, and $2 < m < n$,

$$|f(x) - P_n(x)| \leq \sum_{k=2}^{m} |b_k| (1 - e^{-\rho_k(n)}) y^{\lambda_k} + \sum_{k=m+1}^{\infty} |b_k| y^{\lambda_k},$$

since $0 < 1 - e^{-\rho_k(n)} < 1$. For a fixed y, and $\varepsilon > 0$, we can choose m such that $\sum_{k=m+1}^{\infty} |b_k| y^{\lambda_k} < \varepsilon$. Then, in view of (9), we can choose N such that $N > m$ and $1 - e^{-\rho_k(n)} < \varepsilon\{\sum_{k=2}^{\infty} |b_k| y^{\lambda_k}\}^{-1}$ for $k \leq m$ and $n \geq N$. Then, for $n \geq N$

$$|f(x) - P_n(x)| \leq \varepsilon + \varepsilon = 2\varepsilon.$$

Thus, for $x \geq 0$

(19)
$$P_n(x) \rightarrow f(x) \quad \text{as} \quad n \rightarrow \infty$$

uniformly on every finite interval. The polynomial $P_n(x)$ has a zero of order n at

$$x = \Lambda_n = (-a_1)^{1/\lambda_1} = \left\{ \prod_{i=2}^{n} \left(1 - \frac{\lambda_1}{\lambda_i}\right) \right\}^{-1/\lambda_1}.$$

Because of condition (1), $\lim_n \Lambda_n = +\infty$.

It is easy to show that $P_n'(x)$ has no zeros in the interval $0 < x < \Lambda_n$. Indeed, $P_n'(x)$ has a zero of order $(n-1)$ at $z = \Lambda_n$. Suppose it also had a zero at y_1 $(0 < y_1 < \Lambda_n)$. The polynomial $(d/dx)(x^{1-\lambda_1}P_n'(x))$ has a zero of order $(n-2)$ at $z = \Lambda_n$, and by Rolle's theorem, it also would have a zero at $y_2(y_1 < y_2 < \Lambda_n)$. By repeated application of Rolle's theorem we would conclude that the polynomial

$$b_n^{(n)}\lambda_n(\lambda_n - \lambda_1) \cdots (\lambda_n - \lambda_{n-1})x^{\lambda_n - \lambda_{n-1} - 1}$$

also had a zero y_n in the interval $0 < y_n < \Lambda_n$ which is impossible.

Thus $P_n'(x)$ has no zeros in the interval $0 < x < \Lambda_n$, and therefore $P_n(x)$ is a decreasing function in the interval $0 \leq x \leq \Lambda_n$. Further, since $P_n(\Lambda_n) = 0$, it follows that $P_n(x) > 0$ for $0 \leq x < \Lambda_n$. Consequently, in every finite interval $f(x)$ is the uniform limit of a sequence of decreasing positive functions, and hence itself is decreasing and nonnegative. Thus $f(x)$ is positive and decreasing for $0 \leq x < \infty$, and tends to a finite limit as $x \to +\infty$. This concludes our proof.

The function $f(z)$ we have constructed has some interesting additional properties. In particular, not only is $f(x) > 0$, and $f'(x) < 0$ for every $x \geq 0$, but the "quasi-derivatives" of f defined recursively by

$$f_0 = f, \qquad f_\nu(z) = z^{1 + \lambda_{\nu-1} - \lambda_\nu} f'_{\nu-1}(z), \qquad \nu = 1, 2, \ldots,$$

are all positive for even ν, and negative for odd ν, for every $x \geq 0$.

In the special case when $\lambda_n = \rho n$,

$$p_n(z) = (1 - z^\rho)^n, \qquad P_n(z) = \left(1 - \frac{z^\rho}{n}\right)^n, \qquad \text{and} \quad f(z) = e^{-z^\rho}.$$

REFERENCES

1. A. J. Macintyre, Asymptotic paths of integral functions with gap power series, *Proc. London Math. Soc.* **2** (1952), 286–296.

(*Received February 13, 1968*)

Functions of Bounded Indices in One and Several Complex Variables

J. Gopala Krishna and S. M. Shah*

Department of Mathematics
University of Kentucky
Lexington, Kentucky 40506
U.S.A.

1. Introduction

The idea of an entire function of bounded index (b.i.) in the complex plane $\mathscr{C}^1$ was introduced by B. Lepson [8] in connection with his study of the differential equations of infinite order. Following his preliminary investigations Fred Gross, S. M. Shah, and others established some further properties of the spaces of such functions, which we shall survey in Section 2. It turns out that entire solutions of a large class of differential equations in $\mathscr{C}^1$ are functions of b.i., and in particular some well-known special functions are of this kind. Also the functions of b.i. are shown to possess specific growth properties, which from the point of view of the differential equations turn out to be the properties of their entire solutions.

In Section 3 of this paper, we generalize the concepts of b.i., and show how these lead to more precise results concerning the solutions of differential equations, and also point out some immediate consequences of the related ideas in the theory of the maximum term and the central index of a power

* The research of this author is supported by the National Science Foundation under Grant GP-7544. Some of these results were presented at the American Mathematical Society annual meeting 1968; abstract *Notices Amer. Math. Soc.* **15** (1968), 152.

series. In Section 4 we try to extend the theory to the case of several complex variables and indicate some open questions, which we also do at other appropriate places. The ideas discussed in this paper, particularly with reference to analytic functions not necessarily entire, are also relevant in connection with the study of the existence and analytic continuation of the local solutions of partial differential equations (see Theorem 4.1 and Lemma 4.1 of this paper; cf. Chapter II of [4]). However in the present work we do not go into any details in this regard except that we follow an indicative scheme in Section 4.

2. Entire Functions of B.I. in $\mathscr{C}^1$

In this section we use $\mathscr{A} = \mathscr{A}(\mathscr{C}^1)$ to stand for the space of all entire functions in $\mathscr{C}^1$ and $I = I^1$ for the set of all nonnegative integers. We define the absolute Taylor coefficients T_n of F by

(2.1) $T_n\{F\} = T_n(z, F) = |F^{(n)}(z)|/n!$ for $z \in \mathscr{C}^1, n \in I,$

where $F^{(n)}$ stands for the nth derivative of $F(F^{(0)} = F)$. We write, as usual, $\mathscr{M}(r) = \mathscr{M}(r, F)$, for the maximum modulus of $F \in \mathscr{A}$ on the circle $z: [z \in \mathscr{C}^1, |z| = r]$.

Following Fred Gross [5] and S. M. Shah [11], we say that an $F \in \mathscr{A}$ is of bounded index iff (if and only if) there exists an $N \in I$ such that

(2.2) $T_m(z, F) \leq \max_{n \in I, n \leq N} T_n(z, F)$ for $z \in \mathscr{C}^1, m \in I,$

and call the least among such N the index of F, and denote the space of all such functions as $\mathscr{B} = \mathscr{B}(\mathscr{C}^1)$.

According to Lepson [8] an $F \in \mathscr{A}$ is of bounded index, iff there exists an $N \in I$ such that (2.2) holds with $<$ instead of $\leq$. Let us say that all such functions together with the zero function of $\mathscr{A}$ constitute the space $\mathscr{L} = \mathscr{L}(\mathscr{C}^1)$. It is obvious that $\mathscr{L} \subseteq \mathscr{B}$, and the question arises whether the results of Fred Gross, Shah, and others, which conclude that certain functions ($\not\equiv 0$) are of b.i., remain valid when one chooses to interpret "bounded index" in the sense of Lepson. With minor changes in the available discussions, it is easy to see that the answer is in the affirmative in the case of all such results proved about $\mathscr{B}$ so far. For example, improving some theorems of Fred Gross [5, Theorems 1, 2] and one of his own (see [11]) Shah proved

Theorem 2.1. (Shah [12]). *Let $G \in \mathscr{A}$ and satisfy a differential equation of the form*

$$\sum_{n \in I, n \leq p} B_n(z)G^{(n)}(z) = F(z)   \text{for}   z \in \mathscr{C}^1,$$

where B_n ($n \in I, n \leq p \in I$) and F are polynomials of $\mathscr{A}$ and B_p is not the zero function of $\mathscr{A}$ and the degree of B_p is the largest among the degrees of the

polynomials B_n ($n \leqq p$). Then $G \in \mathscr{B}$. Further, the theorem, without the hypothesis on the degrees of the polynomials B_n ($n \leqq p$), would be false.

As mentioned above it is easily seen from Shah's proof itself that the
theorem holds with "$G \in \mathscr{L}$" instead of "$G \in \mathscr{B}$" in its conclusion. It
follows, in particular, that the exponential and trigonometric polynomials in
$\mathscr{C}^1$, the Bessel functions of the first kind, and the confluent hypergeometric
functions, all belong to $\mathscr{L}$. Improving a result of Fred Gross, Shah [12]
further showed that the Mittag-Leffler functions F_s ($s \in I, s > 0$) defined by

$$F_s(z) = \sum_{n \in I} z^n/(ns)! \quad \text{for} \quad z \in \mathscr{C}^1$$

also belong to $\mathscr{B}$ (to $\mathscr{L}$).

Turning to the growth results we have

Theorem 2.2. (Shah [11]). *Let $F \in \mathscr{B}$ and have index N. Then F is an entire
function of exponential type $N + 1$ (as defined in [1, Chapter II]). Further the
upper bound obtained for the growth of F is actually attained by the exponential
function (which belongs to $\mathscr{L}$ as well).*

Theorem 2.2 implies, in particular, that the F of Theorem 2.1 is of
exponential type (cf. [13, 14]). But there can be no analogue of the theorem,
which provides a "lower estimate" for the growth of a function of $\mathscr{B}$ or $\mathscr{L}$,
as we have

Theorem 2.3. (Lee and Shah*). *Let $\{\psi_n, n \in I\}$ be a real-valued sequence
$\to +\infty$ as $n \to +\infty$. Let $G \in \mathscr{A}$, be transcendental and be such that $G(0) \neq 0$.
Then there exists an $F \in \mathscr{B}$ and with index 1 such that*

$$\psi_n |F^{(n)}(z)| \leqq \max(|F(z)|, |F^{(1)}(z)|) \leqq \mathscr{M}(|z|, G) \quad \text{for} \quad z \in \mathscr{C}^1, n \in I, \text{ but } > 1.$$

One might think of a characterization of $\mathscr{B}$, of $\mathscr{L}$ or of some of their
subspaces. For example, Theorems 2.1 and 2.2 imply

Corollary 2.1. (Fred Gross [5]). *Let $F \in \mathscr{A}$ and have only a finite set of
zeros. Then $F \in \mathscr{B}$ ($F \in \mathscr{L}$) iff there exists a polynomial $P \in \mathscr{A}$ and an $a \in \mathscr{C}^1$
such that*

$$F(z) = P(z)\exp(az) \quad \text{for} \quad z \in \mathscr{C}^1.$$

However there cannot be a characterization of either $\mathscr{B}$ or $\mathscr{L}$ based
entirely on the growth of the maximum moduli of their functions, as we have

---

* B. S. Lee and S. M. Shah, on the growth of entire functions of bounded index, *J. Math.
Mech.*, to appear.

Theorem 2.4. (Pugh and Shah [10], cf. [11, 8]). *To each transcendental* $G \in \mathscr{A}$, *corresponds an* $F \in \mathscr{A}$ *but not* $\in \mathscr{B}$ *such that* $\log \mathscr{M}(r, F) \sim \log \mathscr{M}(r, G)$, *as* $r \to +\infty$.

We finally indicate some investigations leading to an estimation of the growth of the entire solutions of linear differential equations considered in Theorem 2.1, in terms of the coefficients of the equations. Shah ([11, 12]) gave some bounds for the indices of such solutions and subsequently showed in collaborations with Marić and with Lee that more sharp estimations are possible under stronger hypotheses (see [7, 9]).

3. Generalized Concepts and the Theory in $\mathscr{C}^1$

In this section we generalize the concepts of Section 2 and indicate the contribution of this work to the special case of $\mathscr{C}^1$ and point out certain additional complications in the case of several complex variables.

We start with some notation and conventions to which we shall be adhering hereafter. We write $\mathscr{C}^k$ for the Cartesian product of the complex plane with itself k-times, Ω for a domain (nonempty connected open set) in $\mathscr{C}^k$, $\mathscr{A} = \mathscr{A}(\Omega)$ for the space of functions analytic in Ω, Ω^+ for the set of all points of Ω with positive real coordinates, and $I = I^k$ for the set of all points of $\mathscr{C}^k$ with non-negative integral coordinates. When it is easily understood from the context we refer to the points $(z_1, \ldots, z_k)$, $(|z_1|, \ldots, |z_k|)$, $(n_1, \ldots, n_k)$, etc. of $\mathscr{C}^k$ by their unsuffixed symbols z, $|z|$, n, etc., and sometimes omit specifications such as "$\ldots \in I$".

In the case of any $x, y \in \mathscr{C}^k$ and with real coordinates we say that $x \leqq y$ or $y \geqq x$ iff $x_j \leqq y_j$ for $1 \leqq j \leqq k$, that $x < y$ or $y > x$ if $x \leqq y$ but $x \neq y$, and that $x \ll y$ or $y \gg x$ iff $x_j < y_j$ for $1 \leqq j \leqq k$. Corresponding to any $\delta \in \mathscr{C}^{k+}$ we write

$$\Omega(\delta) = [z: z \in \mathscr{C}^k, |z - a| \ll \delta \text{ for some } a \in \Omega].$$

Corresponding to any $F \in \mathscr{A}$ and any $n \in I$ we write $F^{(n)}$ for the nth partial derivative of $F(F^{(0, \ldots, 0)} = F)$ and define $T_n = T_n(F)$ by

$$T_n = T_n(F) = T_n(z, F) = |F^{(n)}(z)|/n! \quad \text{for} \quad z \in \Omega,$$

where $n! = n_1! n_2! \ldots n_k!$.

We generalize the concepts of $\mathscr{B}(\mathscr{C}^1)$ and $\mathscr{L}(\mathscr{C}^1)$ of Section 2. We say that $F \in \mathscr{B}(\Omega, \alpha)$, for $\alpha \in \mathscr{C}^{k+}$, iff $F \in \mathscr{A}(\Omega)$ and there exists an $N = N(\alpha, F) \in I$ such that

$$(3.1) \qquad T_m(z, F)\alpha^m \leqq \max_{\substack{n \leqq N \\ n \in I}} T_n(z, F)\alpha^n \qquad \text{for} \quad z \in \Omega, m \in I,$$

where $\alpha^m = \alpha_1^{m_1} \alpha_2^{m_2} \ldots \alpha_k^{m_k}$, in which case we refer to any such N as a $\mathscr{B}(\Omega, \alpha)$-index of F. We say that $F \in \mathscr{L}(\Omega, \alpha)$, for $\alpha \in \mathscr{C}^{k+}$, iff $F \in \mathscr{A}(\Omega)$ and either (i) is the zero function of $\mathscr{A}$, in which case we say that any $N \in I$ is an $\mathscr{L}(\Omega, \alpha)$-index of F, or (ii) there exists an $N = N(\alpha, F) \in I$ such that

(3.2) $T_m(z, F)\alpha^m < \max\limits_{n \leq N} T_n(z, F)\alpha^n$ for $z \in \Omega, m$ not $\leq N (m \in I)$,

in which case we refer to any such N as an $\mathscr{L}(\Omega, \alpha)$-index of F.

Our discussions (Theorem 4.5 together with Theorem 4.1 of Section 4) show that entire solutions of a larger class of differential equations than those considered in Theorem 2.1 belong to $\mathscr{L}(\mathscr{C}^1, \alpha)$ for every $\alpha \in \mathscr{C}^{1+}$, while we have

Theorem 3.1. *Let $\alpha \in \mathscr{C}^{1+}$ and $F \in \mathscr{B}(\mathscr{C}^1, \alpha)$ with $N = N(\alpha, F)$ as the minimal $\mathscr{B}(\mathscr{C}^1, \alpha)$-index. Then F is of exponential type $(N + 1)/\alpha$.*

Proof. This theorem is one among such easily deduced generalizations of the results of Section 2, and is an obvious consequence of Theorem 2.2 and the fact that if G is the function defined by $G(z) = F(\alpha z)$ for $z \in \mathscr{C}^1$, then $G \in \mathscr{B}(\mathscr{C}^1, 1)$ with $N = N(\alpha, F)$ as the minimal $\mathscr{B}(\mathscr{C}^1, 1)$-index.

Thus to get an upper estimate for the exponential type of an entire solution F of a differential equation of the kind referred above, one might obtain a $\mathscr{B}(\mathscr{C}^1, \alpha)$-index, say $B(\alpha)$ of F, as small as possible, and take

$$\inf\{(B(\alpha) + 1)/\alpha : \alpha \in \mathscr{C}^{1+}\}$$

instead of taking $B(1) + 1$ (cf. the techniques referred in Section 2 for estimating the index, for example, with reference to the equation $2F^{(1)} + F = 0$).

In extending Theorem 2.1 and its generalizations in $\mathscr{C}^1$ to $\mathscr{C}^k$, we find it desirable to consider the spaces $\mathscr{B}_w(\Omega, \alpha)$ and $\mathscr{L}_w(\Omega, \alpha)$ which, respectively, reduce to $\mathscr{B}(\Omega, \alpha)$ and $\mathscr{L}(\Omega, \alpha)$ when $k = 1$. We introduce these spaces along with the corresponding indices of their functions as we did in introducing $\mathscr{B}(\Omega, \alpha)$ and $\mathscr{L}(\Omega, \alpha)$, except that we modify (3.1) and (3.2), respectively, as

(3.3) $T_m(z, F)\alpha^m \leq \max\limits_{n < m} T_n(z, F)\alpha^n$ for $z \in \Omega, m > N$,

and

(3.4) $T_m(z, F)\alpha^m < \max\limits_{n < m} T_n(z, F)\alpha^n$ for $z \in \Omega, m > N$.

Before concluding this section let us observe some properties of $\mathscr{B}(\Omega, \alpha)$ and $\mathscr{L}(\Omega, \alpha)$, when $k = 1$, which are simple consequences of the theory of the maximum term and the central index (see [14]) of a power series. The maximum term $\mu = \mu(F)$ and the central index $v = v(F)$ of the Taylor series of an $F \in \mathscr{A}(\Omega)$ $(\Omega \subseteq \mathscr{C}^1)$ may be defined by

$$\mu = \mu(F) = \mu(\alpha, z, F) = \max_{n \in I^1} T_n(z, F)\alpha^n$$

and

$$v = v(F) = v(\alpha, z, F) = \max[n : T_n(z, F)\alpha_n = \mu(\alpha, z, F)],$$

for $z \in \Omega$, $\alpha \in \Omega_z = [\alpha : \alpha \in \mathscr{C}^{1+}$, the disk $[\xi : |\xi - z| \leq \alpha] \subseteq \Omega]$. From the elementary properties of the central index we get

Theorem 3.2. *Let $\alpha \in \mathscr{C}^{1+}$, $\Omega \subseteq \mathscr{C}^1$ and $F \in \mathscr{A}(\Omega)$. Then (i) $F \in \mathscr{L}(\Omega, \alpha)$, iff*

$$\sup_{z \in \Omega} v(\alpha, z, F) < +\infty,$$

in which case the sup is the minimal $\mathscr{L}(\Omega, \alpha)$-index of F, and (ii) $F \in \mathscr{B}(\Omega, \alpha)$, iff

$$\sup_{z \in \Omega} v(\alpha - 0, z, F) < +\infty.$$

in which case the sup is the minimal $\mathscr{B}(\Omega, \alpha)$-index of F.

Corollary 3.1. *Let $\alpha, \beta \in \mathscr{C}^{1+}$, $\alpha < \beta$ and let $\Omega \subseteq \mathscr{C}^1$. Then*

$$\mathscr{B}(\Omega, \alpha) \supseteq \mathscr{L}(\Omega, \alpha) \supseteq \mathscr{B}(\Omega, \beta) \supseteq \mathscr{L}(\Omega, \beta).$$

We finally observe

Theorem 3.3. *Let $\alpha \in \mathscr{C}^{1+}$, $p \in I$, $\Omega \subseteq \mathscr{C}^1$, $F \in \mathscr{A}(\Omega)$, and $F^{(p)} \in \mathscr{L}(\Omega, \alpha)$ $(\in \mathscr{B}(\Omega, \alpha))$ with N as an $\mathscr{L}(\Omega, \alpha)$-index (a $\mathscr{B}(\Omega, \alpha)$-index). Then $F \in \mathscr{L}(\Omega, \alpha)$ $(\in \mathscr{B}(\Omega, \alpha))$ with $N + p$ as an $\mathscr{L}(\Omega, \alpha)$-index (a $\mathscr{B}(\Omega, \alpha)$-index).*

Proof. The theorem is an obvious consequence of (3.4) and the essentially known (see [13, Chapter II])

Lemma 3.1. *Let p, Ω, F be as in Theorem 3.3, $z \in \Omega$, and let $\alpha \in \Omega_z$ (see the definition of v). Then*

$$v(\alpha, z, F) \leq v(\alpha, z, F^{(p)}) + p.$$

4. The Theory in $\mathscr{C}^k$

In this section we follow the notation and conventions mentioned in Section 3 without further clarification. Our present attempts are mainly oriented toward discussing some properties of the solutions of linear partial differential equations in $\mathscr{C}^k$. In trying to conclude that a solution F, for example, $\in \mathscr{L}(\Omega, \alpha)$ we can often ignore a bounded part of Ω because of (cf. [8, Theorem 25, Section 5])

Theorem 4.1. *Let Ω be bounded in $\mathscr{C}^k$. Let $\delta \in \mathscr{C}^{k+}$, $F \in \mathscr{A}(\Omega)$, and let F admit an analytic continuation (not necessarily single valued) to $\Omega(\delta)$. Then $F \in \mathscr{L}(\Omega, \alpha)$ for every $\alpha \ll \delta$ ($\alpha \in \mathscr{C}^{k+}$).*

Proof. Let us assume that the conclusion is false. This leads to the existence of an $\alpha \in \mathscr{C}^{k+}$ and $\ll \delta$ such that $F \notin \mathscr{L}(\Omega, \alpha)$, which implies the existence of a net $(z(N), N \in I)$ in Ω and a net $(q(N), N \in I)$ in I, both corresponding to the directed set $(I, \geqq)$ (see [6]) such that

$$(4.1) \quad \begin{cases} q(N) \quad \text{ is not } \leqq N \text{ or } N \in I, \ , \\ T_{q(N)}(z(N), F)\alpha^{q(N)} \geqq \max_{n \leqq N} T_n(z(N), F)\alpha^n \quad \text{for} \quad N \in I. \end{cases}$$

Since Ω is bounded the net $(z(N), N \in I)$ admits a limit point, say a, in the closure of Ω. Let $\beta \in \mathscr{C}^{k+}$ and be such that $\alpha \ll \beta \ll \delta$. By hypothesis there exists an analytic continuation G of F into the closed polydisk $D = [z: z \in \mathscr{C}^k, |z - a| \leqq \beta]$.

Let us fix an $m \in I$ and let $h \in \mathscr{C}^{1+}$. Now there exists a $\gamma \in \mathscr{C}^{k+}$ such that $\alpha + 2\gamma \ll \beta$ and an infinite subset J of I such that for $N \in J$,

$$m \leqq N, \ |z(N) - a| < \gamma \quad \text{and} \quad T_m(a, G) - h \leqq T_m(z(N), F).$$

The last expression is, by (4.1),

$$\leqq T_{q(N)}(z(N), G).$$

which, by Cauchy's formula (see [3, Chapter 1]),

$$\leqq M\alpha^{q(N)}/(\alpha + \gamma)^{q(N)},$$

where $M = \max_{z \in D} |G(z)|$. Thus

$$T_m(a, G) - h \leqq M\theta^{\left(\Sigma_{j=1}^k q_j(N)\right)},$$

for all $N \in J$, where $\theta = \max_{1 \leqq j < k} \alpha_j/(\alpha_j + \gamma_j) < 1$. Hence by (4.1) and the arbitrariness of $h \in \mathscr{C}^{1+}$ and $m \in I$, it follows that G is identically zero in D and hence that F is identically zero in Ω, which contradicts our assumption. Hence the theorem.

We are now in a position to prove (cf. Theorem 3.2 of Section 3)

Theorem 4.2. *Let $\alpha, \beta \in \mathscr{C}^{k+}$ and let $\alpha \ll \beta$. Then (i) $\mathscr{B}(\Omega, \alpha) \supseteq \mathscr{L}(\Omega, \alpha)$ and (ii) $\mathscr{L}(\Omega, \alpha) \supseteq \mathscr{B}(\Omega, \beta)$ if Ω is bounded.*

Proof. The first part is obvious and (ii) follows from Theorem 4.1 and

Lemma 4.1. *Let* $\delta \in \mathscr{C}^{k+}$, $S \subseteq \mathscr{C}^{k+}$, δ *be a point or a limit point of S and let $F \in \mathscr{B}(\Omega, \alpha)$ for every $\alpha \in S$. Then F admits an analytic continuation into $\Omega(\delta)$ (which may not be single valued).*

Proof. By Taylor's theorem and the Abel-property of the power series (see [3, Theorem 3.3]) it follows that F admits analytic continuation into $\Omega(\alpha)$ for any $\alpha \in S$ and this implies the lemma.

We need some relations between the spaces $\mathscr{L}_w(\Omega, \alpha)$ and $\mathscr{L}(\Omega, \alpha)$. It is obvious that $\mathscr{L}_w(\Omega, \alpha) \supseteq \mathscr{L}(\Omega, \alpha)$ and in the other direction we prove

Theorem 4.3. *Let* $\alpha \in \mathscr{C}^{k+}$, $N \in I^k$ *and let $N_{(j)}$ $(1 \leqq j \leqq k)$ denote the point of I^k whose jth coordinate is N_j and whose other coordinates are all zeros. Let $F \in \mathscr{L}_w(\Omega, \alpha)$ with each $N_{(j)}$ $(1 \leqq j \leqq k)$ as an $\mathscr{L}_w(\Omega, \alpha)$-index. Then $F \in \mathscr{L}(\Omega, \alpha)$ with N as an $\mathscr{L}(\Omega, \alpha)$-index.*

Proof. Let us fix a $z \in \Omega$ and assume that the conclusion is false, that is, that (3.2) is false for some m not $\leqq N$; choose a minimal m for which this is the case. Since m is not $\leqq N$, there exists a j $(1 \leqq j \leqq k)$ such that $m > N_{(j)}$. Hence by the hypothesis,

$$T_m(z, F)\alpha^m < \max_{n < m} T_n(z, F)\alpha^n,$$

which, by the minimality of m, implies (3.2) for the particular z and m contrary to the implication of our assumption. Hence the theorem.

Remark 4.1 Using the minimality principal, as in Theorem 4.3, it is easy to, get some criteria for a function to belong to either $\mathscr{B}_w(\Omega, \alpha)$ or $\mathscr{L}_w(\Omega, \alpha)$ $(\alpha \in \mathscr{C}^{k+})$, which are more conveniently verified in practice. For example $F \in \mathscr{B}_w(\Omega, \alpha)$ with N as a $\mathscr{B}_w(\Omega, \alpha)$-index, iff $F \in \mathscr{A}(\Omega)$ and

$$T_m(z, F)\alpha^m \leqq \max_{n < m \text{ but not } > N} \{T_n(z, F)\alpha^n\} \quad \text{for} \quad z \in \Omega, m > N.$$

We now proceed to give some conditions under which the solutions of a linear differential equation belong to $\mathscr{L}_w(\Omega, \alpha)$. We first prove

Theorem 4.4. *Let* $\delta \in \mathscr{C}^{k+}$, $p \in I^k$ *and let A_n be functions analytic and bounded in $\Omega(\delta)$ for $n < p$ $(n \in I)$. Let $F \in \mathscr{A}(\Omega)$ and satisfy the differential equation*

$$F^{(p)}(z) + \sum_{n < p} A_n(z)F^{(n)}(z) = 0 \quad \text{for} \quad z \in \Omega.$$

Then $F \in \mathscr{L}_w(\Omega, \alpha)$ for every $\alpha \in \mathscr{C}^{k+}$ and $\ll \delta$.

Proof. Let $\alpha, \beta \in \mathscr{C}^{k+}$ and be such that $\alpha \ll \beta \ll \delta$. Let $h \in I^k$. The hypothesis together with Cauchy's inequality (see [3]) implies the existence of a positive real M such that

$$|A_n^{(t)}(z)| \leq t! \, M/\beta^t \quad \text{for} \quad z \in \Omega, \, t \in I.$$

Also using Leibnitz's formula (see [2, Section 8.13.2]) and the hypothesis on F, we get that for $z \in \Omega$,

$$F^{(p+h)}(z) + \sum_{n<p} \sum_{t \leq h} {}^h C_t \, A_n^{(t)}(z) F^{(n+h-t)}(z) = 0,$$

where (and hereafter)

$$
{}^h C_t = \prod_{j=1}^{k} {}^{h_j} C_{t_j}.
$$

Hence for $z \in \Omega$,

$$T_{p+h}(z, F)\alpha^{p+h} \leq \sum_{n<p} \sum_{t \leq h} \left[{}^h C_t \, \frac{t! \, M \alpha^t}{\beta^t} \, T_{n+h-t}(z, F)\alpha^{n+h-t} \, \frac{(n+h-t)!}{(p+h)!} \, \alpha^{p-n} \right]$$

$$\leq H \sum_{n<p} \sum_{t \leq h} \left[{}^h C_t(t!) \, \frac{(n+h-t)!}{(p+h)!} \, \frac{\alpha^t}{\beta^t} \right],$$

where

$$H = H(z) = M \left[\max_{m \leq p} \alpha^m \right] \left[\max_{m < p+h} T_m(z, F)\alpha^m \right].$$

Let $q(s) \in \mathscr{C}^k$, for $1 \leq s \leq k$ and be defined by

$$q(s)_j = \begin{cases} p_j - 1 & \text{if } j = s, \\ p_j & \text{otherwise.} \end{cases}$$

It now follows that for $z \in \Omega$,

$$T_{p+h}(z, F)\alpha^{p+h} \leq \sum_{t \leq h} H \frac{\alpha^t}{\beta^t} \left[\sum_{s=1}^{k} \sum_{n \leq q(s)} {}^h C_t(t!) \, \frac{(h+n-t)!}{(h+p)!} \right]$$

$$\leq H \sum_{t \leq h} \frac{\alpha^t}{\beta^t} \left[\sum_{s=1}^{k} \frac{1}{h_s + p_s} \, ({}^h C_t)(t!) \, \frac{(h+q(s)-t)!}{(h+q(s))!} \right.$$

$$\left. \times \left\{ \sum_{n \leq q(s)} \frac{(h+n-t)!}{(h+q(s)-t)!} \right\} \right]$$

$$\leq H \sum_{t \leq h} \left[\frac{\alpha^t}{\beta^t} \sum_{s=1}^{k} \frac{1}{h_s + p_s} \, e^k \right]$$

$$\leq H \theta^k \left\{ \prod_{j=1}^{k} \frac{\beta_j}{\beta_j - \alpha_j} \right\} \sum_{s=1}^{k} \frac{1}{h_s + p_s},$$

and this implies the theorem.

We now consider nonhomogeneous equations.

Theorem 4.5. *Let* $\delta \in \mathscr{C}^{k+}$, $q \in I^k$ *and* $B_n \in \mathscr{A}(\Omega(\delta))$, *for* $n \leq q$ $(n \in I)$. *Let* B_q *be nonvanishing in* $\Omega(\delta)$ *and let the quotients* B_n/B_q *for* $n < q$ *and* $B_q^{(1(j))}/B_q$ *for* $1 \leq j \leq k$ *be all bounded in* $\mathscr{A}(\Omega(\delta))$, *where* $1(j)$ *stands for the element of* I *whose jth coordinate is 1 while the others are zeros. Let F be as in Theorem 4.4. Let* $G \in \mathscr{A}(\Omega)$ *and satisfy*

$$\sum_{n \leq q} B_n(z) G^{(n)}(z) = F(z) \qquad \text{for} \quad z \in \Omega.$$

Then $G \in \mathscr{L}_w(\Omega, \alpha)$ *for every* $\alpha \ll \delta (\alpha \in \mathscr{C}^{k+})$.

It is convenient to prove first

Lemma 4.2. *Let* β, $\gamma \in \mathscr{C}^{k+}$ *and let* $\beta \ll \gamma$. *Let* P, $Q \in \mathscr{A}(\Omega(\gamma))$. *Let* Q *be nonvanishing in* $\Omega(\gamma)$ *and let* P/Q *and* $Q^{(1(j))}/Q$ *for some* $1 \leq j \leq k$, *be bounded in* $\Omega(\gamma)$. *Then* $P^{(1(j))}/Q$ *is bounded in* $\Omega(\beta)$.

Proof. We have in $\Omega(\gamma)$

$$P^{(1(j))}/Q = (P/Q)^{(1(j))} + (P/Q)(Q^{(1(j))}/Q).$$

By using Cauchy's formulas for derivatives (see [3]) it is easily seen that $(P/Q)^{(1(j))}$ is bounded in $\Omega(\beta)$ and the lemma follows.

Proof of Theorem 4.5. Let $\alpha \in \mathscr{C}^{k+}$. Let $\beta^{(t)} \in \mathscr{C}^{k+}$ for $t \in I^1$ and be such that $\beta^{(t+1)} \ll \beta^{(t)} \ll \delta (t \in I^1)$ and further be such that $\beta^{(t)} \to \beta \gg \alpha$ as $t \to +\infty$. By a repeated application of Lemma 4.2 we get that the functions $B_n^{(m)}/B_q$ for $n \leq q$, $m \leq p$ $(m, n \in I^k)$ are all bounded in $\Omega(\beta)$. From the relations satisfied by F and G, using Leibnitz's formula (as in the proof of Theorem 4.4) we get that in Ω(with $A_p \equiv 1$)

$$(4.2) \qquad \sum_{n \leq q} \sum_{m \leq p} \sum_{t \leq m} {}^m C_t A_m B_n^{(t)} G^{(n+m-t)} = 0.$$

Now by virtue of our observation about the functions $B_n^{(m)}/B_q$, the equation reduces to one of the form

$$G^{(p+q)} + \sum_{n < p+q} D_n G^{(n)} = 0,$$

where $D_n (n < p+q)$ are functions analytic and bounded in $\Omega(\beta)$. Hence by Theorem 4.4 $G \in \mathscr{L}(\Omega, \alpha)$.

The coefficient function F of the equation of Theorem 4.5 can perhaps be more general, at least in some specialized equations, as is illustrated by an extension of Theorem 3.3, viz.,

Theorem 4.6. *Let* $\alpha \in \mathscr{C}^{k+}$, $p \in I^k$, $G \in \mathscr{A}(\Omega)$ *and let*

$$G^{(p)} = F \in \mathscr{B}_w(\Omega, \alpha)(\in \mathscr{L}_w(\Omega, \alpha))$$

with N *as a* $\mathscr{B}_w(\Omega, \alpha)$*-index (an* $\mathscr{L}_w(\Omega, \alpha)$*-index). Then* $G \in \mathscr{B}_w(\Omega, \alpha)(\in \mathscr{L}_w(\Omega, \alpha))$
with $N + p$ *as a* $\mathscr{B}_w(\Omega, \alpha)$*-index (an* $\mathscr{L}_w(\Omega, \alpha)$*-index).*

Proof. For any $m \geq N + p$ $(m \in I)$ and $z \in \Omega$, we have that

$$T_m(z, G)\alpha^m = \frac{(m-p)!\,\alpha^p}{m!} T_{m-p}(z, F)\alpha^{m-p}$$

$$\leqq (<) \frac{(m-p)!\,\alpha^p}{m!} \max_{n < m-p} T_n(z, F)\alpha^n$$

$$= \max_{n < m-p} \left[\frac{(m-p)!\,(n+p)!}{m!\,n!} T_{n+p}(z, G)\alpha^{n+p} \right]$$

$$\leqq \max_{u < m-p} T_{n+p}(z, G)\alpha^{n+p}$$

$$\leqq \max_{n < m} T_n(z, G)\alpha^n,$$

and hence the theorem.

The analogues of Theorem 4.6 for the space $\mathscr{B}(\Omega, \alpha)$ and $\mathscr{L}(\Omega, \alpha)$ are not true when $k > 1$ (see Example 4.2).

When the coefficients and partials of the equations of Theorems 4.4 and 4.5 are independent of some of the variables, we can give a more precise information about their solutions, which turns out to be a step towards considering whether the solutions of a system of such equations belong to $\mathscr{L}(\Omega, \alpha)$.

Theorem 4.7. *Let the hypothesis of Theorem 4.5 hold. Let J be a nonempty subset of the set of all positive integers $\leq k$. Let $A_n(z)$ $(n < p)$, $B_n(z)$ $(n \leqq q)$ depend only on $(z_j : j \in J)$ for $z \in \Omega$. Let $p_j = q_j = 0$ for $j \notin J$. Then $G \in \mathscr{L}_w(\Omega, \alpha)$ and admits an $\mathscr{L}_w(\Omega, \alpha)$-index $N = N(\alpha, G)$, which is such that $N_j = 0$ if $j \notin J$, for every $\alpha \ll \delta (\alpha \in \mathscr{C}^{k+})$.*

Proof. It is sufficient to consider the case when $F \equiv 0$ and $B_q \equiv 1$ in Ω, because of Equation 4.2, and this we do by essentially refining the proof of Theorem 4.4. We start with α, $\beta \in \mathscr{C}^{k+}$, such that $\alpha \ll \beta \ll \delta$. Since now $B_n^{(t)} \equiv 0$ in Ω $(n < q)$ whenever t is such that $t_j > 0$ for some $j \notin J$, we get as in the case of Theorem 4.4 that for $z \in \Omega$, $h \in I^k$

$$(4.3) \qquad T_{q+h}(z, G)\alpha^{q+h} \leqq K \sum_{n<q} \sum_{\substack{t \leqq h \\ t_j = 0 \text{ for } j \notin J}} \left[{}^h C_t(t!) \frac{(n+h-t)!\,\alpha^t}{(p+h)!} \frac{\alpha^t}{\beta^t} \right],$$

where $K = L[\max_{m < q + h} T_m(z, G)\alpha^m]$, $L = L(\alpha)$ being a positive real number. But the right side of (4.3), in the present context, reduces to

$$K \sum_{\substack{t \leq h \\ t_j = 0 \text{ for } j \notin J}} \frac{\alpha^t}{\beta^t} \left[\sum_{n < q} \prod_{j \in J} \left\{ {}^{h_j}C_{t_j}(t_j!) \frac{(h_j + n_j - t_j)!}{(h_j + q_j)!} \right\} \right],$$

which as in the proof of Theorem 4.4 is

$$\leq K \Delta \left[\sum_{s \in J} \frac{1}{h_s + q_s} \right],$$

where $\Delta = \Delta(\alpha, \beta)$ is a positive real number, and hence the theorem.

Remark 4.2 When $k > 1$, it appears in general necessary to consider a system of differential equations with more than one equation, to be able to say that all the solutions in Ω belong to $\mathscr{L}(\Omega, \alpha)$ (see also Example 4.2). Theorems 4.3 and 4.7, in particular, suggest some conditions under which one can arrive at such a conclusion. However these conditions limit the considerations to very specialized systems, and it is not clear whether it is possible to consider significantly more general systems.

Remark 4.3 We now indicate some more open problems. (i) It appears possible to extend the growth results, particularly Theorem 3.1, to the spaces $\mathscr{B}(\mathscr{C}^k, \alpha)$ and $\mathscr{B}_w(\mathscr{C}^k, \alpha)$, and in the case of the spaces $\mathscr{B}_w(\mathscr{C}^k, \alpha)$ one should only expect "partial growth results" (see Example 4.2). (ii) Theorem 4.6 suggests the question whether a derivative of a function of any of the spaces considered, belongs to the same space or some one of the spaces. The question is of interest particularly when $\Omega = \mathscr{C}^k$ and in certain other cases the answer is "no" (see Example 4.1). (iii) It is not apparent whether any of the spaces under consideration is an algebra or even a vector space over the complex number field, with the usual function-space-operations. However when $\Omega \neq \mathscr{C}^k$, the spaces $\mathscr{B}(\Omega, \alpha)$ and $\mathscr{L}(\Omega, \alpha)$ need not be algebras (see Example 4.1).

We finally give the examples to which we were referring at different places.

Example 4.1 Let $\Omega = [z : z \in \mathscr{C}^k$, real part of $z_j < 1$ for $1 \leq j \leq k]$. Let $\alpha = (1, \ldots, 1) \in \mathscr{C}^{k+}$. Let $q \in \mathscr{C}^{k+}$ and be with rational integral coordinates. Let $G = G_{(q)} \in \mathscr{A}(\Omega)$ and be defined by $G(z) = \prod_{j=1}^{k} (z_j - 2)^{-q_j}$. We then have for any $z \in \Omega$ and any $n \in I^k$

$$T_n(z, G)\alpha^n = \prod_{j=1}^{k} \frac{|q_j(q_j + 1) \cdots (q_j + n_j - 1)|}{n_j! \, |z_j - 2|^{q_j + n_j}}.$$

Now $G \in \mathscr{L}(\Omega, \alpha)$, if $q = \alpha$, particularly since $|z_j - 2| > 1$ for $1 \leq j \leq k$,

$z \in \Omega$. But G does not belong even to $\mathcal{B}_w(\Omega, \alpha)$, if $q > \alpha$. For otherwise there would exist an $N \in I^k$ such that

$$T_m(z, G) \leqq \max_{n < m} T_n(z, G) \qquad \text{for} \quad z \in \Omega, \, m > N,$$

which would lead to an absurdity, on taking limits as $z \to \alpha$ $(z \in \Omega)$.

Thus while $G_{(\alpha)} \in \mathcal{L}(\Omega, \alpha)$, none of its derivatives nor the product $G_{(\alpha)} G_{(\alpha)}$ belongs even to $\mathcal{B}_w(\Omega, \alpha)$ (see Remark 4.3). The function $G_{(2\alpha)}$ restricted to the polydisk $\Lambda = [z : z \in \mathscr{C}^k, |z| < \alpha]$ does not belong to $\mathcal{L}(\Lambda, \beta)$ for any β not $\ll \alpha (\beta \in \mathscr{C}^{k+})$ (cf. Theorem 4.1).

Example 4.2 Let $s \in I^1$ and be less than k. Let $G \in \mathcal{A}(\mathscr{C}^s)$ but not to $\mathcal{B}(\mathscr{C}^s, \beta)$ for any $\beta \in \mathscr{C}^{s+}$ (see Theorem 3.1). Let $F \in \mathcal{A}(\mathscr{C}^k)$ and be such that $F(z) = G(z_1, \ldots, z_s)$ for $z \in \mathscr{C}^k$. Then F does not belong to $\mathcal{L}(\mathscr{C}^k, \alpha)$ for any $\alpha \in \mathscr{C}^{k+}$, although it satisfies $F^{(p)}(z) = 0$ for $z \in \mathscr{C}^k$ and for any $p \in I^k$, with $p_k > 0$ (see Remarks 4.2, 4.3).

REFERENCES

1. R. P. Boas, Jr., *Entire functions*, Academic Press, New York, 1954.
2. J. Dieudonné, *Foundations of modern analysis*, Academic Press, New York, 1960.
3. B. A. Fuks, *Introduction to the theory of analytic functions of several complex variables*, Providence, Amer. Math. Soc., 1963.
4. E. Goursat, *A course in mathematical analysis*, translated by Hedrick and Dunkel, Vol. II, Part II, Dover, New York, 1959.
5. Fred Gross, Entire functions of bounded index, *Proc. Amer. Math. Soc.* **18** (1967) 974–980.
6. J. L. Kelley, *General topology*, Van Nostrand, Princeton, 1955.
7. B. S. Lee and S. M. Shah, An inequality involving the Bessel function and its derivatives, *J. Math. Anal. and Appl.*, **30** (1970), 144-156.
8. B. Lepson, Differential equations of infinite order, hyperdirichlet series and entire functions of bounded index, Lecture, Notes, Summer Institute on Entire Functions. Univ. of California, La Jolla, Calif., 1966.
9. V. Marić and S. M. Shah, Entire functions defined by gap power series and satisfying a differential equation, *Tōhoku Math. J.*, to appear.
10. W. J. Pugh and S. M. Shah, Entire functions with widely spaced zeros and of bounded index, *Pacific J. Math.*, to appear; abstract *Notices Amer. Math. Soc.* **19** (1968), 490.
11. S. M. Shah, Entire functions of bounded index, *Proc. Amer. Math. Soc.* **19** (1968), 1017–1022.
12. ———, Entire functions satisfying a linear differential equation, *J. Math. Mech.* **18** (1968–1969), 131–136.
13. G. Valiron, *Lectures on the general theory of integral functions*, Imprimerie et Librairie Édouard Privat, Librairie de l'Universite, Toulouse, 1923; Chelsea, New York, 1949.
14. H. Wittich, *Neuere untersuchungen über eindeutige analytische funktionen*, Springer-Verlag, Berlin, 1955.

(Received May 28, 1968)

The Minimum Modulus of Functions of Slow Growth in the Unit Disk

C. N. Linden

Department of Pure Mathematics
University College of Swansea
Wales

1. Introduction

If $f(z)$ is regular in the unit disk $C = \{z : |z| < 1\}$, let

$$\mu(r, f) = \min_{|z|=r} |f(z)|, \qquad M(r, f) = \max_{|z|=r} |f(z)|.$$

The aim of this paper is to examine the behavior of $\mu(r, f)$ in those cases where $M(r, f)$ is unbounded but does not increase very rapidly as $r \to 1 - 0$. To be more specific, and to put the problem in an appropriate context, we first recall some known results.

Let

$$(1.1) \qquad \alpha(f) = \limsup_{r \to 1-0} \frac{\log^+ \log^+ M(r, f)}{-\log(1 - r)},$$

where $\log^+ x$ denotes $\max(0, \log x)$. Then we have the following theorems due to Heins [5] and the author [6], respectively.

Theorem A. *If $f(z)$ is regular, nonconstant, and bounded in C, there exists a constant K such that*

(1.2)
$$\log \mu(r,f) > - \frac{K}{1-r},$$

for a sequence of numbers r increasing to 1.

Theorem B. If $f(z)$ is regular and nonconstant in C, and $\alpha(f) > 1$, then there exists a constant K such that

(1.3)
$$\log \mu(r,f) > - K \log M(r,f) \log \log M(r,f),$$

for a sequence of numbers r increasing to 1.

The inequality (1.2) of Theorem A is best possible, as is exemplified by the function $f(z) = \exp[1/(z-1)]$, which is regular in C and satisfies the conditions

$$\log \mu(r,f) = - \frac{1}{1-r}, \qquad \log M(r,f) = O(1) \qquad \text{as } r \to 1 - 0.$$

Nevertheless the theorem can be readily extended to apply (i) to functions which are of bounded characteristic in C, and (ii) to those functions which are regular and nonzero in C, and satisfy the condition $\alpha(f) < 1$. In particular the conclusion based on the hypothesis (i) follows from Theorem A and the standard factorization of functions of bounded characteristic, while the conclusion based on (ii) is an immediate consequence of a theorem of Cartwright [3]. We note further that conclusion (1.3) of Theorem B is also best possible in general when $\alpha(f) > 1$, although it can be sharpened when $\alpha(f)$ is infinite and $M(r,f)$ increases sufficiently rapidly as $r \to 1 - 0$ (see [6, 7]).

In this paper we consider the behavior of $\mu(r,f)$ when $\alpha(f) < 1$, and we prove the following theorem as our main result.

Theorem 1. Let $f(z)$ be regular and nonconstant in C, and suppose that $\alpha(f) < 1$. Then there exist positive constants K and L, and a number ρ_0 in $(0, 1)$, such that if $\rho \in (\rho_0, 1)$ the interval $(\rho, \frac{1}{2}(1 + \rho))$ contains a set of values r of measure at least $L(1 - \rho)$ for which

(1.4)
$$\log \mu(r, f) > - \frac{K}{1-r} \log\left(\frac{1}{1-r}\right).$$

The constant L need not be less than $\frac{1}{4}$.

In comparing Theorem A, under the alternative hypothesis (i), with Theorem 1, we first remark that the statement that $f(z)$ is of bounded characteristic in C is sufficient, but not necessary, for $\alpha(f)$ to belong to $[0, 1]$. Further, inequality (1.4) is not as sharp as inequality (1.2), and although we leave open the question as to whether (1.4) is best possible under the hypotheses of Theorem 1, we give improvements of the minimum modulus inequality represented by (1.4), by proving Theorems 3 and 4 under hypotheses which are more restrictive than those given in Theorem 1. Finally we note that Theorem 1 gives a stronger result concerning the set on which $\log \mu(r, f)$ has the ascribed lower bound than does Theorem A. However, under the particular hypothesis (ii), (1.2) is valid for all numbers r in some interval $(r_0, 1)$ of positive length.

Theorem 1 follows immediately from Theorem 2. For convenience we shall prove our main result in this latter form, thereby noting how the constant K of (1.4) depends on $f(z)$.

Theorem 2. *Suppose that $0 \leqq \alpha < 1$. Let $f(z)$ be regular in C, and suppose that $f(0) = 1$ and*

$$(1.5) \qquad \log M(r, f) < A(1 - r)^{-\alpha}, \qquad 0 \leqq r < 1.$$

Then there is a constant ρ_1 in $(0, 1)$ depending on α, and a constant K depending on A and α, such that if $\rho \in (\rho_1, 1)$ the interval $(\rho, \frac{1}{2}(1 + \rho))$ contains a set of numbers r of measure at least $\frac{1}{4}(1 - \rho)$ such that

$$\log \mu(r, f) > -\frac{K}{1 - r} \log \frac{1}{1 - r}.$$

Henceforth the symbol K will be used to denote positive constants. The value of K will depend on parameters which are indicated in the text, and the different values of K which arise will be distinguished by suffixes.

2. Preliminaries

The proof of Theorem 2 is based on some results of Hayman [4], but before stating these we must introduce some relevant notation. Let $F(w)$ be regular for $|w| \leq 1$, and let $n(w, t, F)$ denote the number of zeros of $F(\zeta)$ in $\{\zeta : |w - \zeta| \leqq t\}$, multiple zeros being counted according to multiplicity. Then set

$$N(w, h, F) = \int_0^h \frac{n(w, t, F)}{t} \, dt, \qquad U(w, h) = \log |F(w)| + N(w, h, F).$$

Hayman's results are stated as follows.

Lemma A. *Suppose that $\sigma > 0$ and $h > 0$. Let $F(w)$ be regular for $|w| \leqq \sigma + h$, and suppose that $n(w, h, F) \leqq n_0$ for $|w| = \sigma$. Then there exists an absolute constant K_1 and a set of numbers s of measure of least $\frac{1}{4}h$ such that $\sigma < s < \sigma + \frac{1}{2}h$ and*

$$N(w, \tfrac{1}{2}h, F) \leqq n_0 \log \frac{K_1(\sigma + h)}{h},$$

for $|w| = s$.

Lemma B. *Suppose that $F(w)$ is regular for $|w| \leqq 1$, $0 < s < 1$, $0 < \delta < \tfrac{1}{2}$, and $h = \delta(1 - s)$. Then if $F(0) \neq 0$ there exists a number K_2 depending on δ such that*

$$|U(s, h)| \leqq \frac{K_2}{1 - s} \int_0^{2\pi} \left\{ \log^+ |F(e^{i\psi})| + \log^+ \frac{1}{|F(0)|} \right\} d\psi,$$

$$n(s, h, F) \leqq \frac{1 + 2\delta}{2\pi(1 - s)} \int_0^{2\pi} \left\{ \log^+ |F(e^{i\psi})| + \log^+ \frac{1}{|F(0)|} \right\} d\psi.$$

Lemma A, and Lemma B in the case where $F(0) = 1$, have been proved by Hayman. The generalization of the latter lemma which appears here is obtained immediately from Hayman's result.

3. A Study of Auxiliary Functions

We will consider a mapping of C defined by

(3.1) $w = z - D(1 - z)^\beta$, $0 < D \leqq \tfrac{1}{12}$, $1 < \beta < 3$,

where we take that branch of $(1 - z)^\beta$ which is positive when $0 < z < 1$. This mapping has been applied elsewhere [8], where it has been established that any function $f(z)$ which is regular in C is transformed by (3.1) into a function $F(w)$ which is regular for $|w| \leqq 1$ except possibly at $w = 1$. If $f(z)$ satisfies the hypothesis of Theorem 2 then we have also

(3.2) $\displaystyle\int_0^{2\pi} \log^+ |F(\sigma e^{i\psi})| \, d\psi < K_3(\alpha, \beta, D, A),$

when $\alpha\beta < 1$, $0 < \sigma \leqq 1$.

In the following lemma we compare the general behavior of the function $f(z)$ with that of

(3.3) $F(w) = F(z - D(1 - z)^\beta) = f(z)$.

Lemma 1. *If $f(z)$ and $F(w)$ are related by (3.3), where $0 < D \leqq \tfrac{1}{12}$, $1 < \beta < 3$, set*

(3.4) $u(z, h) = \log |f(z)| + N(z, h, f)$, $U(w, h) = \log |F(w)| + N(w, h, F)$.

Then there exists a constant r_1, depending on D and β, such that $0 < r_1 < 1$ and

(3.5) $$n(s, \tfrac{1}{4}h, F) \leq n(r, \tfrac{1}{2}h, f) \leq n(s, h, F),$$

(3.6) $$U(s, \tfrac{1}{4}h) \leq u(r, \tfrac{1}{2}h) \leq U(s, h),$$

when $r_1 < r < 1$, $0 < h < 1 - r$, and

(3.7) $$s = r - D(1 - r)^{\beta}.$$

Let $w = z - D(1 - z)^{\beta}$. Then

(3.8) $$w - s = z - r + D\{(1 - r)^{\beta} - (1 - z)^{\beta}\}.$$

Now if $|z - r| = \tfrac{1}{2}t < \tfrac{1}{2}(1 - r)$, we have

$$(1 - r)^{\beta} - (1 - z)^{\beta} = \int_0^1 \beta\{(1 - r) + u(r - z)\}^{\beta - 1}(z - r)\, du,$$

so that

$$|(1 - r)^{\beta} - (1 - z)^{\beta}| \leq \beta|z - r|(\tfrac{3}{2}(1 - r))^{\beta - 1}.$$

Therefore it follows from (3.8) that

$$|w - s| = |z - r|\{1 + H(r, z)\},$$

where $|H(r, z)| \leq \beta D(\tfrac{3}{2}(1 - r))^{\beta - 1} \to 0$ as $r \to 1 - 0$. Hence there exists a constant $r_1(\beta, D)$ in $(0, 1)$ such that

(3.9) $$\tfrac{1}{4}t < |w - s| < t,$$

whenever $|z - r| = \tfrac{1}{2}t$, $r_1(\beta, D) < r < 1$, and $0 < t < 1 - r$.
It follows from (3.9) that

(3.10) $$n(s, \tfrac{1}{4}t, F) \leq n(r, \tfrac{1}{2}t, f) \leq n(s, t, F),$$

and (3.5) follows by replacing t by h. Dividing the terms of (3.10) by t and integrating we deduce

$$N(s, \tfrac{1}{4}h, F) \leq N(s, \tfrac{1}{2}h, f) \leq N(s, h, F),$$

and inequalities (3.5) follow from (3.4) and the fact that $f(r) = F(s)$.

In the proof of Theorem 2 it will be convenient to make certain additional assumptions about the function $f(z)$ which are not stated in the hypothesis of that theorem. The following lemma shows that the required assumptions can be made later without any loss of generality.

Lemma 2. *Suppose $0 < \alpha < 1$. Let $f(z)$ be regular in C, $f(0) = 1$, and let*

$$\log |f(z)| < \frac{A}{(1 - |z|)^\alpha}, \qquad |z| < 1.$$

If $r_0 \in (0, 1)$ let $a_1, a_2, \ldots, a_m$ denote the set of zeros of $f(z)$ in

$$\{z : |z| \le \tfrac{1}{2}(1 + r_0)\},$$

each zero being counted according to the appropriate multiplicity. Then

(3.11)
$$f(z) = f_1(z) \prod_{n=1}^{m} \left(1 - \frac{z}{a_n}\right),$$

where $f_1(z)$ is regular in C, $f_1(0) = 1$, and

(3.12)
$$\log |f_1(z)| < \frac{AK_4(r_0, \alpha)}{(1 - |z|)^\alpha}, \qquad |z| < 1,$$

(3.13)
$$|f_1(z)| > AK_5(r_0, \alpha), \qquad |z| \le r_0,$$

(3.14) $\log \mu(r, f) > AK_6(r_0, \alpha) + \log \mu(r, f_1), \quad \tfrac{1}{4}(3 + r_0) \le |z| < 1.$

The regularity of $f_1(z)$ and the condition that $f_1(0) = 1$ are immediate.

By a standard deduction from Jensen's formula (see [2], Theorem 5), we have

(3.15)
$$m \le \frac{A(\tfrac{1}{4}(1 - r_0))^{-\alpha}}{\log\left(\dfrac{3 + r_0}{2 + 2r_0}\right)} = AK_7(r_0, \alpha).$$

Therefore if $\tfrac{1}{4}(3 + r_0) \le |z| < 1$, we obtain

(3.16)
$$\prod_{n=1}^{m} \left|1 - \frac{z}{a_n}\right| \ge (\tfrac{1}{4}(1 - r_0))^m,$$

and (3.12) follows from (3.11),

Since $f_1(z)$ has no zeros in $\{z : |z| < \tfrac{1}{2}(1 + r_0)\}$ we can deduce (3.13), for example, by application of Theorem II, [3] to $\log |f_1(z)|$.

Finally we note that (3.14) is a consequence of (3.11), (3.15), and (3.16).

4. Proof of Theorem 2

For the proof of Theorem 2 we set $D = \frac{1}{12}$ and $\beta = \min(2, \tfrac{1}{2}(1 + 1/\alpha))$. Let $z = r_0$ be that zero of $z - D(1 - z)^\beta$ which lies in $(0, 1)$. Then r_0 depends only on α, and $z = r_0$ is mapped by (3.1) onto the point $w = 0$.

Lemma 2 shows that in proving Theorem 2 we may assume without any loss of generality that

(4.1)
$$\log |f(z)| < \frac{AK_4(r_0, \alpha)}{(1 - |z|)^\alpha}, \qquad |z| < 1,$$

(4.2)
$$|f(z)| > AK_5(r_0, \alpha), \qquad |z| \le r_0.$$

Let $f(z)$ be transformed by (3.3) into $F(w)$. Then (3.2) takes the form

$$\int_0^{2\pi} \log^+ |F(\sigma e^{i\psi})| \, d\psi < K_8(\alpha, A),$$

for $0 < \sigma \le 1$. Since $F(w)$ is continuous in $|w| \le 1$, except possibly at the single point $w = 1$, we can apply Lemma B with $\delta = \frac{1}{2}$. Noting that inequality (4.2) implies $|F(0)| = |f(r_0)| > K_5(r_0, \alpha, A)$, we obtain

$$|U(s, h)| \le \frac{K_9(\alpha, A)}{1 - s}, \qquad n(s, h, F) \le \frac{K_{10}(\alpha, A)}{1 - s},$$

for $h = \frac{1}{2}(1 - s)$, $s = r - D(1 - r)^\beta$. Clearly we have $1 - s > 1 - r$ and, for any θ in $[0, 2\pi)$, $f(ze^{i\theta})$ satisfies exactly the same hypothesis as $f(z)$. Hence, by applying Lemma 1, we obtain

(4.3)
$$u(re^{i\theta}, \tfrac{1}{2}k) \ge \frac{-K_9(\alpha, A)}{1 - r}, \qquad n(re^{i\theta}, \tfrac{1}{2}k, f) \le \frac{K_{10}(\alpha, A)}{1 - r},$$

for $k = \frac{1}{2}(1 - r)$, $r_1 < r < 1$.

Since $n(re^{i\theta}, \frac{1}{2}k, f)$ denotes the number of zeros of $f(z)$ in $\{z : |z - re^{i\theta}| \le \frac{1}{2}k\}$ we can, by applying the second of the inequalities (4.3) at a suitable number of points, cover $\{z : |z - re^{i\theta}| < k\}$ with a set of circles in each of which $f(z)$ has at most $3K_{10}(\alpha, A)/(1 - r)$ zeros. The required number of such circles is independent of both r and θ, so that we obtain

(4.4)
$$n(re^{i\theta}, k, f) \le \frac{K_{11}(\alpha, A)}{1 - r}, \qquad \tfrac{1}{2}(1 + r_1) < r < 1.$$

The application of Lemma A now shows that if $\rho \in (\frac{1}{2}(r_1 + 1), 1)$ then

(4.5)
$$N(re^{i\theta}, \tfrac{1}{2}k, f) \le \frac{K_{12}(\alpha, A)}{1 - r} \log \frac{1}{1 - r}, \qquad 0 \le \theta < 2\pi,$$

on a set of r in $(\rho, \frac{1}{2}(\rho + 1))$ of measure at least $\frac{1}{4}(1 - \rho)$. The proof of Theorem 2 is now completed with reference to (3.4) and (4.3), and Theorem 1 follows.

As a consequence of Theorem 1, we note the following corollary which can be easily deduced.

Corollary. *Let $\mathcal{E}_\rho$ denote the set of values r on $(0, \rho)$ such that (1.4) holds for the value of K determined in the proof of Theorem 1. Then*

(4.6)
$$\liminf_{\rho \to 1-0} \left\{\frac{\int_{\mathcal{E}_\rho} \frac{1}{1-t}\,dt}{-\log(1-\rho)}\right\} \geqq \frac{1}{2}\,.$$

For the proof of the corollary let m be a fixed positive number such that $2^m(1 - \rho_0) > 1$, and let

$$E_n = [1 - 2^{1-n}, 1 - 2^{-n}], \quad F_n = [1 - 2^{1-n}, 1 - 3.2^{-1-n}].$$

Then if $N > m$ and $\rho \in E_{N+1}$, we have

$$\int_{\mathcal{E}_\rho} \frac{1}{1-t}\,dt \geqq \sum_{n=m+1}^{N} \int_{F_n} \frac{1}{1-t}\,dt$$

$$\geqq \tfrac{1}{2}\sum_{n=m}^{N-1} \int_{E_n} \frac{1}{1-t}\,dt$$

$$= \tfrac{1}{2}\log 2^{N-m}$$

$$> \tfrac{1}{2}\log\left(\frac{2^{-m-1}}{1-\rho}\right).$$

Hence (4.6) is valid.

5. Further Remarks

In Section 1 it was demonstrated that inequality (1.2) could not be improved in general, even for bounded regular functions. It is therefore of interest to consider under what conditions $\log|f(z)|$ has a lower bound which is $O(1/(1 - |z|))$ on a set of z which approaches the boundary of the disk C.

We now present two further theorems. The first of these concerns a generalization of hypothesis (ii). Inequality (1.4) can be improved in Theorem 1 only if the bound to $N(re^{i\theta}, \tfrac{1}{2}h, f)$ given by (4.5) can be improved. This could be achieved by replacing (4.4) by an appropriate assumption concerning $n(re^{i\theta}, \tfrac{1}{2}h, f)$. Thus the proof of Theorem 1 can be readily amended to give the following theorem.

Theorem 3. *Let $f(z)$ be regular and nonconstant in C, and suppose that $\alpha(f) < 1$. If there is a number r_0 in $(0, 1)$ and a constant B such that*

(5.1)
$$n(re^{i\theta}, \tfrac{1}{2}(1 - r), f) < \frac{-B}{(1 - r)\log(1 - r)}, \qquad 0 \leqq \theta < 2\pi, \ r_0 \leqq r < 1,$$

there are positive constants K and L, and a number ρ_0 in $(0, 1)$, such that if $\rho \in (\rho_0, 1)$ then the interval $(\rho, \frac{1}{2}(1 + \rho))$ contains a set of values r of measure at least $L(1 - \rho)$ for which

$$(5.2) \qquad \qquad \log \mu(r, f) > -\frac{K}{1 - r}.$$

The constant L need not be less than $\frac{1}{4}$.

It is clear that hypothesis (ii) of Section 1 implies that $n(re^{i\theta}, \frac{1}{2}(1 - r), f)$ is equal to zero when $0 \leqq \theta < 2\pi, 0 < r < 1$ so that (5.2) is satisfied if (ii) holds. However the assumption (5.1) requires a rather detailed knowledge of the distribution of the zeros of $f(z)$. On one hand (5.1) is implied by

$$n(0, r, f) < \frac{-\frac{1}{2}B}{(1 - r) \log (1 - r)}, \qquad r_0 \leqq r < 1.$$

On the other hand, although (5.1) is not true in general, for bounded regular functions, it is possible for (5.1) and

$$n(0, r, f) \neq O((1 - r)^{-\beta}), \qquad r \to 1 - 0,$$

to hold simultaneously for any constant β in $(0, 2)$.

Finally, we state a minimum-modulus theorem for those function-values that are taken on chords of C.

Theorem 4. *Let $f(z)$ be regular and nonconstant in C, and let $\alpha(f) < 1$. If d is a line in C with an end point at $e^{i\theta}$ then for some constant K*

$$\limsup_{z \in d, z \to e^{i\theta}} (1 - |z|)\log |f(z)| > -K.$$

Theorem 4 could be proved, for example, by mapping C by means of the function

$$(5.3) \qquad \qquad w = (1 - ze^{-i\theta})^{-\beta},$$

where β is a number in $(\alpha(f), 1)$ which is chosen so that (5.3) maps d onto a straight line in the right half of the w-plane. For some positive number R_0 the function

$$\Phi(w) = f(1 - w^{-\beta}) = f(z)$$

is regular in the set

$$\{w : |w| > R_0, |\arg w| \leqq \tfrac{1}{2}\pi\},$$

and $|\Phi(w)| < K|w|^{\alpha(f)/\beta}$. The application of a well-known type of minimum-modulus theorem for functions regular in an angle (see, for example, [1], Theorem 10.4.1) yields a property of $\Phi(w)$ which implies Theorem 4.

REFERENCES

1. R. P. Boas, Jr., *Entire functions*, Academic Press, New York, 1954.
2. M. L. Cartwright, *Integral functions*, Cambridge University Press, Cambridge, England, 1956.
3. ———, On analytic functions regular in the unit circle, *Quart. J. Math. Oxford* **4** (1933), 246–257.
4. W. K. Hayman, The minimum modulus of large integral functions, *Proc. London Math. Soc.* (3) **2** (1952), 469–512.
5. M. Heins, The minimum modulus of a bounded analytic function, *Duke Math. J.* **14** (1947), 179–215.
6. C. N. Linden, The minimum modulus of functions regular and of finite order in the unit circle, *Quart. J. Math. Oxford Ser* 2 **7** (1956), 196–216.
7. ———, Minimum modulus theorems for functions regular in the unit circle, *Quart. J. Math. Oxford Ser.* 2 **12** (1961), 1–16.
8. ———, The distribution of the zeros of regular functions, *Proc. London Math. Soc.* (3) **15** (1965), 301–322.

(Received October 15, 1967)

On the "Pits Effect" for Standard Integral Functions of Finite Nonzero Order

J. E. Littlewood

University of Wisconsin
Madison, Wisconsin 53706
U.S.A.
and
Trinity College
Cambridge,
England

1.

Littlewood and Offord showed [1] that if, starting with any integral function of finite nonzero order, we introduce a random factor ± 1, then "most" of the functions $f = \sum \pm c_n z^n$ are exponentially large except in exponentially small circles around the zeros. If an ordinate $|f(z)|$ is erected at the point z the resulting surface is an exponentially rapidly rising bowl, with exponentially narrow pits going down to the bottom. They described this as the "pits effect."

Y. M. Chen and I [2] have made a study of what factors $\exp[i\Lambda(n)]$, with $\Lambda(n) \geqslant n \log n$, produce a pits effect when applied to smooth functions $f_0 = \exp[-\lambda(n)]z^n$ of various kinds. Our results are as follows.

Functions $f_0(z)$

1. $\sum e^{n^a} z^n \ (0 < a < 1)$: function in $|z| < 1$ of high order.
2. $\sum \exp(-n \log \log n) z^n$
3. $\sum \exp(-n \log^a n) z^n \quad (0 < a < 1)$ $\rho = \infty$
4. $\sum \exp(-n \log n / \log \log n) z^n$
5. $\sum \exp(-n \log n / \rho) z^n \quad (\rho > \frac{1}{2})$
   (The reason for the $\rho > \frac{1}{2}$ will appear below.)

$\Lambda(n)$

(a) $cn \log n$ $(0 < c < \infty)$
(b) $n \log^\beta n$ $(\beta > 1)$
(c) n^{1+b} $(0 < b < \tfrac{1}{2})$

We find the pits effect in every combination of (1) to (5) with (a), (b), (c), except that (i) in the combination (c), (1) we have to assume $b < \tfrac{1}{2}a$, (ii) pits definitely do not exist in the marginal combination (5), (a) unless $c > c_0(\rho)$.

I have proved [3] further that the factor $\exp(i\alpha\pi n^2)$, with $\alpha = \dfrac{1}{1 +} \dfrac{1}{1 +} \cdots =$ $\tfrac{1}{2}(\sqrt{5} - 1)$, produces a pits effect when applied to any smooth enough integral function with positive coefficients, of whatever order (zero, finite, or infinite). (An earlier paper [4] by Nassif proved this for $\rho < \rho_0$, where ρ_0 is a certain absolute constant, $\rho_0 > 1$.) The gap $n^{1+b}(b < \tfrac{1}{2})$ to $\alpha\pi n^2$ is very interesting.

Chen's and my method (much the same in all cases) depends on a number of conditions being satisfied, and this is the reason for some of the gaps. One vital condition was

(1.1) $$\Lambda(n) \succ \lambda(n).$$

Now Watson [5] has given asymptotic formulas for the function

(1.2) $$\sum \frac{z^n}{\Gamma(1 + n\lambda)}, \qquad \lambda = k + i\beta, \qquad k = 1/\rho,$$

which is effectively

$$\sum \frac{z^n}{\Gamma(1 + n/\rho)} e^{-i\beta n \log n}.$$

(He considers functions with a further factor $\phi(n)(n + a)^{-b}$, where $\phi(n)$ is analytic and satisfies certain asymptotic conditions; and Maitland-Wright [6] considers still further elaborations. The simplest case $\phi(n) = 1$, $b = 0$ will serve our purpose.)

Watson remarks that his formulas should be useful for estimating the zeros of f, but he does not carry this out.

Watson's results show at once that when $\beta^2 + k^2 < 2k$, for example, when ρ is large and β is not, $f(z) = O(1)$ in a finite proportion of the plane, so that no question of a pits effect arises. This fits in with the $\Lambda(n) \succ \lambda(n)$ condition (1.1); we are in the marginal case $\Lambda(n) \asymp \lambda(n)$, and the ratio $\Lambda(n)/\lambda(n)$ is too small for a pits effect.

What I do in the present note is to evaluate the zeros of (1.2), and to show that there is a pits effect if

(1.3) $$\beta^2 + k^2 > 2k$$

(we have seen that there is none when $\beta^2 + k^2 < 2k$, and we will ignore the marginal case $\beta^2 + k^2 = 2k$). In particular, if $\beta = 0$, there is one if $\rho < \frac{1}{2}$, and this could be extended to a wide class of smooth integral functions of order $\rho < \frac{1}{2}$. Thus for such functions there is a pits effect without any randomizing factor being necessary, a fact which complicates the story.

It is necessary for these $f(z)$ of order $\rho < \frac{1}{2}$ to be smooth; I construct a pathological function of zero order with highly multiple zeros which shows no pits effect.

2.

We consider

(2.1) $$f(z) = \sum \frac{z^n}{\Gamma(1 + \lambda n)}, \qquad \lambda = k + i\beta, \quad k = 1/\rho.$$

Let

(2.2) $$P = \frac{\beta^2 + k^2}{2k} = \frac{|\lambda|}{2k} > 1,$$

a condition equivalent to Watson's form $|\lambda - 1|^2 > 1$.

Let B's denote positive constants depending only on β and k, and let A's denote positive absolute constants. The A's or B's with suffixes retain their identity, those without are not in general the same from one occurrence to the next.

Watson's result for his general form reduces for (2.1) and $P > 1$ (after much tracing and particularization of parameters) to the following very simple asymptotic formula.

(2.3) $$\begin{cases} f(z) = \sum_{m=-p}^{p} f_m(z), \\ p = \text{greatest integer} < P \text{ (so } p \geq 1), \\ f_m(z) = \exp(z^{1/\lambda} e^{2m\pi i/\lambda}) \cdot [1 + O(r^{-K})], \end{cases}$$

where K is an arbitrarily large positive constant. In what follows K will always have this meaning, but K's will vary from one occurrence to another (this convention will avoid certain minor complications).

Formula (2.3) is valid for *all* (large) $z = re^{\theta i}$, but with a particular (and unusual) determination of θ and $z^{1/\lambda}$, namely

$$z^{1/\lambda} = \exp[(\log r + i\theta)/\lambda],$$

(2.4)

$$\theta = \sigma \log r + \phi, \qquad \sigma = \beta/k, \qquad -\pi < \phi \leq \pi.$$

Let

(2.5) $$\phi_m = \phi + 2m\pi, \qquad \psi_m = \phi_m/2P.$$

Then calculation gives

(2.6) $$\left\{ \begin{array}{l} \log f_m = r^\rho e^{\sigma\psi_m + i\psi_m} + O(r^{-K}), \\ \log |f_m| = u_m = r^\rho e^{\sigma\psi_m} \cos \psi_m + O(r^{-K}), \\ \arg f_m = r^\rho e^{\sigma\psi_m} \sin \psi_m + O(r^{-K}). \end{array} \right\} \qquad (|m| \leq p)$$

We shall find that in general, and roughly speaking, one u_m dominates the rest and then f is approximately the corresponding f_m. For f not to be exponentially large, in particular for z to be a zero, there must be a pair of u_m nearly equal and dominating the other u_m.

3.

Let

(3.1) $$g(t) = e^{\sigma t} \cos t,$$

where t is confined to the range to which all ψ_m belong, namely

(3.2) $$\phi/2P - p\pi/P \leq t \leq \phi/2P + p\pi/P.$$

This is contained in

$$-\frac{(2p+1)}{2P} \pi \leq t \leq \frac{2p+1}{2P} \pi,$$

and since $p < P$, $P > 1$, it is further contained in

(3.3) $$-\tfrac{3}{2}\pi + B < t < \tfrac{3}{2}\pi - B.$$

Let

(3.4) $$\tan t^* = \sigma, \qquad 0 \leq t^* < \tfrac{1}{2}\pi.$$

We have

$$g'(t)/e^{\sigma t} = \sigma \cos t - \sin t.$$

The only zeros of g' in (3.2) are t^* and $t^* - \pi$, and in this range g has a maximum $e^{\sigma t^*} \cos t$ at $t = t^*$, a negative minimum at $t = t^* - \pi$, and is negative for the part of the range for which $t < -\frac{1}{2}\pi$, in particular for $t \leq t^* - \pi$, and finally $g(t)\downarrow$ from t^* to $t^* - \pi$.

Let μ be the greatest m for which $\psi_m = \phi_m/2P \leq t^*$, so that

$$(3.5) \qquad \phi_\mu/2P \leq t^*, \qquad \phi_{\mu+1}/2P > t^*.$$

Note that μ depends on ϕ but not on r.

If $m < \mu$ then either

$$t^* - \pi \quad \leq \psi_m \leq \quad \psi_\mu - \pi/P,$$

or

$$\psi_m \leq t^* - \pi.$$

In the latter case $\log |f_m| \leq r^p g(\psi_m) + O(r^{-K}) < 0$, and $f_m = O(1)$. In the former,

$$\log |f_m/f_\mu| = r^p[g(\psi_\mu - \pi/P) - g(\psi_\mu)] + O(r^{-K}) < -Br^p,$$

$$|f_m/f_\mu| < \exp(-Br^p).$$

If $m > \mu + 1$ then, similarly, $|f_m|/|f_{\mu+1}| < \exp(-Br^p)$.

We are aiming at proving that f is generally exponentially large, and look for conditions that this should *not* happen. Let us assume till further notice that

$$(3.6) \qquad |f(z)| < |f_\mu(z)|/r$$

(which of course includes the case when z is a zero of f), and develop the consequences, which we hope will be conditions *not* satisfied except in exponentially small circles.

After (3.6) we must, by what precedes, have

$$(3.7) \qquad \mathscr{R} \log f_{\mu+1} = \mathscr{R} \log f_\mu + o(1)$$

and

$$(3.8) \qquad \mathscr{I} \log f_{\mu+1} - \mathscr{I} \log f_\mu = \pi + o(1) \qquad (\text{mod } 2\pi),$$

and then f_m other than f_μ, $f_{\mu+1}$ have moduli $< \exp(-Br^p)$ times those of $f_\mu, f_{\mu+1}$.

Equations (3.7) and (3.8) are, by (2.6),

$$(3.9) \qquad e^{\sigma\psi_{\mu+1}} \cos\psi_{\mu+1} = e^{\sigma\psi_\mu} \cos\psi_\mu + o(1),$$

$$(3.10) \qquad r^p(e^{\sigma\psi_{\mu+1}} \sin\psi_{\mu+1} - e^{\sigma\psi_\mu} \sin\psi_\mu) = (2M+1)\pi + o(1),$$

where M is some positive integer.

The first reduces to

(3.11)
$$\begin{cases} \psi_\mu = c_1 + o(1), \quad \text{where} \\ \tan c_1 = \cot \tau - e^{-\sigma\tau} \operatorname{cosec} \tau, \qquad \tau = \pi/P, \\ |c_1| < \tfrac{1}{2}\pi - B. \end{cases}$$

We have further,

(3.12)
$$\begin{cases} f_\mu(z) = \exp[\{b + o((1)\}r^\rho], \text{ where} \\ b = e^{\sigma c_1} \cos c_1. \end{cases}$$

Since $\psi_\mu = (\phi + 2\mu\pi)/2P$, $P > 1$, and $|\phi| \le \pi$, (3.11) determines both μ, exactly, and ϕ to $o(1)$, as constants depending only on β, k. Also, b depends only on β, k.

Continuing consequences of (3.6) we have, combining (3.9) and (3.10),

$$r^\rho e^{\sigma\psi_{\mu+1}}[\sin \psi_{\mu+1} - \sin \psi_\mu \cos \psi_{\mu+1}/\cos \psi_\mu] \equiv \pi + o(1),$$

and so

(3.13) $$r^\rho/c_2 = (2M + 1)\pi + o(1),$$

(3.14) $$c_2 = e^{-\sigma(c_1+\pi)} \sin \tau \sec c_1 \ne 0,$$

c_2 depending only on β, k.

If we retrace the preceding argument with the stronger hypothesis that z is a zero of f, we find, recalling that the formula for f_m has a factor $1 + O(r^{-K})$, that we get the same conclusions as above, but with $O(r^{-K})$'s replacing $o(1)$'s.

Summing up, we have

Lemma 1. *We have* $|f(z)| \ge |f_\mu(z)|/r$ *and* $|f(z)| > \exp[br^\rho + o(r^\rho)]$, *where* $b = e^{\sigma c_1} \cos c_1$, *unless the following conditions are satisfied:*
   (i) $r^\rho = (2M + 1)c_2\pi + o(1)$,
   (ii) $\mathscr{R} \log f_{\mu+1} = \mathscr{R} \log f_\mu + o(1)$,
   (iii) $|f_\mu(z)| = \exp[br^\rho + o(r^\rho)]$,
   (iv) $\psi_\mu = c_1 + o(1)$.
   If z is a zero of f the $o(1)$'s are replaced by $O(r^{-K})$'s.

4.

   I digress to discuss a curious point that arises here. The foregoing argument shows that (3.8) must imply

$$\psi_\mu \le t^* < \psi_{\mu+1}.$$

Since

$$\sin \pi/P \cdot \tan t^* = \sigma \tan \pi/P,$$

it must be the case, by (3.9), that

(4.1) $$\cos \pi/P - \sigma \sin \pi/P - e^{-\sigma\pi/P} \leqq 0.$$

In this P is any number >1, and, since β/k is of weight 0 in β, k and P of weight -1, σ can have any value in $0 \leqq \sigma < \infty$ for given P. Hence it must be true that

(4.2) $$F(x) = \cos x - \sigma \sin x - e^{-\sigma x} \leqq 0$$

for all $0 < x < \pi$, $0 \leqq \sigma < \infty$.

An independent proof is as follows.

$F(x)$ is initially decreasing from $x = 0$, so that $F < 0$ at $x = \delta$, say. In (δ, π), $F < 0$ if $\cos x \leqq e^{-\sigma x}$, and $F'(x) = -\sin x - \sigma \cos x + \sigma e^{-\sigma x} < 0$ if $\cos x > e^{-\sigma x}$. Thus in (δ, π) either $F < 0$ or $F' < 0$. If F is ever >0 there is a ξ with $F(\xi) = 0$, $F(x) < 0$ $(\delta < x < \xi)$. Clearly $F'(\xi) \geqq 0$, contrary to the above.

5.

Return to the second part of Lemma 1. Let $z_M = r_M e^{i\theta_M}$, where, using (3.13), we take

(5.1) $$r_M = [(2M + 1)c_2 \pi]^{1/\rho},$$

and, by using (2.6), with $\phi + 2\mu\pi = 2Pc_1$,

(5.2) $$\theta_M = \sigma \log r_M + 2c_1 P - 2\mu\pi.$$

Then we have

(5.3) $$f_{\mu+1}(z_M)/f_\mu(z_M) = -1 + O(r_M^{-K}).$$

Let C_M be the ζ-circle

$$C_M : |\zeta - z_M| = R = r_M^{-K_1},$$

where K_1 is arbitrarily large but fixed.

On C_M, with $K > 2K_1$, say,

(5.4) $$\log[f_m(z_M + \zeta)/f_m(z_M)] = [(z_M + \zeta)^{1/\lambda} - z_M^{1/\lambda}]e^{2m\pi i/\lambda} + O(r_M^{-K})$$

$$= z_M^{1/\lambda} e^{2m\pi i/\lambda}\left[\frac{\zeta}{\lambda z_M} + O(r_M^{-K_1})\right].$$

In particular

(5.5) $$f_\mu(z_M + \zeta)/f_\mu(z_M) = \exp[Q\zeta + O(r_M^{-K})]$$

$$= 1 + Q\zeta + O(r_M^{-K}), \qquad Q = z_M^{1/\lambda}/\lambda z_M.$$

Similarly

(5.6) $$f_{\mu+1}(z_M + \zeta)/f_{\mu+1}(z_M) = 1 + Qe^{2\pi i/\lambda}\zeta + O(r_M^{-K}),$$

and combining this with (5.3), we get

(5.7) $$f_{\mu+1}(z_M + \zeta)/f_{\mu}(z_M) = -1 - Qe^{2\pi i/\lambda}\zeta + O(r_M^{-K}).$$

Since finally

$$\left|\left[\sum_{m \neq \mu, \mu+1} f_m(z)\right]\Big/f_\mu(z_M)\right| = O[\exp(-Br_M^\rho)],$$

and so, by (5.4),

$$\sum_{m \neq \mu, \mu+1} f_m(z_M + \zeta)/f_\mu(z_M) = O(r_M^{-K}),$$

we have from this, (5.5), and (5.7),

(5.8) $$\frac{f(z_M + \zeta)}{f_\mu(z_M)} = Q(i - e^{2\pi i/\lambda})\zeta + O(r_M^{-K}).$$

It follows that the increment in the argument of $f(z_M + \zeta)$ around C_M is approximately and so exactly 2π.

There is a single and simple zero $\mathscr{Z}_M$ in C_M, which, recall, is

(5.9) $$|z - z_M| = r_M^{-K_1} = R_1.$$

Further, by (2.6), we have on C_M,

(5.10) $$|f(z_M + \zeta)| > B|Q\zeta||f_\mu(z_M)| > \exp[br^\rho + o(r^\rho)].$$

6.

By (5.1),

(6.1) $$r_{M+1} - r_M \sim B_1 M^{1/\rho - 1}.$$

Let C_M' be

(6.2) $$|z| = r_M + \tfrac{1}{2}B_1 M^{1/\rho - 1}.$$

Since (i) of Lemma 1 is not satisfied on C_M' we have

(6.3) $$|f| > \exp[br^\rho + o(r^\rho)] \quad \text{on} \quad C_M'.$$

Since this is true also on the C_n with $n \leq M$, which are all inside C_M', and since there are no zeros in $C_M' - \sum_{n \leq M} C_n$, $|f|$ attains its minimum on the boundary of $C_M' - \sum_{n \leq M} C_n$, and (6.3) holds everywhere except inside the C_M.

7.

We can now establish the pits effect. Let $\mathscr{Z}_M$ be the zero in C_M and let $|\mathscr{Z}_M| = R_m$. We may suppose that $\mathscr{Z}_M$ lies in the inner half of C_M, since otherwise we could have taken $K_1 + 1$ instead of K_1. Let

$$F(z) = \frac{f(z)}{z - \mathscr{Z}_M}.$$

On C_M

$$|F| > \exp[bR_M^\rho + o(R_M^\rho)]/\max_{(C_M)} |z - \mathscr{Z}_M|$$
$$> \exp[bR_M^\rho + o(R_M^\rho)],$$

and in particular, since F has no zero inside the circumference C_M,

(7.1) $$|f'(\mathscr{Z}_M)| = |F(\mathscr{Z}_M)| \geqq \min_{(C_M)}[F] > \exp[bR_M^\rho + o(R_M^\rho)].$$

Next, we have on the circumference C_M,

$$|F| \leqq \max_{(C_M)} |f|/\min_{(C_M)} |z - \mathscr{Z}_M|$$
$$\leqq \exp[bR_M^\rho + o(R_M^\rho)]/\tfrac{1}{3}R_M^{-K_1}$$

(since $r_M - R_M = o(1)$)

(7.2) $$< \exp[bR_M^\rho + o(R_M^\rho)].$$

Hence

$$\left|\frac{f^{(n)}(\mathscr{Z}_M)}{n!}\right| = \left|\frac{1}{2\pi}\int_{C_M} \frac{f(z)}{(z - \mathscr{Z}_m)^{n+1}}\,dz\right|$$

(7.3) $$\leqq \max_{(C_M)} |f(z)|/(\tfrac{1}{3}R_M)^{-K_1 n}$$
$$\leqq \exp[bR_m^\rho + o(R_m^\rho)]R_m^{-K_1 n}.$$

Let now γ_M be the circle

(7.4) $$|z - \mathscr{Z}_M| = d = \exp(-\tfrac{1}{2}bR_M^\rho).$$

Then on γ_M

$$|f| \geqq d|f'(\mathscr{Z}_M)| - \sum_2^\infty \left|\frac{f^{(n)}(z_M)}{n!}\right| d^n$$

$$\geqq d\left[\exp[bR_M^\rho + o(R_M^\rho)] - \sum_{n=1}^\infty d^n R_M^{K_1 n}\exp[bR_M^\rho + o(R_M^\rho)]\right]$$

Since $dR_M^{K_1} < \exp(-\tfrac{1}{4}bR_M^\rho) < \tfrac{1}{2}$ for large M, this gives, on γ_M,

$$|f| \geqq \exp[\tfrac{1}{2}bR_M^\rho + o(R_M^\rho)],$$

and so, for z of γ_M,

(7.5) $$|f| \geqq \exp[\tfrac{1}{2}br^\rho + o(r^\rho)].$$

Since this is true also on C_M, and $C_M - \gamma_M$ has no zeros, it is clearly true in $C_M - \gamma_M$, and, being true outside C_M's, it is true everywhere except in the γ_M. This combined with (7.4) is a case of the pits effect.

To see that some degree of smoothness is necessary, consider the function

$$f(z) = \prod \left(1 - \frac{z}{5^{n^2}}\right)^{4^n},$$

and a circle C_N,

$$|z - 5^{N^2}| = \rho_N = 5^{N^2/3}.$$

At the point ρ_N, or $r = 5^{N^2} - \rho_N$ of C_N we have

$$|f| \leqq \left(\prod_{n=1}^{N-1} r^{4^n}\right)\left(\frac{\rho_N}{5^{N^2}}\right)^{4^N} \cdot 1$$

$$\leqq (5^{N^2})^{4^N/3}(\rho_N/5^{N^2})^{4^N} = (\rho_N 5^{-2N^2/3})^{4^N}$$

$$= (5^{-N^2/3})^{4^N}$$

It is easily seen that ρ_N is a very large distance from any zero of f, and f, which is of "small" zero order, is small at ρ_N.

REFERENCES

1. J. E. Littlewood and A. C. Offord, On the distribution of the zeros and α-values of a random integral function. I, *J. London Math. Soc.* **20** (1945), 130–136; II, *Ann. of Math.* **49** (1948), 885–952.
2. Y. M. Chen and J. E. Littlewood, Some new properties of power series, *Indian J. Math.* **9** (1967), 289–324.
3. J. E. Littlewood, A "pits effect" for all smooth enough integral functions with a coefficient factor $\exp(n^2\alpha\pi i), \alpha = \tfrac{1}{2}(\sqrt{5} - 1)$, *J. London Math. Soc.* **43** (1968), 79–92.
4. M. Nassif, On the behavior of the function $\sum e^{\sqrt{2\pi i n^2}} \dfrac{z^{2n}}{n!}$, *Proc. London Math. Soc.* **54** (1950), 201–216.
5. G. N. Watson, A class of integral functions defined by Taylor Series, *Trans. Cambridge Philos. Soc.* **XXII** (II), (1912), 9–37.
6. E. Maitland-Wright, The asymptotic expansions of integral functions defined by Taylor Series, *Philos. Trans. Roy. Soc. London Ser. A* (1939–1940), 432–451.

(Received August 10, 1967)

A Distortion Theorem for a Class of Conformal Mappings

A. J. Lohwater

Department of Mathematics
Case Western Reserve University
Cleveland, Ohio, 44106
U.S.A.

Frank Ryan

Department of Mathematics
Case Western Reserve University
Cleveland, Ohio 44106
U.S.A.

1. Introduction

Our interest in the conformal mapping of Smirnov and non-Smirnov domains stems from a question raised by P. L. Duren. It turns out that Duren's question can be answered rather easily, and, after recalling the fundamental concepts, we give a solution to the problem. In the second section we pose and solve a problem on the distortion of a conformal mapping of the unit disk onto a non-Smirnov domain; the proof of the main theorem of this paper (Theorem 3) suggests certain problems concerning the uniqueness and boundary behavior of harmonic functions, and we give one such theorem (Theorem 4) as a corollary to Theorem 3. Further extensions of Theorem 4 will be published elsewhere.

If D is a domain in the complex plane bounded by a rectifiable Jordan curve, and if $f(z)$ is a schlicht analytic function mapping $|z| < 1$ conformally onto D, it is known [4, p.78] that the derivative $f'(z)$ has a representation of the form

(1)
$$f'(z) = e^{i\gamma} \exp\left\{\frac{1}{2\pi} \int_0^{2\pi} \frac{e^{it} + z}{e^{it} - z} \, dg(t)\right\},$$

where $g(t)$ is of bounded variation on $0 \leqq t \leqq 2\pi$. It is also known that, in the Lebesgue decomposition

(2)
$$g(t) = \mu(t) + v(t)$$

of $g(t)$ into its singular part $\mu(t)$ and absolutely continuous part $v(t)$, the function $\mu(t)$ is monotone nonincreasing and continuous with $\mu'(t) = 0$ almost everywhere, and that $g'(t) = v'(t) = \log|f'(e^{it})|$ almost everywhere. The domain D is called a *Smirnov domain* if the singular function does not appear in (2); the presence of a singular function $\mu(t)$ in the representation (1) is independent of the choice of mapping function.

It is known that not every rectifiably bounded Jordan domain is a Smirnov domain; for a discussion of this problem, together with a characterization of the singular measures associated with such domains, we refer to a recent paper of Duren, Shapiro, and Shields [1]. We remark only that, for one of the known examples of a non-Smirnov domain, the derivative of the mapping function has the form

(3)
$$f'(z) = \exp\left\{\frac{1}{2\pi} \int_0^{2\pi} \frac{e^{it} + z}{e^{it} - z} \, d\mu(t)\right\},$$

where $\mu(t)$ is nonincreasing and continuous with $\mu'(t) = 0$ almost everywhere. The function appearing on the right-hand side of (3) belongs to a class of functions first pointed out by R. Nevanlinna [3] and extensively studied by W. Seidel [5]. Using terminology which is now classical, we say that a function $f(z)$ is of *class* (A) in $|z| < 1$ if $f(z)$ is analytic and bounded, $|f(z)| < 1$, in $|z| < 1$, and if the radial limits of $f(z)$ are of modulus 1 almost everywhere on $|z| = 1$. (Such functions are sometimes referred to as *inner functions*.) It is known [5] that a function of class (A) has a representation of the form

(4)
$$f(z) = e^{i\gamma} B(z) \exp\left\{\frac{1}{2\pi} \int_0^{2\pi} \frac{e^{it} + z}{e^{it} - z} \, dg(t)\right\},$$

where $g(t)$ is a monotone nonincreasing function with $g'(t) = 0$ almost everywhere, and $B(z)$ is a Blaschke product extended over the zeros of $f(z)$. Except in the case that $f(z)$ in (4) reduces to a finite Blaschke product—or a constant—the boundary behavior of a function of class (A) described in [5] is pathological in a well-defined sense; in particular, every value of modulus less than 1 is assumed infinitely often except possibly for an extremely small set of values.

The example (3) shows, however, that functions of class (A) can actually appear in the representation of the derivative $f'(z)$ of a schlicht function, and the question asked by Duren is whether there exists a schlicht function of the form

(5)
$$f(z) = z \exp\left\{\frac{1}{2\pi} \int_0^{2\pi} \frac{e^{it} + z}{e^{it} - z} \, d\mu(t)\right\}$$

$$\times \exp\left\{\frac{1}{2\pi} \int_0^{2\pi} \frac{e^{it} + z}{e^{it} - z} \, dv(t)\right\},$$

where $\mu(t)$ is a nonincreasing singular function and $v(t)$ is absolutely continuous, in which $\mu(t)$ is not identically constant. We shall show first that, if $f(z)$ is schlicht, then $\mu(t)$ is identically constant, and we shall see that the method of proof suggests a relatively simple sufficient condition that the domain D mentioned above be a Smirnov domain. We shall need a trivial fact, which, for the purposes of reference, we state as a lemma.

Lemma. *If $f(z)$ is analytic and schlicht in $|z| < 1$ with $f(0) = 0$, then* $\lim \inf_{|z| \to 1} |f(z)| > 0$.

We omit the proof of the Lemma and proceed directly to the answer to Duren's question.

Theorem 1. *If $f(z)$ is analytic and schlicht in $|z| < 1$, then no representation of the type (5) for $f(z)$ can contain as a factor a function of class (A), i.e., $\mu(t)$ is identically constant.*

Proof. Let us assume, to the contrary, that $\mu(t)$ is not identically constant and force a contradiction. We may write the absolutely continuous function $v(t)$ in (5) as the difference of two nondecreasing absolutely continuous functions,

$$v(t) = v_1(t) - v_2(t),$$

and form the function

(6)
$$F(z) = \exp\left\{\frac{1}{2\pi} \int_0^{2\pi} \frac{e^{it} + z}{e^{it} - z} (d\mu(t) + dv_1(t))\right\}.$$

Then $f(z)$ has the form $zF(z)G(z)$, where the function

$$G(z) = \exp\left\{\frac{1}{2\pi} \int_0^{2\pi} \frac{e^{it} + z}{e^{it} - z} (-dv_2(t))\right\}$$

is bounded, $|G(z)| < 1$, in $|z| < 1$. We assert now that $\lim \inf_{|z| < 1} |F(z)| = 0$.

Suppose that there exists a number $m > 0$ such that $\liminf_{|z|<1} |F(z)| \geqq m$. Then the function

$$\phi(z) = \frac{1}{F(z)} = \exp\left\{\frac{1}{2\pi} \int_0^{2\pi} \frac{e^{it} + z}{e^{it} - z} (-d\mu(t) - dv_1(t))\right\}$$

is analytic and bounded, $|\phi(z)| < 1/m$, in $|z| < 1$, where it is now to be remarked that $-\mu(t)$ is a nondecreasing function and $-v_1(t)$ is a nonincreasing function. Since $\mu'(t) = 0$ almost everywhere and since $-v'(t) \leqq 0$ almost everywhere, it follows that the radial limits of $|\phi(z)|$ are almost everywhere of modulus no greater than 1, so that $|\phi(z)| \leqq 1$ everywhere in $|z| < 1$. Since $\phi(z) \neq 0$ in $|z| < 1$, $\log|\phi(z)| = u(r, \theta)$ is harmonic and nonpositive in $|z| < 1$, so that in the representation

$$u(r, \theta) = \frac{1}{2\pi} \int_0^{2\pi} \frac{1 - r^2}{1 + r^2 - 2r \cos(\theta - t)} (-d\mu(t) - dv_1(t))$$

the function $-\mu - v_1$ is nonincreasing. This implies that the singular part $-\mu$ must also be nonincreasing. But if $-\mu$ is both nonincreasing and nondecreasing, then μ must be identically constant. Hence, if μ is not identically constant, it follows that $\liminf_{|z|<1} |F(z)| = 0$, or, what is the same thing, there exists a sequence $\{z_n\}$ tending to some point on $|z| = 1$ such that $\lim_{n \to \infty} F(z_n) = 0$.

To complete the proof, it suffices to observe that, since $G(z)$ is bounded, $|G(z)| < 1$, in $|z| < 1$, it now follows that $\lim_{n \to \infty} f(z_n) = \lim_{n \to \infty} z_n F(z_n) G(z_n) = 0$, so that $\liminf_{|z| \to 1} |f(z)| = 0$. Because this contradicts the Lemma, it follows that the factor arising from a singular measure cannot be present in the representation (5). This answers the question raised by Duren.

2. The Main Theorem

We turn now to the question of a rectifiably bounded Jordan domain D which is not a Smirnov domain, i.e., a domain D for which the representation (1) of the mapping function actually contains a factor of class (A) arising from a nonconstant singular function $\mu(t)$ in the Lebesgue decomposition (2). We have shown in the proof of Theorem 1 that, if the singular measure $d\mu(t)$ is actually present in the expression for $F(z)$ in (6) above, then the value 0 is in the cluster set $C(F, P)$ for some point P on $|z| = 1$, that is, there exists a sequence $\{z_n\}$, $\lim_{n \to \infty} z_n = P$, such that $\lim_{n \to \infty} F(z_n) = 0$. If we use the expression (1) for $f'(z)$ and form the function (6) as before, where $v_1(t)$ is the increasing part of $v(t)$ appearing in (1), then, as in the proof of Theorem 1, there is a sequence $\{z_n\}$ tending to a point P on $|z| = 1$ such that

$$\lim_{n \to \infty} f'(z_n) = 0.$$

We may state this fact as follows.

Theorem 2. *Let D be a domain in the complex plane bounded by a rectifiable Jordan curve, and let $f(z)$ be any schlicht analytic function mapping $|z| < 1$ conformally onto D. If the cluster set $C(f', P)$ of $f'(z)$ does not contain the value 0 for any point P on $|z| = 1$, then D is a Smirnov domain.*

We remark that the example (3) has the property that $f'(z)$ has 0 as a radial limit on a nondenumerable set of radii. To see this, we write

$$\log |f'(z)| = u(r, \theta) = \frac{1}{2\pi} \int_0^{2\pi} \frac{1 - r^2}{1 + r^2 - 2r \cos(\theta - t)} \, d\mu(t),$$

and observe that a continuous nonincreasing singular function which is not identically constant must have the property that $\mu'(t) = -\infty$ on a non-denumerable set of points. At any such point θ_0, $\lim_{r \to 1} u(r, \theta_0) = -\infty$, whence $\lim_{r \to 1} |f'(re^{i\theta})| = 0$. Thus 0 is not only a cluster value of $f'(z)$ at a nondenumerable set of points P on $|z| = 1$, the value 0 is *an asymptotic value—in this case, a radial limit value—at a nondenumerable set of points on $|z| = 1$.*

Our main theorem states that this last property is characteristic of non-Smirnov domains. We have

Theorem 3. *Let $f(z)$ be a schlicht analytic mapping of $|z| < 1$ onto a domain D bounded by a rectifiable Jordan curve. If D is not a Smirnov domain, then $f'(z)$ admits 0 nondenumerably often as a radial limit value.*

Proof. Since $f'(z) \neq 0$ in $|z| < 1$, the function

$$(7) \qquad u(r, \theta) = \log |f'(z)| = \frac{1}{2\pi} \int_0^{2\pi} \frac{1 - r^2}{1 + r^2 - 2r \cos(\theta - t)} \, dg(t)$$

obtained from the integral representation (1) is harmonic in $|z| < 1$. Because D is not a Smirnov domain, a singular term $\mu(t)$ is actually present in the decomposition (2), and we denote by E the nondenumerable set of points t in $0 \leq t \leq 2\pi$ such that $\mu'(t) = -\infty$. Since E is Borel measurable, we may consider the decomposition

$$(8) \qquad\qquad\qquad G(E) = M(E) + N(E),$$

where G, M, and N are the set functions or signed measures corresponding, respectively, to g, μ, and ν. Since N is absolutely continuous, and since E is of measure zero, we have $N(E) = 0$, whence $G(E) = M(E) > 0$. This implies that $g'(t) = -\infty$ on a nondenumerable subset of E, so that, on this non-denumerable subset $\lim_{r \to 1} u(r, \theta) = -\infty$, or, what is equivalent,

$$\lim_{r \to 1} |f'(re^{i\theta})| = 0,$$

and the theorem is proved.

3. A Class of Harmonic Functions

Let $u(r, \theta)$ be harmonic in $|z| < 1$ and let $u(r, \theta)$ satisfy the condition

$$(9) \qquad \int_0^{2\pi} |u(r, \theta)| \, d\theta < M,$$

where M is a finite constant independent of r. It is known [2] that if

$$(10) \qquad \lim_{r \to 1} u(r, \theta) = 0$$

for almost all θ in $0 \leq \theta \leq 2\pi$ and if, on the remaining set (of measure zero), whenever $\lim_{r \to 1} u(r, \theta)$ exists, it is finite, then $u(r, \theta)$ is identically zero in $|z| < 1$. Our method of proving Theorem 3 suggests a theorem in the theory of harmonic functions.

Theorem 4. *Let $u(r, \theta)$ satisfy* (9) *and let us assume that*

$$(11) \qquad \lim_{r \to 1} u(r, \theta) \geq 0$$

for almost all θ in $0 \leq \theta \leq 2\pi$. If we assume that there is no radius θ on which $\lim_{r \to 1} u(r, \theta)$ is $-\infty$, then $u(r, \theta)$ is nonnegative in $|z| < 1$.

Second, we remark that if we drop condition (9) but retain (11), our method of proving Theorem 3 may be used to show that if $-\infty$ is not an asymptotic value of $u(r, \theta)$, then $u(r, \theta)$ is nonnegative in $|z| < 1$; the proof is long and will appear elsewhere.

REFERENCES

1. P. L. Duren, H. S. Shapiro, and A. L. Shields, Singular measures and domains not of Smirnov type, *Duke Math. J.* **33** (1966), 247–254.
2. A. J. Lohwater, The boundary values of a class of meromorphic functions, *Duke Math. J.* **19** (1952), 243–252.
3. R. Nevalinna, Über beschränkte analytische Funktionen, *Ann. Acad. Sci. Fenn. Ser. A* **32** (1929), 1–75.
4. I. I. Priwalow, *Randeigenschaften analytischer Funktionen*, VEB Deutscher Verlag der Wissenschaften, Berlin, 1956.
5. W. Seidel, On the distribution of values of bounded analytic functions, *Trans. Amer. Math. Soc.* **36** (1934), 201–226.

(*Received February 9, 1968*)

Some Examples in the Theory of Groups

I. D. Macdonald

Department of Mathematics
University of Queensland
St. Lucia, Brisbane
Australia

The intent of this note is to illustrate certain problems in group theory, which have perhaps little in common apart from their " nonlinear " nature, by means of three mildly surprising examples. The arguments used are straightforward though the construction of the examples is tedious; the calculations have been set forth not in every detail but merely with enough indications for the reader to recover them, should he feel willing and able. Notation is either standard or to be found in [4], and we draw the reader's attention to the usefulness of Lemma 2 of that paper in organizing the calculations.

Example 1. Let $R_n(G)$ denote the set of nth-right-Engel elements of the group G. Kappe has shown in [3] that $R_2(G)$ is always a subgroup of G, and some results on similar lines are known for $R_3(G)$. For instance, Heineken ([2], Lemma 12) proved that if a in $R_3(G)$ has odd order and if $gp\{a, x\}$ is nilpotent for each element x of G then $a^{-1} \in R_3(G)$. We shall prove

There exists a finite 2-group E_1 and an element a in E_1 such that

$$\{a, a^{-1}, a^2\} \cap R_3(E_1) = \{a\}.$$

We start by presenting an infinite group G. This is to have

$$\{a, b, c, d_1, d_2, e, f\}$$

as a set of generators. The list of defining relations is the following: firstly we want

$$e^8 = 1, \qquad f^2 = 1;$$

secondly we want the commutator relations specified in the following table:

	b	c	d_1	d_2	e	f
a	c^{-1}	d_1	e^6f	e^5f	e^4	1
b		d_2	e	1	f	1
c			e^4	f	1	1
d_1				1	1	1
d_2					1	1
e						1

(In this table $[x, y]$ is to be found at the intersection of the column labeled x and the row labeled y.) The presentation contains many redundant generators and relations, but we prefer matters so for the sake of the subsequent construction.

It is easy to check that the relations imply that $G = gp\{a, b\}$ and that G has class 5. To show that $a \in R_3(G)$ is equivalent to showing that $[a, 3xy] = 1$ where $x = a^\alpha b^\beta$ and $y \in \delta(G)$. We indicate the relevant steps:

$$[a, 3xy] = [a, 3x][a, x, x, y][a, x, y, x][a, y, x, x].$$

$$[a, 3x] = e^{4\alpha\beta(\alpha+\beta)} f^{\alpha\beta(\alpha+\beta)},$$

$$[a, x, y, x] = 1,$$

$$[a, x, x, y][a, y, x, x] = 1.$$

Details may be filled in by using the relations. Therefore $a \in R_3(G)$.

Because G has class 5, we now have

$$[a^{-1}, 3xy] = [a, xy, a, xy, xy] = [a, x, a, x, x],$$

$$[a^2, 3xy] = [a, xy, a, xy, xy] = [a, x, a, x, x];$$

and it follows that neither a^{-1} nor a^2 lies in $R_3(G)$ if $f \neq 1$. At this stage, construction of the group G becomes necessary in order to prove that the defining relations do not imply $f = 1$. One way to do this is by means of successive splitting extensions, and in any case the details are for the reader to complete. Therefore $\{a, a^{-1}, a^2\} \cap R_3(G) = \{a\}$.

Finally we assert that one can easily show that G is residually a finite 2-group, and so there exists a finite factor group which has the properties required of E_1. We take this factor group to be E_1.

Example 2. Let $L_n(G)$ denote the set of nth-left-Engel elements in G, that is

$$L_n(G) = \{a: [x, na] = 1 \quad \text{for all } x \in G\}.$$

Heineken has proved very simply in [1] a result which states (in our left-normed notation) that $R_n(G) \subseteq L_{n+1}(G)^{-1}$. Since it is well-known (see [3]) that $R_2(G) \subseteq L_2(G) = L_2(G)^{-1}$ we are led to seek an improvement on Heineken's result when $n > 2$. Our example will show that such an improvement cannot usually be found.

Let n be any integer greater than 2. For each prime divisor p of n there exists a finite metabelian p-group E_2 such that both $R_n(E_2) \subseteq L_n(E_2)$ and $R_n(E_2) \subseteq L_n(E_2)^{-1}$ are false.

Once again we start by writing down a group G, which will be constructed in due course and of which E_2 will be a suitable factor group. Generators are to be a, b and the elements $c(i, j)$ where $0 \le i$, $0 \le j$, $i + j \le n - 1$, and if $i + j = n - 1$ then $j = 0$ or $p - 1$. The defining relations are to be such as are necessary to make the subgroup generated by the $c(i, j)$ elementary abelian of order $p^{2+n(n-1)/2}$, together with the following:

$$[a, b] = c(0, 0);$$
$$[c(i, j), a)] = c(i + 1, j) \quad \text{for } 0 \le i + j < n + 2,$$
$$[c(i, j), b] = c(i, j + 1) \quad \text{for } 0 \le i + j < n + 2;$$
$$[c(n - 2, 0), a] = c(n - 1, 0),$$
$$[c(n - p, p - 2), b] = c(n - p, p - 1),$$
$$[c(n - j - 2, j), a] = 1 \quad \text{for } j \neq 0,$$
$$[c(n - j - 2, j), b] = 1 \quad \text{for } j \neq p - 2,$$
$$[c(n - 1, 0), a] = [c(n - 1, 0), b] = 1,$$
$$[c(n - p, p - 1), a] = [c(n - p, p - 1), b] = 1;$$
$$c(n - 1, 0)c(n - p, p - 1)^{\binom{n-1}{p-1}} = 1.$$

This presentation makes it plain that G is metabelian and is nilpotent of class $n + 1$; further $G = gp\{a, b\}$. We consider $[a, nxy]$ where $x = a^\alpha b^\beta$ and $y \in \delta(G)$ with the aim of showing that $a \in R_n(G)$. Since G has class $n + 1$ we have

$$[a, nxy] = [a, na^\alpha b^\beta],$$

and the latter commutator may be evaluated by the usual rules for commutator expansion in metabelian groups. Thus we find that

$$[a, nxy] = c(n-1, 0)^{\alpha^{n-1}\beta} c(n-p, p-1)^{\binom{n-1}{p-1}\alpha^{n-p}\beta^p}$$

But $\alpha^p \equiv \alpha$ modulo p and each $c(i, j)$ has order p; so $[a, nxy] = 1$ and $a \in R_n(G)$.

Construction of G, by cyclic extensions or otherwise, is now necessary to verify that the defining relations do not imply that $c(n-1, 0) = 1$. Again we omit the details, remarking only that the fact that $\binom{n-1}{p-1}$ is not divisible by p is relevant here. Since $[b, na] = c(n-1, 0)^{-1}$, we at once deduce that $a \notin L_n(G) \cup L_n(G)^{-1}$.

Finally, we note that a much smaller effort will convince the reader that G is residually a finite p-group. We may therefore take for E_2 a factor group of G which has the required properties.

Example 3. The Hughes subgroup $H_p(G)$ of the finite p-group G is the subgroup generated by all elements that do not have order p, and of course in most of the familiar p-groups we have $|G : H_p(G)| = 1$ or p unless $H_p(G) = 1$. However Wall has constructed [5] a finite 5-group W with $|W : H_5(W)| = 25$ and $H_5(W) \neq 1$. If W were a two-generator group then every element of order 25 would lie in $\Phi(W)$ (since any pair of generators would be elements of order 5); in fact W is not a two-generator group and there is an element of order 25 outside $\Phi(W)$. Thus there is some interest attached to finite p-groups G whose elements of maximal order all lie in $\Phi(G)$.

We define $H_4(G)$ to be that subgroup of the finite 2-group G generated by all elements whose order does not divide 4 (with the convention $gp\{\phi\} = 1$), and we shall prove:

There exists a finite 2-group E_3 such that $|E_3 : H_4(E_3)| = 8$ with $H_4(E_3) \neq 1$, and $H_4(E_3) \leqq \Phi(E_3)$.

We present E_3 as follows. The generating set is to be

$$\{a, b, c, d_1, d_2, e_1, e_2, e_3, f_1, f_2, z\}.$$

The defining relations are to be

$$a^4 = 1, \quad b^4 = 1,$$
$$c^2 = e_1 e_2 e_3 f_2 z, \quad d_1^2 = f_1, \quad d_2^2 = f_2,$$
$$e_1^2 = e_2^2 = e_3^2 = z,$$
$$f_1^2 = f_2^2 = z^2 = 1,$$

together with those contained in the following table:

	b	c	d_1	d_2	e_1	e_2	e_3	f_1	f_2	z
a	c^{-1}	d_1	e_1	$e_2f_1f_2z$	f_1	f_2	f_1f_2z	z	z	1
b		d_2	e_2	e_3	f_1f_2z	f_1	f_2	z	z	1
c			f_1	f_2	z	1	z	1	1	1
d_1				z	1	1	1	1	1	1
d_2					1	1	1	1	1	1
e_1						1	1	1	1	1
e_2							1	1	1	1
e_3								1	1	1
f_1									1	1
f_2										1

These relations imply that $E_3 = gp\{a, b\}$ and that E_3 is a finite 2-group of class 6. Further,

$$\gamma_2{}^4 = \gamma_4{}^2 = \gamma_6,$$
$$\gamma_3{}^4 = 1,$$
$$\gamma_5{}^2 = 1,$$

where γ_n denotes the nth term of the lower central series of E_3.

The next step is to use the relations to show that $y^4 = 1$ if $y \notin \Phi(E_3)$. In preparation, one may show that, because each element of $\Phi(E_3)$ is the product of squares, one has identities such as the following:

$$[g_1, y_1]^4 = 1,$$
$$[g_1, y_1, g_2]^2 = 1,$$
$$[g_1, y_1, g_2, g_3] = 1,$$
$$[g_1, y_1, y_2, y_3, g_2] = [g_1, y_1, y_2, g_2, y_3] = [g_1, y_1, g_2, y_2, y_3] = 1$$

where y_1, y_2, y_3 are arbitrary elements of E_3 and g_1, g_2, g_3 are arbitrary elements of $\Phi(E_3)$.

Now an arbitrary element not in $\Phi(E_3)$ can be written as xg where $x = a$ or b or ab and $g \in \Phi(E_3)$. A standard commutator collection gives

$$(xg)^4 = x^4 g^4 [g, x]^2 [g, 3x][g, 2x, g].$$

Another collection process shows that $x^4 = 1$ when $x = ab$. What we therefore want to prove is that

(1) $$g^4 = [g, x]^2 [g, 3x][g, 2x, g]$$

for $x = a$ or b or ab and $g \in \Phi(E_3)$.

Now suppose (1) holds for $g = g_1$ and g_2, with x fixed; does it also hold for $g = g_1 g_2$? Further calculation shows that this is the case if and only if

(2) $$[g_1, g_2]^2 = [g_1, g_2, x^2][g_1, x; g_2, x].$$

We therefore turn to (2), which is to be verified case by case with $x = a$ or b or ab and $g_1, g_2 = a^2$ or b^2 or c, the remaining possibilities being trivially easy. The truth of (2) makes it sufficient to establish (1) when $x = a$ or b or ab and g is in a fixed set of generators for $\Phi(E_3)$, say

$$\{a^2, b^2, c, d_1, d_2, e_1, e_2, e_3\}.$$

(In fact e_1, e_2, e_3 are redundant.) This again is done case by case, perhaps the longest calculation in the present proof but in the end proving (1) and the fact that $y^4 = 1$ if $y \notin \Phi(E_3)$.

Thus all elements of E_3 which have order 8 lie in $\Phi(E_3)$. To prove that elements of order 8 exist in E_3, a construction is necessary. The recommended way to do this, though admittedly laborious, is by successive cyclic extensions, with particular care given by proving that the automorphisms corresponding to a and b each have order 4; the details are omitted. We conclude that the defining relations for E_3 do not imply that $z = 1$, and that E_3 has order 2^{13}. Note that $c^4 = z$, which implies that c has order 8.

It remains to prove that $|E_3 : H_4(E_3)| = 8$. We have already seen that $\delta(E_3) \leqq H_4(E_3) \leqq \Phi(E_3)$, and we intend to show that $H_4(E_3) = gp\{a^2b^2, \delta(E_3)\}$. First we assert that $(a^2 b^2)^4 = z$ so that $a^2 b^2$ has order 8, leaving the details to the reader. Next it has to be shown that $(a^2 g)^4 = 1$ and $(b^2 g)^4 = 1$ for all $g \in \delta(G)$, and a commutator collection process will accomplish this. It follows that $|E_3 : H_4(E_3)| = 8$ and indeed that $E_3 / H_4(E_3) \cong Z_4 \times Z_2$, where Z_n denotes the cyclic group of order n.

The factor group $E_3/\gamma_6(E_3)$ is in fact the Burnside group of exponent 4 on two generators. The original search for E_3 began by making use of close acquaintance with the Burnside group, and after all the computation is done one can extract from the above working a proof of the known fact that its order is 2^{12}.

A second remark may also be in order: *if G is a finite 2-group on two generators then $|G : H_4(G)| \leqq 8$ unless $H_4(G) = 1$.* For if we take a normal subgroup N of G having index 2 in $H_4(G)$ then

$$|G : H_4(G)| \leqq |G/N : H_4(G/N)|,$$

and it may be proved on the basis of the results about E_3 that $|G/N : H_4(G/N)| \leqq 8$.

REFERENCES

1. Hermann Heineken, Eine Bemerkung über engelsche Elemente, *Archiv Math. (Basel)* **11** (1960), 321.
2. ———, Engelsche Elemente der Länge drei, *Illinois J. Math.* **5** (1961), 681–707.
3. Wolfgang Kappe, Die A-Norm einer Gruppe, *Illinois J. Math.* **5** (1961), 187–197.
4. I. D. Macdonald and B. H. Neumann, A third-Engel 5-group, *J. Austral. Math. Soc.* **7** (1967), 555–569.
5. G. E. Wall, On Hughes' H_p problem, in *Proceedings of the International Conference on the Theory of Groups held at the Australian National University, Canberra, 10–20 August 1965*, Ed. by L. G. Kovács and B. H. Neumann, Gordon and Breach, New York 1967, pp. 357–362.

(Received February 20, 1968)

Exceptional Values of $f^{(n)}(z)$, Asymptotic Values of $f(z)$, and Linearly Accessible Asymptotic Values

Gerald R. MacLane*

Division of Mathematical Sciences
Purdue University
Layfayette, Indiana 47907
U.S.A.

1. Introduction

Let $f(z)$ be holomorphic in a domain D bounded by a Jordan curve C. Let Γ be an arc which is contained in D except for an end point $\zeta \in C$. If $f(z) \to a$ as $z \to \zeta$, $z \in \Gamma$, then f has the asymptotic value a on Γ, where a can be finite or ∞. We say that $f \in \mathscr{A}[D]$ provided f is nonconstant and it has asymptotic values at each point of a set of points $A \subset C$ with A dense on C.

Each asymptotic path, such as Γ above, is associated with an asymptotic tract $\mathscr{T} = \{D(\varepsilon, a)\}$: $D(\varepsilon, a)$ is one component of the open set

$$\{z \mid z \in D, |f(z) - a| < \varepsilon\}$$

with the following properties: (1) there is one nonvoid $D(\varepsilon, a)$ for each $\varepsilon > 0$; (2) $\bigcap_{\varepsilon > 0} D(\varepsilon, a) = \varnothing$; and (3) $\varepsilon_1 > \varepsilon_2 > 0 \Rightarrow D(\varepsilon_1) \supset D(\varepsilon_2)$. In case $a = \infty$, replace $|f(z) - a| < \varepsilon$ by $|f(z)| > 1/\varepsilon$. The end of the tract, $\bigcap \overline{D(\varepsilon, a)}$, is

* Supported in part by the U.S. Air Force Office of Scientific Research.

271

either a point ζ of C or an arc on C. The tract is called, respectively, a point-tract or an arc-tract. For $f \in \mathscr{A}[D]$, arc-tracts are possible only for the value $a = \infty$. Details of these matters will be found in [6].

In case $D = \{|z| < 1\}$, we write simply $\mathscr{A}$. Let $A^*[f]$ denote the set of points $\zeta \in C$ at which f has finite asymptotic values, and $A^\infty[f]$ the set at which f has the asymptotic value ∞. If a is a finite asymptotic value along an arc Γ such that $w = f(z)$ maps Γ one-one onto a linear segment, then we say that this asymptotic value (or tract) is linearly accessible. The set of points ζ at which f has linearly accessible finite asymptotic values is denoted by $A_l^*[f]$. Similarly, if $w = f(z)$ maps Γ one-one onto a ray to ∞, we say that that asymptotic value ∞ is linearly accessible; we denote the set of points on $|z| = 1$ where such occur by $A_l^\infty[f]$, and we set $A_l[f] = A_l^*[f] \cup A_l^\infty[f]$.

If $w = f(z)$ maps Γ onto a curve of finite length, then we will say that the asymptotic value is finitely accessible. The set of points on $|z| = 1$ at which f has finitely accessible asymptotic values will be denoted by $A_f^*[f]$.

In [6] we considered various basic facts about the class $\mathscr{A}$. Among other things, we showed the following, which will be used in this paper.

Theorem [6, Theorem 11]. *Let $f \in \mathscr{A}$ and let γ be an arc on $|z| = 1$ such that $A^\infty[f] \cap \gamma = \varnothing$. Then $\mathrm{meas}(A^*[f] \cap \gamma) > 0$.*

Theorem [6, Theorem 14]. *Let $f(z)$ be holomorphic and nonconstant in $|z| < 1$. If its maximum modulus satisfies*

(III) $$\int_0^1 (1 - r)\log^+ M(r)\, dr < \infty,$$

then $f \in \mathscr{A}$.

Theorem [6, Theorem 17]. $\mathscr{N} \subset \mathscr{A}$ *and the inclusion is proper. Also, if $f \in \mathscr{N}$, then*

1. *Given ζ, $|\zeta| = 1$, f has at most one asymptotic value at ζ. If f has the asymptotic value a at ζ, then f has the angular limit a at ζ.*

2. *f has no arc-tracts.*

Here $\mathscr{N}$ denotes the class of all nonconstant holomorphic functions in $|z| < 1$ which are normal in the sense of Lehto and Virtanen [4].

[6, Theorem 17], as well as the special case of [6, Theorem 11] (and also [6, Theorem 9]) where $f \in \mathscr{N}$, were first proved by Bagemihl and Seidel [1, 2]. The present author was ignorant of this fact when [6] was written.

In Section 2 we prove a fundamental lemma relating asymptotic values of a linear combination, $\phi(z)$, of the derivatives of f to asymptotic values of f.

In Section 3 we show that under suitable conditions, $\phi(z) \in \mathscr{A} \Rightarrow f \in \mathscr{A}$ and $A^*[f]$ dense.

Various results on linearly accessible asymptotic values are collected in Section 4.

In Section 5 the results of Section 3 are strengthened for the particular case where $f'(z) \neq 0$. These results depend on some of those in Section 4, as well as Section 3.

In Section 6 we consider the question of when $f \in \mathscr{A} \Rightarrow e^f \in \mathscr{A}$. This result is not true without supplementary hypotheses. The section concludes with Theorem 13 which shows that, under the special conditions $f \neq 0$ and $f' \neq 0$, the condition (III) may be weakened to

(IV)
$$\int_0^1 (1 - r)\log^+ \log^+ M(r)\, dr < \infty.$$

Finally, Section 7 contains several examples which illuminate some of the preceding results.

2. Fundamental Lemma

Let D be a domain in the finite z-plane and let $f(z)$ be holomorphic in D. Let $b \neq \infty$ be a boundary point of D, and let $p_0(z), p_1(z), \ldots, p_{n-1}(z), q(z)$ be given functions, holomorphic in some disk $\Delta_0 = \{|z - b| < r_0\}$. Let $\Gamma: z = \psi(u)$, $0 \leq u \leq 1$, be a continuous curve such that $\psi(1) = b$ and

$$\Gamma - \psi(1) \subset D \cap \Delta_0 = D_0.$$

If Γ has the following three properties:
1. $\Gamma_s : z = \psi(u)$, $0 \leq u \leq s$, is rectifiable for any $s < 1$.
2. The function

(1)
$$\phi(z) \equiv f^{(n)}(z) + \sum_{m=0}^{n-1} p_m(z) f^{(m)}(z) + q(z) \qquad (z \in D_0)$$

satisfies $\phi(\psi(u)) \to \lambda \neq \infty$ as $u \uparrow 1$.
3. Either (a) Γ is rectifiable
      or (b) $\phi_1(\psi(u)) = \phi(\psi(u)) - q(\psi(u)) \in BV[0, 1]$.
Then f has a finite asymptotic value on Γ.

Remarks. Briefly, but inaccurately, if $\phi(z)$ has a finite asymptotic value on Γ, then so does $f(z)$. One should bear in mind the particular cases $\phi(z) = f^{(n)}(z)$ for any positive integer, in particular $n = 1$.

In 3b it is understood that $\phi(\psi(1)) = \lambda$.

Proof. Consider the differential equation.

$$(2) \qquad w^{(n)}(z) + \sum_{m=0}^{n-1} p_m(z) w^{(m)}(z) = \phi(z) - q(z) \qquad (z \in D_0),$$

where $\phi(z)$ is defined by (1). Then $f(z)$ is *one* solution of (2) in D_0. Actually, our domain of operation will be D_1, that component of D_0 which contains $\Gamma - \psi(1)$. Let $g_1(z), \ldots, g_n(z)$ be a set of linearly independent solutions of the homogeneous equation associated with (2). The functions g_i are holomorphic in Δ_0. We use the classical fact (variation of parameters) that the solution f of (2) is given by

$$(3) \qquad f(z) = \sum_{m=1}^{n} \left\{ \alpha_m + \int_a^z h_m(t)[\phi(t) - q(t)] \, dt \right\} g_m(z)$$

for $z \in D_1$, where $a = \psi(0)$ and the α_m are constants. The functions h_m are holomorphic in Δ_0. Since g_m is holomorphic in Δ_0, the lemma follows if we show that each

$$(4) \qquad \beta_m(z) = \int_a^z h_m(t)[\phi(t) - q(t)] \, dt = \int_a^z h_m(t)\phi_1(t) \, dt$$

has a finite asymptotic value on Γ.

In case 3a this follows immediately since the integrand is continuous on Γ, and β_m will have the symptotic value

$$\beta_{m0} = \int_0^1 h_m(\psi(u))[\phi(\psi(u)) - q(\psi(u))] \, d\psi(u).$$

In case 3b, let $H_m'(z) = h_m(z)$ in Δ_0. Then

$$\beta_m(z) = \int_a^z \phi_1(t) \, dH_m(t) = \phi_1(z)H_m(z) - \phi_1(a)H_m(a) - \int_a^z H_m(t) \, d\phi_1(t).$$

Clearly each of the first two terms has a finite asymptotic value on Γ. Also the last term, since

$$\int_a^{\psi(s)} H_m \, d\phi_1 = \int_0^s H_m(\psi(u)) \, d\phi_1(\psi(u)) \to \int_0^1 H_m(\psi(u)) \, d\phi_1(\psi(u));$$

for $H_m(\psi(u))$ is continuous on $[0, 1]$ and $\phi_1(\psi(u))$ is both continuous and of bounded variation there.

3. Relation of Fundamental Lemma to Class $\mathscr{A}$

Theorem 1. *Let $f(z)$ be holomorphic and nonconstant in $|z| < 1$. Let there be given*

    1. *A disk $D = \{|z - 1| < \rho\}$. Set $\Delta = D \cap \{|z| < 1\}$.*

    2. *An integer n and functions $p_0(z), \ldots, p_{n-1}(z), q(z)$ holomorphic in D.*

Let

$$(5) \qquad \phi(z) = f^{(n)}(z) + \sum_{m=0}^{n-1} p_m(z) f^{(m)}(z) + q(z) \qquad (z \in \Delta).$$

If $\phi \in \mathscr{A}[\Delta]$ and if there is a constant c such that $\phi_1(z) = \phi(z) - q(z) \neq c$, then $f \in \mathscr{A}[\Delta]$, and more strongly

$$(6) \qquad A^*[f] \text{ is dense on } \gamma = D \cap \{|z| = 1\}.$$

Remarks. D need not be centered at 1; any point of $|z| = 1$ will do. If there are a finite number of such disks D, covering $|z| = 1$, then the conclusion becomes $f \in \mathscr{A}$ and $A^*[f]$ dense on $|z| = 1$. In this situation, the integer n and the functions p and q can vary from disk to disk, as can the constant c. In elementary cases, such as $\phi(z) = f^{(n)}(z) + q(z)$ where q is a constant or holomorphic on $|z| \leq 1$, one disk will, of course, do.

Proof. Since D could be replaced by a smaller disk, contained in D and centered on $|z| = 1$, it will suffice to show that f possesses a finite asymptotic value at one point of γ.

Define $E \subset \{|z| = 1\}$ as follows: $\zeta \in E \Leftrightarrow |\zeta| = 1$, there exists a neighborhood $U_\zeta = \{|z - \zeta| < h\} \cap \{|z| < 1\}$, and a Jordan arc J_ζ such that $\phi(z)$ maps U_ζ into the complement of J_ζ; i.e., in U_ζ, ϕ assumes no value of J_ζ. Clearly E is open. Similarly, we define E_1, using ϕ_1 instead of ϕ. Set $E_2 = E \cup E_1$.

Consider the case where γ contains points of E_2. By shrinking D we may assume that $\gamma \subset E_2$. It follows from simple generalizations of the Fatou and Riesz-Riesz theorems that both ϕ and ϕ_1 have finite angular limits almost everywhere on γ. In the proof of this one uses the fact that $\phi(z) - \phi_1(z) = q(z)$ has angular limits everywhere on γ. Considering (3) and (4) it is easily seen that the $\beta_m(z)$, and hence f also, have finite angular limits almost everywhere on γ.

It follows from a theorem of Priwalow [8, p. 210] that the asymptotic values assumed by $f(z)$ on any interval γ contained in E_2 form a set containing a closed set of positive harmonic measure. So, in particular, *this set must be infinite.*

If E_2 is dense on γ, then there is nothing more to prove. So let us assume that $\gamma \subset E_2'$. Since $\phi \in \mathscr{A}[\Delta]$, also $\phi_1 \in \mathscr{A}[\Delta]$. It follows that ϕ_1 has arbitrarily large asymptotic values densely on γ. Before showing this, we note that all asymptotic tracts of ϕ_1 must be point-tracts [6, p. 6] because ϕ_1 omits the value c [6, Theorem 7]. Now if the asymptotic values of ϕ_1 were bounded by M, take two distinct points ζ_1 and ζ_2 on γ at which ϕ_1 has asymptotic values and join ζ_1 and ζ_2 by a curve $\Gamma \subset \Delta$ which is an asymptotic path at both ζ_1 and ζ_2. Then ϕ_1 is bounded on Γ, say by B. Let G be the domain bounded by Γ and part of γ. If ϕ_1 were bounded in G, then clearly we would have $\zeta_1\zeta_2 \leq E_2$, a contradiction. Pick a value w_0 such that $\phi_1(z_0) = w_0$, $z_0 \in G$, and $|w_0| > \max(B, M)$. Then by the standard lifting argument [6, p. 13, second section], ϕ_1 has an asymptotic value a at some boundary point of G on $|z| = 1$, satisfying

$$\infty \geq a > |w_0| > M : \text{contradiction}.$$

Thus we now assume that Γ is such that the two asymptotic values of ϕ_1 at its end are $> 2|c|$ in absolute value, where c is the omitted value of ϕ_1. Hence $|\phi_1 - c| \geq \delta > 0$ on Γ. Since ϕ_1 omits few values in G (less than a Jordan arc full), there exists $z_0 \in G$ such that $|\phi_1(z_0) - c| < \delta$. Now the standard lifting argument shows that there exists $\Gamma_1 \subset G$ and ending at a point of γ such that not only does ϕ_1 have a finite asymptotic value b on Γ, $|b - c| < \delta$, but actually $\phi_1(z)$ maps Γ, one to one, onto a linear segment ending at b. Now use the fundamental lemma, case 3b, and we see that f has a finite asymptotic value on Γ_1. This completes the proof of Theorem 1.

Considering the two parts of the proof, one thinks of drawing certain conclusions about the number of distinct (numerically) asymptotic values of f locally. However, such results are true in general for $f \in \mathcal{A}$:

Theorem 2. *Let $f \in \mathcal{A}$, and let γ be any interval on $|z| = 1$. Then either $A^\infty[f]$ is dense on γ or f assumes infinitely many numerically distinct asymptotic values on γ.*

Remark. This is a trivial corollary of [6, Theorem 11 and its corollary]. However, it will prove useful to us here, and its proof is much easier than that of [6, Theorem 11].

Proof. If $A_\infty \cap \gamma = \varnothing$, then either f is locally bounded at some arc $\gamma_1 \subset \gamma$ (in which case use Riesz-Riesz theorem), or f is unbounded and hence has arbitrarily large, but still finite, asymptotic values on γ.

4. On Linear Accessibility

We now turn our attention to linearly accessible asymptotic values. We first prove some general facts and then, in Section 5, connect linear accessibility with Theorem 1. The next two theorems are rather specialized. Theorem 4 will be used in the proof of Theorem 5. Theorem 4 is an interesting complement to the theorem of Lohwater and Piranian [5]; they produce a schlicht function, mapping $|z| < 1$ onto a Jordan domain, such that the linearly accessible asymptotic values occur at the points of a set of measure zero on $|z| = 1$. Theorem 4 is of course pretty obvious for schlicht functions.

We shall say that $f(z)$ is *finitely valent densely* (fvd) provided there is a set of points $\{w_n\}$ dense in the w-plane and a set of nonnegative integers $\{p_n\}$ such that $f(z)$ assumes the value w_n no more than p_n times in $|z| < 1$. Here $\sup p_n = \infty$ is allowable. As usual, f will be said to be p-valent provided it assumes no value more than p times.

Theorem 3. *Let $f(z)$, holomorphic and nonconstant in $|z| < 1$, be finitely valent densely. Then $f \in \mathcal{N} \subset \mathcal{A}$, and $A_l^*[f]$ is dense on $|z| = 1$. More strongly, the set of numerically distinct finite linear asymptotic values occurring at points of any arc $\gamma \subset \{|z| = 1\}$ is infinite.*

Theorem 4. *Let $f(z)$, holomorphic and nonconstant in $|z| < 1$, be p-valent. Then $f \in \mathcal{N} \subset \mathcal{A}$, and the set of numerically distinct linearly accessible asymptotic values occurring on any arc $\gamma \subset \{|z| = 1\}$ has the power of the continuum.*

Proof of Theorem 3

Since $f(z)$ assumes the values w_1 and w_2 only a finite number of times, it is elementary to prove $f \in \mathcal{N}$. Here in considering asymptotic tracts it will prove convenient to use squares rather than disks in defining them. Let $Q(\varepsilon, a)$ be the square with center a, and sides, parallel to the real and imaginary axes, of length ε. The domains $D(\varepsilon, a)$ making up an asymptotic tract for the finite value a will be domains in which $f(z)$ takes values inside $Q(\varepsilon, a)$; if $a = \infty$, domains in which f takes values outside $Q(1/\varepsilon, 0)$.

Now let γ be any arc of $|z| = 1$ and let a be a finite asymptotic value of f at an interior point of γ. By choosing ε small enough, the associated $D(\varepsilon, a)$ will be bounded by curves in $|z| < 1$ on which f takes values on $Q(\varepsilon, a)$ and interior points of γ. We can at the same time, since f is fvd, choose ε so that a point w_n lies on $Q(\varepsilon, a)$. Let $w = f(z)$ map $|z| < 1$ onto the Riemann surface $\mathscr{F}$ over the w-plane, and let $\mathscr{D}(\varepsilon, a)$ correspond to $D(\varepsilon, a)$. That part of the boundary of $\mathscr{D}(\varepsilon, a)$ which lies over $Q(\varepsilon, a)$ can contain no closed curves, as otherwise $D(\varepsilon, a)$ would be relatively compact in $|z| < 1$. Hence each boundary component over $Q(\varepsilon, a)$ will be an open polygonal arc which, because of the presence of w_n on $Q(\varepsilon, a)$ will contain only a *finite* number of segments. The last segment of any such polygon obviously produces a linearly accessible asymptotic value of f at some interior point of γ.

If the only asymptotic values of f on γ are ∞, the proof is slightly different. By [6, Theorem 17], f has no arc tracts; hence all boundary points of $D(\varepsilon, \infty)$ [values of f outside $Q(1/\varepsilon, 0)$] will be interior to $|z| < 1$ or interior points of γ, if ε is small enough. Also we choose ε so that w_n is on $Q(1/\varepsilon, 0)$ for some n. In this case $D(\varepsilon, \infty)$ may be multiply connected and *some* boundary components of $\mathscr{D}(\varepsilon, \infty)$ may be closed polygons over $Q(1/\varepsilon, 0)$. If *all* boundary components were closed for each ε in a permissible sequence (compatible with $\{w_n\}$) of ε's decreasing to zero, then it is clear that f would possess a global tract for ∞ [6, p. 6]. But f has no arc-tracts, and hence for some ε, $\mathscr{D}(\varepsilon, \infty)$ will have an open finite sided boundary component over $Q(1/\varepsilon, 0)$, and the argument is finished as before.

Proof of Theorem 4

The proof of Theorem 4 requires only a slight change in the proof of Theorem 3: now *all* ε sufficiently small may be used, and each one leads to at least one linearly accessible asymptotic value.

Remarks. The hypothesis of Theorem 4 can be weakened without altering the conclusion. For a function f let $p(f, w)$ denote the number of times that f assumes the value w in $|z| < 1$. *The hypotheses p-valent may be replaced by the following*: $p(f, w) < \infty$ *for all w. Or, even weaker: each neighborhood in the w-plane contains a subset A, of the power of the continuum, on which* $p(f, w) < \infty$. The proof of Theorem 4 remains unaltered, except that in the last case the squares $Q(\varepsilon, a)$ used may on occasion have to be rotated about a.

Theorem 5. *Let $f \in \mathscr{A}$ and $f(z) \neq c$, where c is some constant. Then $A_l^*[f]$ is dense on $|z| = 1$.*

Proof. We may assume $c = 0$ without loss of generality. Consider an arbitrary arc γ of $|z| = 1$; we shall produce a linear asymptotic value at some point of γ. Suppose first that there is an interior point ζ of γ such that $\lim \inf_{z \to \zeta} |f(z)| = 0$. By virtue of Theorem 2 we can find a cross-cut Γ of $|z| < 1$ from $\zeta_1 \in \gamma$ to $\zeta_2 \in \gamma$ with the following properties: f has nonzero asymptotic values on Γ at both ζ_1 and ζ_2; if $\gamma_1 \subset \gamma$ is the open arc from ζ_1 to ζ_2, then $\zeta \in \gamma_1$. Let Δ be the domain bounded by Γ and γ_1. Now $|f(z)| \geq m > 0$ on Γ. Pick a point $z_0 \in \Delta$ such that $|f(z_0)| < m$. The usual argument of lifting the ray $[f(z_0), 0)$ produces the desired linear asymptotic value at a point of γ_1.

If there is no such point ζ on γ, then (shorten γ if necessary) there is a neighborhood U of γ such that

$$(7) \qquad\qquad |f(z)| \geq m > 0 \qquad (z \in U).$$

By the Riesz-Riesz theorem, $f(z)$ has finite asymptotic values at a dense set of points of γ. Thus there exists a cross-cut Γ of U with the following properties. (1) Γ ends at distinct points ζ_1 and ζ_2 of γ and f has finite asymptotic values on Γ at both ends: $f(z) \to a_i$ as $z \to \zeta_i$ on Γ, $i = 1, 2$. (2) The image of Γ by $w = f(z)$ will be a polygonal curve $\tilde{\pi}$ on the Riemann surface $\mathscr{F}_1$ onto which f maps U; in general, it will have an infinite number of sides in either direction; we will assume so, for if not, the proof of the Theorem is complete. Let π be the projection of $\tilde{\pi}$ into the w-plane. The lengths of the sides of π tend to zero and the sides tend to a_1 and a_2.

Let $\gamma_1 \subset \gamma$ be the arc between ζ_1 and ζ_2 and let Δ be the domain bounded by Γ and γ_1. Let $w = f(z)$ map Δ onto the Riemann surface $\mathscr{F}$. Now it is easy enough to produce a linear asymptotic value of f in Δ. Just pick a point on $\mathscr{F}$ and head along a line to some point of $|w| = m$. But we must be sure to produce an asymptotic value corresponding to a point of γ_1. The simple argument just stated might lead us to a point of Γ, and that would not serve.

If f satisfies the condition that each neighborhood in the w-plane contains a subset A, of the power of the continuum, on which $p(f(z), w) < \infty$ for $z \in \Delta$, then we are done. (See Remarks following proof of Theorem 4.) If not, let w_0 be some value which $f(z)$ assumes ∞ many times in Δ and such that $w_0 \neq a_1, a_2$. Pick any ray R from w_0 to a point $w_1 = me^{i\alpha}$, where α is picked so that (1) R is a positive distance from both a_1 and a_2 and R is not parallel to any side of π; (2) there are no branch points of $\mathscr{F}$ over R. Consider all the liftings of R into $\mathscr{F}$, starting at each of the infinitely many points over w_0; these will all be unique and disjoint because of 2. Some of the liftings may be halted at points of $\tilde{\pi}$, giving asymptotic values at points of Γ. But clearly two distinct liftings cannot be halted at the same point of $\tilde{\pi}$. By 1, R intersects π in only a finite number of points. Hence this total lifting process produces a countable infinity of linear asymptotic values, a finite number (possibly none) corresponding to points of Γ, all the rest associated with points of γ_1. This completes the proof of Theorem 5.

Remarks. The set of permissible R in the above paragraph has the power of the continuum. Hence we can draw a stronger conclusion than that stated in Theorem 5, along arcs where (7) is satisfied. Actually, (7) may be replaced by something more general, and we have the following.

Corollary 5. *Let $f(z)$ be holomorphic and nonconstant in $|z| < 1$. Let Ω be a set in the finite w-plane with $\operatorname{card}(\Omega) = \omega \geqq 2$. Suppose $f(z)$ assumes no value in Ω. Then $f \in \mathscr{N} \subset \mathscr{A}$.*

For an arbitrary $\gamma \subset \{|z| < 1\}$, let $\eta(\gamma)$ denote the cardinal number of numerically distinct finite linear asymptotic values associated with points of γ. Then $\eta(\gamma) \geqq \omega$.

In case $\omega \leqq \eta(\gamma) < \infty$, then each of at least ω of the finite linear asymptotic values is associated with an infinity of distinct tracts ending at distinct points of γ.

Proof. The proof follows, in essence, the second part of the proof of Theorem 5, starting with (7), which is now replaced by our knowledge of Ω. We can't use the Riesz-Riesz theorem, but a_1 or $a_2 = \infty$ will do. The only important point here is that if $\omega < \infty$, then by Theorem 2 we can ensure a_1, $a_2 \notin \Omega$, for π mustn't condense on any point of Ω.

To replace the last paragraph of the proof of Theorem 5: For the first part, use the weakest version of Theorem 4 (see Remarks following proof of Theorem 4). Then for the final argument (lifting R): there exists a neighborhood V in the w-plane such that $p(f(z), w) = \infty$ for $z \in \Delta$ and $w \in V - E$ where card$(E) \leq \mathfrak{c}$, the cardinal of the continuum. In case $\omega < \infty$, it is easily seen that there exists $w_0 \in V - E$ such that the points of Ω lie on ω distinct rays from w_0. For the case $\omega \not< \infty$, pick 3 noncollinear points of $V - E$ and use the easily proved

Lemma. *Let Ω be an infinite set in the finite plane, card $\Omega = \omega$. Let $a \notin \Omega$. Define* asp$(\Omega, a) = $ card$\{\theta \mid \theta = \arg(w - a), w \in \Omega, -\pi < \theta \leq \pi\}$*. If $a, b, c \notin \Omega$ are noncollinear, then*

$$\omega = \max\{\text{asp}(\Omega, a), \text{asp}(\Omega, b), \text{asp}(\Omega, c)\}.$$

So we can always find $w_0 \in V - E$ so that the points of Ω lie on ω distinct rays from w_0. Consider all possible liftings of these rays into $\mathscr{F}$, etc.

Theorem 6. *Let $f \in \mathscr{A}$ and let γ be an arc on $|z| = 1$. If $A^\infty \cap \gamma = \varnothing$, then the set of numerically distinct linearly accessible finite asymptotic values associated with points of γ has the power of the continuum.*

Remark. These asymptotic values also occur at *distinct* points of γ [6, p. 7, (2.6)].

Proof. As we know [6, Theorem 11 and Corollary], $f(z)$ has infinitely many distinct asymptotic values at points of γ, enough to include a closed set of positive harmonic measure. But that result says nothing about linear accessibility.

Let Γ be a cross-cut of $|z| < 1$, ending at ζ_1 and $\zeta_2 \in \gamma$ such that f has a finite asymptotic value along each end of γ. Let $\gamma_1 \subset \gamma$ be the arc from ζ_1 to ζ_2 and let Δ be the domain bounded by Γ and γ_1. If f is bounded in Δ, then our conclusion follows from Corollary 5.

Otherwise, let $|f| \leq M$ on Γ and choose $z_0 \in \Delta$ such that $|f(z_0)| > M$. There is a continuum of distinct rays from $f(z_0)$ to ∞ which do not meet $|w| = M$. Lifting each of these rays obviously leads to completion of the proof.

If we consider the more general question of linear accessibility, including the linear accessibility of ∞, then we obtain the following result.

Theorem 7. *Let $f \in \mathscr{A}$ have no arc-tracts. Then $A_l[f] = A_l^*[f] \cup A_l^\infty[f]$ is dense on $|z| = 1$.*

Proof. Let γ be any open arc on $|z| = 1$. If $A^{\infty} \cap \gamma = \varnothing$, then Theorem 6 settles the matter. If not, then f has a point-tract $\{D(\varepsilon_n)\}$ for ∞ at a point $\zeta \in \gamma$. Choose ε_n so that $D(\varepsilon_n)$ satisfies

$$\overline{D(\varepsilon_n)} \cap \{|z| = 1\} \subset \gamma.$$

Pick a point $z_0 \in D(\varepsilon_n)$ and lift the ray $w = tf(z_0)$, $1 \le t < \infty$, etc.

5. $\mathscr{A}$ and Exceptional Values of $f'(z)$

We start with the case of primary interest, namely, $f'(z) \neq 0$.

Theorem 8. *Let $f \in \mathscr{A}$ and suppose $f'(z) \neq 0$. Let a be a finite asymptotic value of f at $\zeta \in \{|z| = 1\}$, with tract $\mathscr{T} = \{D(\varepsilon)\}$. Then either a is a linearly accessible asymptotic value, in $\mathscr{T}$, or else there is an infinite sequence of numerically distinct linear asymptotic values $\{a_n\}$ occurring at points*

$$\zeta_n \in \{|z| = 1\}$$

such that $a_n \to a$ and $\zeta_n \to \zeta$. The analogous result holds for $a = \infty$, provided f has no arc-tracts.

Remark. If f has an arc-tract for ∞, the theorem remains true in some sense, as will become clear from the proof. Namely, either ∞ is linearly accessible in the tract, although perhaps only along a curve which tends to an arc, not a point, of $|z| = 1$; or we have the distinct finite asymptotic values a_n with $\zeta_n \to \zeta$ where ζ is one end point of the arc which is the end of the tract.

In this connection, if $f \in \mathscr{A}$ and $f' \neq 0$, then f cannot have a global tract. See [6, Theorem 8]; since $f' \neq 0$, the $\Delta(r, n_0)$ with one fixed center would all be schlicht disks, which is impossible. It seems reasonable to guess that that $f' \neq 0$ and $f \in \mathscr{A}$ rules out all arc-tracts, but we have been unable to prove this.

Proof. Consider $a \neq \infty$, and let U be a neighborhood, in $|z| < 1$, of ζ. Choose ε small enough so that $D(\varepsilon) \subset U$. Let $\mathscr{D}(\varepsilon)$ be the part of the Riemann surface of f^{-1} corresponding to $D(\varepsilon)$. Pick a point Q of $\mathscr{D}(\varepsilon)$ over $|w - a| = \varepsilon/2$. Such points certainly exist if ε is small enough. Consider the maximal schlicht disk which can be imbedded in $\mathscr{D}(\varepsilon)$ with center Q. If the radius ρ of this disk is less than $\varepsilon/2$, then some radius of this disk leads to a "boundary" point of $\mathscr{D}(\varepsilon)$ which is *not* over a or $|w - a| = \varepsilon$, and we get a linear asymptotic value $a_1 \neq a$, occurring at a boundary point of U on $|z| = 1$. It is right here that the lack of branch-points in the Riemann surface is really used. Otherwise the maximal schlicht disk could be determined by a branch point rather than a "boundary" point. If for each ε there is a maximal disk of radius $\rho < \varepsilon/2$, then we have our a_n and ζ_n. Note that there is nothing to show that the ζ_n are distinct.

If $\rho = \varepsilon/2$ for all small ε and all choices Q, then it is easily seen that $\mathcal{D}(\varepsilon)$ (which must be simply connected) is either (1) an m-sheeted covering of $|w - a| < \varepsilon$ with a single branch-point (of order $m - 1$) over $w = a$; or (2) the universal covering of $0 < |w - a| < \varepsilon$. The first case, 1, is ruled out since it is incompatible with our original asymptotic value a. In case 2, which can occur, we clearly have the linear accessibility of a in $\mathcal{T}$. Note that we have proved somewhat more than the theorem explicitly states. Namely, *either a_n, ζ_n exist as stated, or $\mathcal{T}$ corresponds to a logarithmic branch point.*

The proof in case $a = \infty$ is similar, and the italicized remark just above holds here also.

Finally, we return to Theorem 1, to prove

Theorem 9. *Let $f' \in \mathcal{A}$ and $f'(z) \neq 0$. Then $f \in \mathcal{A}$ and on any arc $\gamma \subset \{|z| = 1\}$ f possesses at least 3 numerically distinct asymptotic values.*

Remarks. Combining Theorems 8 and 9: *under the hypothesis of Theorem 9, any open arc $\gamma \subset \{|z| = 1\}$ either has:* (1) *infinitely many numerically distinct linear asymptotic values associated with it;* or (2) *each of the (finitely many, numerically) asymptotic values associated with γ corresponds to a logarithmic branch point of f^{-1}.*

The number 3 cannot be improved; consider the modular function.

One of course raises the question of whether $f' \in \mathcal{A} \Rightarrow f \in \mathcal{A}$; to this we have no answer without some extra hypothesis such as $f' \neq 0$.

Note added in Proof. J. E. McMillan [*Duke Math. J.* **36** (1969), 567–570] has proved a closely related theorem: *Let $f \in \mathcal{A}$ have no arc-tracts and let $f' \neq 0$. Then on any arc $\gamma \subset \{|z| = 1\}$ f possesses at least 3 numerically distinct asymptotic values.* This theorem is actually equivalent to Theorem 9. That the hypotheses of Theorem 9 imply those of McMillan's theorem is immediately clear from Theorem 9 itself. The converse follows from the theorem: *Let $f \in \mathcal{A}$ have no arc-tracts and let $f' \neq 0$. Then $f' \in \mathcal{A}$.* This theorem, whose proof is too lengthy to insert in this note, was unknown to both McMillan and the present author. Cogitations set off by the preparation of this note led to its discovery.

Proof. It follows immediately from Theorem 1 that $f \in \mathcal{A}$ and $A^*[f]$ is dense on $|z| = 1$. From Theorem 8 it then follows that $A_l^*[f]$ is dense on $|z| = 1$.

Now to prove the existence of 3 numerically distinct asymptotic values locally. The global case ($\gamma = \{|z| = 1\}$) is trivial. If f is bounded, then it follows from Fatou and Riesz-Riesz. Otherwise f will have arbitrarily large asymptotic values; if it had only two numerically distinct ones, one would have to be ∞ and the other we can take to be zero. Then considering the Riemann surface $\mathscr{F}$ over the w-plane onto which $w = f(z)$ maps $|z| < 1$, and remembering that $\mathscr{F}$ has no branch points, it follows readily that $\mathscr{F}$ contains either the Riemann surface of log w or that of $w^{1/m}$, less its branch-points, for some positive integer m. In the first case, $\mathscr{F}$ would have to be precisely the Riemann surface of log w, which is impossible since the latter is parabolic. Similar considerations dispose of the second possibility.

To prove the local case, we may assume that A^{∞} is dense on γ; for otherwise Theorem 6 settles the business. Suppose there were only one finite asymptotic value, which we may take to be zero, on γ. That there is at least one finite asymptotic value, follows since we have already proved $A^*[f]$ dense. By Theorem 8 there are two points ζ_1 and $\zeta_2 \in \gamma$ at which f has the linear asymptotic value 0. Then we can construct Γ, a cross-cut of $|z| < 1$ with ends at ζ_1 and ζ_2 such that the image of Γ under $w = f(z)$ is a polygonal path π, with a finite number of sides, on the Riemann surface $\mathscr{F}$ onto which $w = f(z)$ maps $|z| < 1$. Let $\gamma_1 \subset \gamma$ denote the arc from ζ_1 to ζ_2, and let Δ be the domain bounded by Γ and γ_1. Let $w = f(z)$ map Δ onto the Riemann surface $\mathscr{S} \subset \mathscr{F}$.

We show next that f has only a finite number of zeros in Δ. With each point p_0 of $\mathscr{S}$ over $w = 0$ we associate the maximal schlicht open disk D_0, $|w| < \rho$, contained in $\mathscr{S}$ and centered at p_0. First it is clear that $\rho < \infty$. Secondly, since $\mathscr{S}$ has no branch-points, D_0 has at least one radius σ whose end point, $w = w_0$, is not a point of $\mathscr{S}$; that is, f, in Δ, has the asymptotic value w_0 on the arc $f^{-1}(\sigma)$. Now $w_0 \neq 0, \infty$; hence $f^{-1}(\sigma)$ must end at a point of Γ, since the only asymptotic values of f on γ are 0 and ∞. This it is seen that D_0 is either tangent to a segment of π or just touches π at a vertex of π. All disks D_0 are centered over $w = 0$; since $\mathscr{S}$ has no branch-points, distinct disks D_0 are "tangent" to distinct segments of π. Hence there are only a finite number of distinct disks D_0.

Thus we may assume that Δ is chosen "narrow" enough so that $f(z) \neq 0$ in Δ. Then $F(z) = \log f(z)$ is holomorphic in Δ and, since $f \in \mathscr{A}$, it is easily seen that $F \in \mathscr{A}[\Delta]$.

Lemma. $F' = f'/f \neq 0$, and $F' \in \mathscr{A}[\Delta]$.

We postpone the lemma's proof. Now using the first paragraph of the proof of Theorem 9 on the function F in the domain Δ, we see that $A_l^*[F]$ is dense on γ_1. Each finite asymptotic value of F yields a nonzero finite asymptotic value of $f = e^F$. This contradicts the assumption that 0 was the only finite asymptotic value of f at points of γ, and the proof of Theorem 9 is complete, except for the proof of our lemma.

Proof of Lemma. By Theorem 5, there are curves Γ, ending at points ζ dense on $|z| = 1$ such that

$$\Gamma: z = \phi(t), \qquad 0 < t \leq 1 \; (t \to 0, \; \phi(t) \to \zeta),$$
$$f'(\phi(t)) = a + ct \qquad (c \neq 0).$$

Then

$$f(\phi(1)) - f(\phi(\tau)) = \int_\tau^1 (a + ct) \, d\phi(t)$$

$$= (a + c)\phi(1) - (a + c\tau)\phi(\tau) - c \int_\tau^1 \phi(t) \, dt,$$

from which it is clear that f has the asymptotic value $f(\zeta)$ on Γ, where

$$f(\phi(1)) - f(\zeta) = (a + c)\phi(1) - a\zeta - c \int_0^1 \phi(t) \, dt.$$

Hence we obtain

$$f(\phi(\tau)) = f(\zeta) + (a + c\tau)\phi(\tau) - a\zeta - c \int_0^\tau \phi(t) \, dt.$$

Thus both f and f' have finite asymptotic values on Γ. If that of f', namely a, is nonzero, then clearly $F' = f'/f$ has a limit on Γ, either finite or infinite. In case $a = 0$, then

$$\frac{1}{F'(\phi(\tau))} = \frac{f(\phi(\tau))}{f'(\phi(\tau))} = \frac{f(\zeta) + c\tau\phi(\tau) - c \int_0^\tau \phi(t) \, dt}{c\tau}$$

$$= \frac{f(\zeta)}{c\tau} + \zeta + o(1) - \zeta - \frac{1}{\tau} \int_0^\tau (\phi(t) - \zeta) \, dt$$

$$= \frac{f(\zeta)}{c\tau} + o(1) \to \begin{cases} \infty & \text{if } f(\zeta) \neq 0 \\ 0 & \text{if } f(\zeta) = 0. \end{cases}$$

This completes the proof of the Lemma, and hence of Theorem 9.

Other Exceptional Values. Special cases of Theorem 1 show that *if $f^{(n)} \in \mathscr{A}$ and $f^{(n)} \neq c$ then $f \in \mathscr{A}$ and $A^*[f]$ is dense on $|z| = 1$.* But it is only in the case $n = 1$ and $c = 0$, or the case $f \in \mathscr{A}$ and f omits a value (Theorem 5), that we are able to draw conclusions on $A_l^*[f]$.

6. Exponentiation of Functions in $\mathscr{A}$

A question of some interest is the following. Under what conditions does $f \in \mathscr{A} \Rightarrow e^f \in \mathscr{A}$. That some condition other than just $f \in \mathscr{A}$ is necessary, is known [3]. The following is also known.

Theorem 10. *If $f \in \mathscr{A}$ has no arc-tracts, then $e^f \in \mathscr{A}$.*

If $f(z)$ has a *finite* asymptotic value along a curve Γ, then so does e^f. So any condition that yields $A^*[f]$ dense on $|z| = 1$, such as the conditions in Theorem 1, will give $e^f \in \mathscr{A}$. In particular, we have

Theorem 11. *Let $f^{(n)} \in \mathscr{A}$, $f^{(n)} \neq c$, for some constant c. Then each of $A^*[f]$ and $A^*[e^f]$ is dense on $|z| = 1$. In particular, $e^f \in \mathscr{A}$.*

In the case $n = 1$ there is a variant as follows.

Theorem 12. *Let f, $f' \neq 0$ and $f'/f \in \mathscr{A}$. Then f and $e^f \in \mathscr{A}$ and each of $A^*[f]$ and $A^*[e^f]$ is dense on $|z| = 1$.*

Proof. By Theorem 11, $A^*[\log f]$ is dense; the same then holds for $A^*[f]$, $A^*[e^f]$, $A^*[\exp(e^f)]$, ... etc.

The condition (III) for $f \in \mathscr{A}$ may be replaced by a much weaker restriction under special circumstances. In preparation we note that if $F(z)$ satisfies (III), then both $F'(z)$ and $\int F(z)\, dz$ do also. For if $M(r)$, $M_1(r)$ and $M_2(r)$ are the maximum moduli of F, F', and $\int F$, respectively, then $M_1(r) = O[(1 - r)^{-2} \times M(\tfrac{1}{2}(r + 1))]$ and $M_2(r) = O[M(r)]$, etc.

Theorem 13. *Let $f(z)$, holomorphic and nonconstant in $|z| < 1$, satisfy $f \neq 0, f' \neq 0$. Let $M(r)$ denote the maximum modulus of $f(z)$. If*

(IV)
$$\int_0^1 (1 - r)\log^+ \log^+ M(r)\, dr < \infty$$

then $f \in \mathscr{A}$, and also the conclusions of Theorems 5 and 8 hold for f.

Proof. Set $F(z) = \log f(z)$, let $M_1(r)$ denote the maximum modulus of $F(z)$, and set

$$A_1(r) = \max_\theta \mathscr{R}F(re^{i\theta}) = \log M(r).$$

Using the Borel-Carathéodory theorem [9, p. 174] with r and $\tfrac{1}{2}(r + 1)$,

$$M_1(r) \leqq \frac{4r}{1 - r} A_1\left(\frac{r + 1}{2}\right) + \frac{1 + 3r}{1 - r} |F(0)|.$$

from which it readily follows that

$$\log^+ M_1(r) \leqslant \log^+ \log^+ M\left(\frac{r + 1}{2}\right) + \log^+ \frac{1}{1 - r} + O(1);$$

from this and (IV) it follows easily that $M_1(r)$ satisfies (III), and hence $F \in \mathscr{A}$.

From the remark preceding Theorem 13, also $F' \in \mathscr{A}$. Also $F' = f'/f \neq 0$.

Then by Theorem 11, $A^*[e^F] = A^*[f]$ is dense on $|z| = 1$. Now we know that f satisfies the hypotheses of Theorems 5 and 8, and the conclusions of those theorems hold.

Remark. It is clear that we may replace the hypothesis $f \neq 0$ by $f \neq c$ for some constant c.

7. Some Examples

Example 1. Any one of the Schwarz triangle functions $f(z)$, where $w = f(z)$ maps $|z| < 1$ onto the simply connected covering $\mathscr{F}$ of $|w| < \infty$ which is smooth and unbordered, except that all points of $\mathscr{F}$ over $0(1)$ are branch-points of multiplicity $m(n)$, with $1/m + 1/n < 1$. Automatically, all "points" over $w = \infty$ turn out to be logarithmic branch-points.

A detailed look at the well-known triangles in $|z| < 1$ corresponding to the half-sheets of $\mathscr{F}$ makes it clear that $f \in \mathscr{A}$, for f has the point asymptotic value infinity at each of a dense set of points on $|z| = 1$. But there are clearly no finite asymptotic values.

Simple considerations on the growth of f show that both f and f' satisfy condition (III) (see Section 1) and hence also $f' \in \mathscr{A}$. This example is instructive in connection with Theorem 9, for it shows that $f' \neq 0$ plays a vital role.

Example 2. A more complicated construction will show that there exist functions $\Phi \in \mathscr{A}$ such that $A^*[\Phi]$ is dense on $|z| = 1$, but $A_l^*[\Phi] = \varnothing$, and, even worse, $A_f^*[\Phi] = \varnothing$.

We start with the Riemann surface $\mathscr{M}$ of the modular function and replace all the logarithmic neighborhoods of $\mathscr{M}$ over $|z| < \frac{1}{3}$ and $|z - 1| < \frac{1}{3}$ by something else. The replacements over $|z| < \frac{1}{3}$ are built as follows. Let $\mathscr{U}_1, \mathscr{U}_2, \ldots$ be distinct copies of the Riemann surface of $w^{1/3}$ over $|w| < \frac{1}{3}$. For convenience we think of each $\mathscr{U}_n$ as being built of 3 sheets which are connected to each other over $(-\frac{1}{3}, 0]$. Let $\{p_n\}_1^\infty$ be a given sequence of real numbers such that $0 < p_n < \frac{1}{4}$. For each $n \geq 1$, slit $\mathscr{U}_n$ in its top sheet and $\mathscr{U}_{n+1}$ in its bottom sheet, over the real axis from p_n to $\frac{1}{3}$; then connect $\mathscr{U}_n$ and $\mathscr{U}_{n+1}$ along these slits in the familiar fashion, forming a first-order branch-point over $w = p_n$.

It is easily seen that the resulting Riemann surface $\mathscr{P}_0(\{p_n\})$ is simply connected and that it has precisely one component over $\frac{1}{4} < |w| < \frac{1}{3}$, namely the universal covering of that annulus. So we can replace each logarithmic neighborhood over $|w| < \frac{1}{3}$ by some $\mathscr{P}_0(\{p_n\})$. The sequences $\{p_n\}$ may all be the same or may vary from neighborhood to neighborhood. Similarly, we replace all logarithmic neighborhoods over $|w - 1| < \frac{1}{3}$ by surfaces $\mathscr{P}_1(\{q_n\})$, which are built in similar fashion to the $\mathscr{P}_0$. Call the resulting Riemann surface $\mathscr{F}$.

Let $w = \Phi(z)$ map $|z| < R \leqq \infty$ onto $\mathscr{F}$. Since the only zeros and 1-points of Φ are triple and $\frac{1}{3} + \frac{1}{3} < 1$, it follows from Montel's theorem [7, p. 65] that $\Phi(z)$ is normal in the sense of Lehto and Virtanen [4]. Clearly a nonconstant entire function cannot be normal. Hence $R < \infty$ (so we may take $R = 1$) and $\Phi \in \mathscr{N}$. Hence $\Phi \in \mathscr{A}$ and all asymptotic tracts are point tracts.

Now let us assume that every sequence $\{p_n\}$ and $\{q_n\}$ satisfies $p_n \to 0$, $q_n \to 0$, $\Sigma p_n = \Sigma q_n = \infty$. It is then clear that associated with each $\mathscr{P}_0(\mathscr{P}_1)$ we have one asymptotic tract with asymptotic value $0(1)$. But none of these are linearly accessible, and moreover, since $\Sigma p_n = \Sigma q_n = \infty$, none is accessible by a path of finite length.

Comparing the structures of $\mathscr{F}$ and $\mathscr{M}$, it is seen that the points of $|z| = 1$ where Φ has asymptotic values 0 $(1, \infty)$ occur in the same relative order as for the modular function. In particular, each of the three asymptotic values occurs at a dense set on $|z| = 1$. There can be no gaps. For $\Phi \in \mathscr{A}$, with no arc tracts, and between two occurrences of the asymptotic value 0 there must be occurrences of the asymptotic values 1 and ∞.

This example is pertinent to various of the preceding theorems. It shows that the hypotheses in Theorems 5, 6, 8, and 9 are not unnecessary. It also shows that $A_l^\infty[f]$ must be included in the conclusion of Theorem 7. It is also instructive to consider the example obtained if $p_n \nrightarrow 0$ and $q_n \nrightarrow 0$; say all p_n and $q_n = \frac{1}{4}$.

Example 3. Another interesting example is obtained as follows. Start with the universal covering of the sphere less the 4 points $\{0, 1, -1, \infty\}$, call it $\mathscr{G}$. Arrange the logarithmic branch-points over -1 in some sequence $\{P_n\}_1^\infty$ and pick a sequence of positive numbers $\{p_n\}_1^\infty$ such that $p_n \to 0$. Now we distort $\mathscr{G}$ as follows. Displace the logarithmic branch-point P_n so that it lies over $-p_n$ rather than -1. Call the resulting surface $\mathscr{F}$. Now $\mathscr{F}$ has no points over $\{0, 1, \infty\}$ and hence is hyperbolic; also, if $w = f(z)$ maps $|z| < 1$ onto $\mathscr{F}$, then $f \in \mathscr{N} \subset \mathscr{A}$.

Consider any component $\mathscr{D}(r)$ of $\mathscr{F}$ over $|w| < r$, for $0 < r < 1$. $\mathscr{D}$ will contain ∞-many of the displaced branch-points P_n. Clearly if we pick $r_1 < r$ so that at least one of these logarithmic branch-points in $\mathscr{D}$ lies over $-p_k < -r_1$, then $\mathscr{D}(r)$ will contain ∞-many distinct noncompact components of $\mathscr{F}$ over $|w| < r_1$. Proceeding in this fashion, we easily see that f has a (power of the) *continuum* of distinct asymptotic tracts, with asymptotic value 0, associated with any arc $\gamma \subset \{|z| = 1\}$. But clearly there are only a *countable* number of distinct linearly accessible asymptotic tracts for f.

REFERENCES

1. F. Bagemihl and W. Seidel, Behavior of meromorphic functions on boundary paths, with applications to normal functions, *Archiv Math. (Basel)* **11** (1960), 263–269.

2. ———, Koebe arcs and Fatou points of normal functions, *Comment. Math. Helv.* **36** (1961), 9–18.

3. K. F. Barth and W. J. Schneider, Exponentiation of functions in MacLane's class $\mathscr{A}$ *J. Reine Angew. Math.* **236** (1969) 120–130.

4. O. Lehto and K. I. Virtanen, Boundary behavior and normal meromorphic functions, *Acta Math.* **97** (1957), 47–65.

5. A. J. Lohwater and G. Piranian, Linear accessibility of boundary points of a Jordan Region, *Comment. Math. Helv.* **25** (1951), 173–180.

6. G. R. MacLane, Asymptotic values of holomorphic functions, *Rice University Studies*, **49** No. 1 (1963).

7. P. Montel, *Leçons sur les Familles Normales de Fonctions Analytiques et leurs Applications*, Gauthier-Villars, Paris, 1927.

8. I. I. Priwalow, *Randeigenschaften analytischer Funktionen*, VEB Deutscher Verlag der Wissenschaften, Berlin, 1956.

9. E. C. Titchmarsh, *The Theory of Functions*, 2nd ed., Oxford University Press, London, 1939.

(Received May 4, 1968)

The Spectra of Step-Function Potentials

J. B. McLeod

Wadham College
Oxford,
England

1.

In a recent series of papers [5, 6, 7] I considered the spectral theory associated with the equation

(1.1) $$\frac{d^2y}{dx^2} + \{\lambda - q(x)\}y = 0 \qquad (0 \leqq x < \infty)$$

with a homogeneous boundary condition at $x = 0$, in cases where $q(x) = h(x)g(x)$ and, roughly, $h(x)$ is increasing and $g(x)$ periodic; and I succeeded in showing that the spectrum is discrete. The precise conditions on $h(x)$, $g(x)$ are as follows:

   (i) $h(x), g(x)$ are real and have continuous derivatives up to and including the third order;

   (ii) $g(x)$, together with its first three derivatives, is bounded;

   (iii) $h(x)$ is a strictly positive function, but $g(x)$ has zeros at the points $a_0, a_1, \ldots$, say, with $a_0 < a_1 < \cdots$, and these zeros are all simple;

   (iv) $g^2(x) + g'^2(x) \geqq K > 0$ for some positive constant K;

   (v) $h' = O(h), \quad h'' = O(h), \quad h''' = O(h)$;

   (vi) $\displaystyle\int_{a_n}^{a_{n+1}} h^{-1/2}(x)\, dx \to 0 \quad \text{as} \quad n \to \infty,$

where a_n, a_{n+1} are any two successive zeros of $g(x)$.

Without thinking too deeply about the matter, I had supposed that this result about the discrete nature of the spectrum would hold much more widely, and that the detailed assumptions above on the differentiability and order of $h(x)$ and $g(x)$ were necessary only because of the nature of the proof. In particular, I had supposed that if $q(x)$ was a step-function, with steps, say, at the integers, such that $q(x)$ was alternately positive and negative and of increasing modulus, then the discreteness of the spectrum remained valid. (The necessity of some assumptions on the nature of the oscillations of $q(x)$ is shown by Hartman's examples [3], where in a sense any spectral behavior results from wildly oscillating $q(x)$; but the negative oscillations are always of a higher order of magnitude than the positive ones, and the lengths of the positive oscillations tend to 0 as $n \to \infty$.)

A communication from Professor J. E. Littlewood (see [4]) suggested that my supposition must be wrong. He considers

(1.2) $$\frac{d^2y}{dx^2} = q(x)y,$$

where

$$q(x) = \begin{cases} c^2 \omega_n{}^2 & (n - \tfrac{1}{2} \le x \le n), \\ -c^2 \omega_n{}^2 & (n < x < n + \tfrac{1}{2}), \end{cases}$$

n being a positive integer, c real, and $\{\omega_n\}$ a monotonic increasing sequence with $\omega_1 \ge 1$, $\omega_n/\log n \to \infty$. His results are as follows.

Theorem 1. *If $\{\omega_n\}$ satisfies the additional condition that*

(1.3) $$\omega_n < \alpha \, \Omega_{n-1} \qquad (0 < \alpha < \tfrac{1}{2})$$

where $\Omega_n = \sum_{m=1}^{n} \omega_m$, then for almost all c in, say, $1 \le c \le 2$, (1.2) has just one solution (apart from an unimportant multiplicative constant) which is in $L^2(0, \infty)$.

(The theorem goes on to give estimates for the size of the solutions of (1.2), but we shall not be concerned with this aspect. We shall also see later that Theorem 1 is true without the restriction (1.3).)

Theorem 2. *If now $\{\omega_n\}$ satisfies*

$$\omega_n > \beta \, \Omega_{n-1} \qquad (\beta > \tfrac{1}{2}),$$

then for almost all c in, say, $1 \le c \le 2$, (1.2) has no solution in $L^2(0, \infty)$, apart, of course, from the trivial one.

(This theorem, as we shall see, is false, and is in fact replaced by the extended form of Theorem 1.)

Theorem 3. *If, as in Theorem 1, (1.3) holds, and if in addition $\omega_n > n^{1+\gamma}$ for some $\gamma > 0$, then the conclusion of Theorem 1 applies not only to (1.2) but also to (1.1) for any real λ.*

(Once again, the restriction (1.3) is not necessary.)

(It will be noted that in none of Littlewood's theorems does the possibility arise that (1.2) should have two linearly independent solutions (and so all) in $L^2(0, \infty)$, and this is only a particular case of a result due to Hartman [2] that (1.1) cannot have more than one linearly independent $L^2(0, \infty)$ solution if $q(x)$ is bounded below in ranges $[x_0, y_0]$, $[x_1, y_1]$, ..., where

$$\underline{\lim}(y_r - x_r) > 0$$

and $x_r, y_r \to \infty$. This carries with it the corollary that, for such $q(x)$, (1.1) is in the Weyl limit-point case, so that the spectral theory associated with (1.1) and a homogeneous boundary condition at $x = 0$ is well defined; and this will apply to all the cases considered in this paper.)

Now the results of Theorems 1 and 3 are at least consistent with the existence of a discrete spectrum for the corresponding spectral problem, but the result of Theorem 2 is not. Before I realized that Theorem 2 was in fact false, I was therefore led to abandon my supposition of a discrete spectrum for wildly oscillating potentials and in fact to prove the following theorem in [8].

Theorem 4. *Let*

$$q(x) = \begin{cases} u_n^2 & (n - \tfrac{1}{2} \leq x \leq n), \\ -v_n^2 & (n < x < n + \tfrac{1}{2}), \end{cases}$$

where $\{u_n\}$, $\{v_n\}$ are two sequences of positive numbers tending to infinity. Even restricting ourselves to cases in which $u_n = v_n$ and u_n and u_{n+1} are of the same order, we can arrange our choice of u_n so that, given any closed set $\mathscr{S}$ on the real λ-axis, the essential spectrum of (1.1) is precisely $\mathscr{S}$ and $\pm \infty$. Furthermore, given any two closed sets $\mathscr{S}_1$, $\mathscr{S}_2$ and any sequence $\{\chi_n\}$ tending to infinity however slowly, we can find sequences $\{u_{n,1}\}$, $\{u_{n,2}\}$ for which the essential spectra of (1.1) are $\mathscr{S}_1$, $\mathscr{S}_2$ respectively (and $\pm \infty$) and

$$|u_{n,1} - u_{n,2}| \leq \chi_n/u_{n,1}.$$

While we can thus, according to Theorem 4, generate any essential spectrum that we please, it is interesting to ask, led on by the "almost all" in Littlewood's results, whether one essential spectrum is "more probable" than any other. Two considerations lead us to believe that, at least under certain conditions, the most probable essential spectrum will be just $\pm \infty$, i.e., the spectrum will be discrete. The first of these considerations is that the spectrum is certainly discrete in the case described in [5, 6, 7]. The second is that, under the conditions of Theorem 3, we have an $L^2(0, \infty)$ solution for every λ, which strongly indicates that the spectrum should be discrete. In fact, it is the purpose of the present paper to use a lemma due to Littlewood and some of the facts from the proof of Theorem 4 to prove the following theorem.

Theorem 5. *Let $\{u_n\}$, $\{v_n\}$ be two given sequences of positive numbers tending to infinity with the properties that*

(1.4) $$u_n = O(v_n), \quad u_n = O(u_{n+1}), \quad v_n = O(u_{n+1}),$$

(1.5) $$n^{1+\gamma} = O(u_n) \quad \text{for some } \gamma > 0.$$

Then if

(1.6) $$q(x) = \begin{cases} c^2 u_n{}^2 & (n - \tfrac{1}{2} \leq x \leq n), \\ -c^2 v_n{}^2 & (n < x < n + \tfrac{1}{2}), \end{cases}$$

the spectrum associated with (1.1) and a homogeneous boundary condition at $x = 0$ is discrete for almost all c.

2.

Before proceeding to the proof of Theorem 5, we make a few remarks about it.

First, the conditions (1.4) and (1.5) can be relaxed at least to some extent without any significant alteration to the proof, except that various estimates made in the course of it have to be considered more carefully. The conditions given seem to have the two advantages that they are reasonably wide and give a reasonably straightforward proof. It is tempting to imagine that the theorem is true without any restrictions at all on u_n, v_n except for something corresponding to (1.5), but I do not know whether this is true.

It should also be noted that "almost all" in the statement of the theorem cannot be replaced by "all." If it were, then $c = 0$ would give an obvious (and perhaps trivial) contradiction; but Theorem 4 assures us that contradictions can occur with $c \neq 0$.

Theorem 5 says more than Theorem 3, but it also contradicts Theorem 2, since a consequence of Theorem 5 is that, for almost all c, there must be an $L^2(0, \infty)$ solution of (1.1) for every real λ. It is therefore necessary to indicate at what point the proof of Theorem 2 breaks down. This is at equation (23) in [4], where (and I take over Littlewood's notation without explanation, the reader being referred to [4]) Littlewood claims that

$$\tfrac{1}{2}|\eta_N \eta_{N+1} \cdots \eta_{n-1}| < \left| \frac{\lambda_n}{\mu_n} + \kappa_0 \right| < 2|\eta_N \eta_{N+1} \cdots \eta_{n-1}|.$$

What in fact is true (and follows from Littlewood's previous work) is that

$$\tfrac{1}{2}|\eta_N\eta_{N+1}\cdots\eta_n| < \left|\frac{\lambda_n}{\mu_n} + \kappa_0\right| < 2|\eta_N\eta_{N+1}\cdots\eta_n|,$$

and this leads to consequential changes in Sections 15, 16 of [4]. Thus in the equation before (26), we should have

$$A|\eta_n| < |u_n| < A|\eta_n| ;$$

(26) should read

$$Bn^{-A}(\omega_n/\omega_{n+1})e_{n+1}^{-2} < |u_n| < Bn^A e_{n+1}^{-2};$$

(27) should read

$$Bn^{-A}(\omega_n/\omega_{n+1})e_{n+1}^{-1} < \left|\frac{b_{n+1}}{b_n}\right| < Bn^A e_{n+1}^{-1};$$

and (28) should read

$$Bn^{-A}(\omega_n/\omega_{n+1})e_{n+1}^{-1} < \left|\frac{a_{n+1}}{a_n}\right| < Bn^A e_{n+1}^{-1}.$$

By comparison with Littlewood's form of (28), the right-hand half of the inequality is now stronger, so that deductions made from it continue to hold good. In particular, Theorem 1 continues to hold, in fact (as Littlewood's analysis, with the correct form of (28), shows) it holds under wider conditions, i.e. without (1.3). Theorem 3 also holds without (1.3.) But Theorem 2 is a deduction from the left-hand half of (28), and this is now weaker, so that in fact the proof of Theorem 2 breaks down.

3.

To prove Theorem 5, we first establish some notation from [8]. If we look at the behavior of a typical solution of (1.1), with $q(x)$ given by (1.6) and $\lambda = 0$, we can suppose that it is given by

$$(3.1) \quad y = \begin{cases} A_n e^{cu_n(x-n)} + B_n e^{-cu_n(x-n)} & (n - \tfrac{1}{2} \le x \le n), \\ C_n \cos\{cv_n(x-n)\} + D_n \sin\{cv_n(x-n)\} & (n < x < n + \tfrac{1}{2}). \end{cases}$$

Continuity of y and y' at $x = n$ gives

$$(3.2) \qquad\qquad C_n = A_n + B_n, \qquad v_n D_n = u_n(A_n - B_n),$$

while the behavior at $x = n + \tfrac{1}{2}$ gives similarly

$$C_n \cos(\tfrac{1}{2}cv_n) + D_n \sin(\tfrac{1}{2}cv_n) = A_{n+1}e^{-cu_{n+1}/2} + B_{n+1}e^{cu_{n+1}/2},$$

$$v_n\{-C_n \sin(\tfrac{1}{2}cv_n) + D_n \cos(\tfrac{1}{2}cv_n)\} = u_{n+1}\{A_{n+1}e^{-cu_{n+1}/2} - B_{n+1}e^{cu_{n+1}/2}\}.$$

Elimination of C_n, D_n gives

(3.3)
$$\frac{A_{n+1}}{B_{n+1}} = e^{cu_{n+1}} \frac{(\alpha_n - \beta_n)A_n/B_n + \gamma_n - \delta_n}{(\alpha_n + \beta_n)A_n/B_n + \gamma_n + \delta_n},$$

where

(3.4)
$$\alpha_n = \cos(\tfrac{1}{2}cv_n) + \frac{u_n}{v_n}\sin(\tfrac{1}{2}cv_n),$$

(3.5)
$$\beta_n = \frac{v_n}{u_{n+1}}\left\{\sin(\tfrac{1}{2}cv_n) - \frac{u_n}{v_n}\cos(\tfrac{1}{2}cv_n)\right\},$$

(3.6)
$$\gamma_n = \cos(\tfrac{1}{2}cv_n) - \frac{u_n}{v_n}\sin(\tfrac{1}{2}cv_n),$$

(3.7)
$$\delta_n = \frac{v_n}{u_{n+1}}\left\{\sin(\tfrac{1}{2}cv_n) + \frac{u_n}{v_n}\cos(\tfrac{1}{2}cv_n)\right\}.$$

If we now take any other value of λ, then this is equivalent to changing u_n, v_n to $u_{n,\lambda}$, $v_{n,\lambda}$, where

$$u_{n,\lambda}^2 = u_n^2 - \lambda/c^2, \qquad v_{n,\lambda}^2 = v_n^2 + \lambda/c^2,$$

so that

(3.8)
$$u_{n,\lambda} = u_n - \tfrac{1}{2}(\lambda/c^2)u_n^{-1} + O(u_n^{-3}),$$

(3.9)
$$v_{n,\lambda} = v_n + \tfrac{1}{2}(\lambda/c^2)v_n^{-1} + O(v_n^{-3}).$$

(The O-terms do, of course, involve λ and c, but in the proof we shall invariably be concerned with λ in a fixed bounded interval, so that the dependence of the O-terms on λ will be irrelevant. And so far as c is concerned, we shall in fact prove the theorem for almost all c in any fixed interval of the form $c_1 \leqq c \leqq c_2$, where $c_1 > 0$, $c_2 < \infty$, so that again the dependence of the O-terms on c is irrelevant. Having proved the theorem for such c, we can then let $c_1 \to 0$ through a sequence of values and $c_2 \to \infty$ through a sequence of values, and since the union of a countable number of null sets is null, the theorem will then hold for almost all c without restriction.)

Having thus defined $u_{n,\lambda}$, $v_{n,\lambda}$, we can similarly denote by $\alpha_{n,\lambda}$ the value of α_n when u_n, v_n are replaced by $u_{n,\lambda}$, $v_{n,\lambda}$ (so that α_n could otherwise be written $\alpha_{n,0}$). We then see easily that

(3.10)
$$\alpha_{n,\lambda} - \alpha_n = O(u_n^{-1}) = O(n^{-1-\gamma}),$$

and similarly for $\beta_{n,\lambda}$, $\gamma_{n,\lambda}$, $\delta_{n,\lambda}$. (We use here the fact that $u_n = O(v_n)$, $u_n = O(u_{n+1})$, $v_n = O(u_{n+1})$, so that $O(v_n^{-1})$ implies $O(u_n^{-1})$, and so on.)

We now state the lemma, proved by Littlewood in [4], on which the proof depends.

Lemma. *Given any sequence $\{\omega_n\}$ with $\omega_n \geq 1$, and any finite number of sequences $\{\theta_n\}$, we have, for each c of a p.p. set $\mathscr{S}$ contained in $1 \leq c \leq \Omega$,*

$$|\sin(\tfrac{1}{2}c\omega_n - \theta_n)| > Bn^{-1-\delta},$$

where δ is an arbitrary positive constant, and $B = B(c)$ is a positive constant depending on δ, c, $\mathscr{S}$, the sequence $\{\omega_n\}$, and the sequences $\{\theta_n\}$.

(We have already remarked that we shall be concerned with a p.p. set in $c_1 \leq c \leq c_2$ rather than $1 \leq c \leq \Omega$, but this is irrelevant to the proof of the lemma.)

To apply the lemma, we note that, from (3.4),

$$\alpha_n = \sqrt{\left(1 + \frac{u_n^2}{v_n^2}\right)} \sin(\tfrac{1}{2}cv_n - \theta_n),$$

for some suitable sequence $\{\theta_n\}$. Hence by the lemma, with $\omega_n = v_n$, and $\delta = \tfrac{1}{2}\gamma$, where γ is defined in (1.5), we have, for almost all c in $c_1 \leq c \leq c_2$,

(3.11)
$$n^{-1-\gamma} = o(\alpha_n).$$

Similarly, we can arrange that, still for almost all c in $c_1 \leq c \leq c_2$, not only is (3.11) true, but also

(3.12)
$$n^{-1-\gamma} = o(\alpha_n \pm \beta_n),$$

(3.13)
$$n^{-1-\gamma} = o(\sin\{\tfrac{1}{2}cv_n - \tan^{-1}(v_n/u_n)\}).$$

From now on, c will be a value for which (3.11), (3.12), (3.13) all hold, and we shall show that for such c the spectrum is discrete.

To do this, we consider any two fixed values of λ, say λ_1, λ_2, $\lambda_1 < \lambda_2$, and show that, if $N(\lambda, T)$ is the number of zeros in $[0, T]$ of some solution of (1.1), then $N(\lambda_2, T) - N(\lambda_1, T)$ is bounded as $T \to \infty$. For, by a theorem due to Hartman [1], it will then follow that the spectrum must be discrete between λ_1, λ_2, so that, since λ_1, λ_2 are arbitrary, the discreteness of the spectrum is obtained.

The proof that $N(\lambda_2, T) - N(\lambda_1, T)$ is bounded follows the lines of the proof in [8], but since there are various minor changes, it may be clearest to give it in its entirety. It will be sufficient to show that, once n is sufficiently large, there are two solutions, one of (1.1) with $\lambda = \lambda_1$, and one with $\lambda = \lambda_2$, such that, in any interval $n - \tfrac{1}{2} \leq x \leq n$ or $n < x < n + \tfrac{1}{2}$, they have the same number of zeros.

First we note that, because of (3.10) and (3.11), we know that $\alpha_{n,\lambda}$ remains of the same sign for λ in $[\lambda_1, \lambda_2]$ and, uniformly for λ in $[\lambda_1, \lambda_2]$,

(3.14)
$$n^{-1-\gamma} = o(\alpha_{n,\lambda}).$$

Similarly, $\alpha_{n,\lambda} \pm \beta_{n,\lambda}$ remain of the same sign and

(3.15)
$$n^{-1-\gamma} = o(\alpha_{n,\lambda} \pm \beta_{n,\lambda}).$$

The solutions of (1.1) with $\lambda = \lambda_1, \lambda_2$ which we will consider are to have the same value of A_N/B_N for some large N. Just how large N has to be will appear in the course of the proof. We will insist that A_N/B_N satisfies

$$|A_N/B_N| > e^{cu_{N,i}}/N^{1+\gamma} \qquad (i = 1, 2),$$

where, to avoid double suffices, we write $u_{N,i}$ for u_{N,λ_i}, and similarly for $\alpha_{N,i}$, $\beta_{N,i}$, and so forth.

Let $C_{N,i}$, $D_{N,i}$, $A_{N+1,i}$, $B_{N+1,i}$, ... be the subsequent coefficients for the solution for $\lambda = \lambda_i$. Then, for $i = 1, 2$, the dominant term in the numerator of (3.3), with $n = N$ and $\lambda = \lambda_i$, is $(\alpha_{N,i} - \beta_{N,i})A_N/B_N$, in view of (3.15) and the fact, immediately deducible from the definitions of γ_n, δ_n, that $\gamma_{n,i} - \delta_{n,i} = O(1)$. In fact, the numerator is

$$(\alpha_{N,i} - \beta_{N,i})A_N/B_N\{1 + O(N^{2+2\gamma}e^{-cu_N})\}.$$

(We recall that, in view of (3.8), it is immaterial whether we write, in the O-term, $e^{-cu_{N,i}}$ or e^{-cu_N}.)

Similarly, the denominator in (3.3) is

$$(\alpha_{N,i} + \beta_{N,i})A_N/B_N\{1 + O(N^{2+2\gamma}e^{-cu_N})\},$$

and so

(3.16)
$$\frac{A_{N+1,i}}{B_{N+1,i}} = e^{cu_{N+1,i}} \frac{\alpha_{N,i} - \beta_{N,i}}{\alpha_{N,i} + \beta_{N,i}}\{1 + O(N^{2+2\gamma}e^{-cu_N})\}.$$

Hence, if N is sufficiently large, as we may suppose, (3.15) and the fact that $\alpha_{n,i} + \beta_{n,i} = O(1)$ enable us to deduce from (3.16) that

$$|A_{N+1,i}/B_{N+1,i}| > e^{cu_{N+1,i}}/(N+1)^{1+\gamma}.$$

Then, repeating the argument inductively, we have, for any positive integer r,

(3.17)
$$|A_{N+r,i}/B_{N+r,i}| > e^{cu_{N+r,i}}/(N+r)^{1+\gamma}.$$

From (3.2), we now deduce that

$$C_{N+r,i}/D_{N+r,i} = (v_{N+r,i}/u_{N+r,i})\{1 + O(B_{N+r,i}/A_{N+r,i})\}$$
$$= (v_{N+r,i}/u_{N+r,i})\{1 + O((N+r)^{1+\gamma}e^{-cu_{N+r}})\}$$

(3.18)
$$= (v_{N+r}/u_{N+r})\{1 + O(u_{N+r}^{-2})\}.$$

It is now easy to see that the two solutions have the same number of zeros, in the sense described earlier. For in the interval $N + r - \frac{1}{2} \leq x \leq N + r$, both of $A_{N+r,i}/B_{N+r,i}$ have the same sign, that of

$$(\alpha_{N+r-1} - \beta_{N+r-1})/(\alpha_{N+r-1} + \beta_{N+r-1}),$$

by (3.16) (with $N + r$ for $N + 1$) and the sentence including (3.15). If this sign is positive, neither solution has a zero in $N + r - \frac{1}{2} \leq x \leq N + r$. If it is negative, then, referring to (3.1), we see that the solution for $\lambda = \lambda_i$ has one (and only one) zero in $N + r - \frac{1}{2} \leq x \leq N + r$ if and only if

(3.19) $\qquad\qquad -e^{cu_{N+r, i}} \leq A_{N+r, i}/B_{N+r, i} \leq -1.$

In view of (3.17), the right-hand half of (3.19) is certainly satisfied, and so the solution for $\lambda = \lambda_i$ has one or no zero in $N + r - \frac{1}{2} \leq x \leq N + r$ according as

(3.20) $\qquad\qquad e^{-cu_{N+r, i}} A_{N+r, i}/B_{N+r, i} + 1 \geq , < 0.$

But, from the sentence including (3.14), both of $\alpha_{N+r-1, i}$ have the same sign. Let us suppose that this sign is positive. (A similar argument applies if it is negative.) By (3.14),

$$\alpha_{N+r-1, i} > K(N + r - 1)^{-1-\gamma},$$

for any constant K we choose, if N is sufficiently large. Hence

$$(\alpha_{N+r-1, i} + \beta_{N+r-1, i}) + (\alpha_{N+r-1, i} - \beta_{N+r-1, i}) > 2K(N + r - 1)^{-1-\gamma}.$$

Now let us suppose that $\alpha_{N+r-1, i} + \beta_{N+r-1, i} > 0$. (We recall that the sign of $\alpha_{N+r-1, i} + \beta_{N+r-1, i}$ is independent of i, and the argument is essentially the same whatever the sign may be.) Then

(3.21) $\qquad 1 + \dfrac{\alpha_{N+r-1, i} - \beta_{N+r-1, i}}{\alpha_{N+r-1, i} + \beta_{N+r-1, i}} \quad > \quad \dfrac{2K(N + r - 1)^{-1-\gamma}}{\alpha_{N+r-1, i} + \beta_{N+r-1, i}}.$

Using (3.16), with $N + r$ in place of $N + 1$, (3.21) leads to

(3.22) $\quad 1 + e^{-cu_{N+r, i}} A_{N+r, i}/B_{N+r, i}\{1 + O((N + r - 1)^{2+2\gamma} e^{-cu_{N+r-1}})\}$

$$> 2K(N + r - 1)^{-1-\gamma}/(\alpha_{N+r-1, i} + \beta_{N+r-1, i}).$$

Again, (3.16) gives

$$|A_{N+r, i}/B_{N+r, i}| = O\{(N + r - 1)^{1+\gamma} e^{cu_{N+r, i}}\},$$

in view of (3.15) and the fact that $\alpha_{n, i} - \beta_{n, i} = O(1)$. Hence (3.22) reduces to

$$1 + e^{-cu_{N+r, i}} A_{N+r, i}/B_{N+r, i}$$

$$> 2K(N + r - 1)^{-1-\gamma}/(\alpha_{N+r-1, i} + \beta_{N+r-1, i})$$

$$+ O\{(N + r - 1)^{3+3\gamma} e^{-cu_{N+r-1}}\},$$

and since $\alpha_{n, i} + \beta_{n, i} = O(1)$, the first term on the last right-hand side dominates the second, so that the first alternative in (3.20) holds for both values of i and both solutions have just one zero in $N + r - \frac{1}{2} \leq x \leq N + r$.

In an interval $N + r < x < N + r + \frac{1}{2}$, the zeros of the solutions occur at

$$cv_{N+r,\,i}\{x - (N + r)\} = \tan^{-1}(-C_{N+r,\,i}/D_{N+r,\,i}) + k\pi$$

$$= \tan^{-1}(v_{N+r}/u_{N+r}) + O\left\{\frac{v_{N+r}/u_{N+r}^3}{1 + v_{N+r}^2/u_{N+r}^2}\right\} + k\pi$$

by use of the mean value theorem and (3.18), k being integral. We can neglect 1 in the denominator of the O-term, since $u_n = O(v_n)$, and the O-term thus reduces to $O(u_{N+r}^{-1} v_{N+r}^{-1})$. The zeros therefore occur at

$$x - (N + r) = \{\tan^{-1}(v_{N+r}/u_{N+r}) + k\pi\}/cv_{N+r} + O(v_{N+r}^{-2}).$$

These points are approximately the same for both solutions, the difference lying entirely in the O-term, and so one solution can gain a zero at the expense of the other only if a zero occurs within $O(v_{N+r}^{-2})$ of the ends of the interval. But at the end $x = N + r$, the nearest zero is approximately at

$$x - (N + r) = \{\tan^{-1}(v_{N+r}/u_{N+r})\}/cv_{N+r},$$

which is not within $O(v_{N+r}^{-2})$, while at the other end the occurrence of a zero within $O(v_{N+r}^{-2})$ of the end would imply that

$$\tfrac{1}{2} = \{\tan^{-1}(v_{N+r}/u_{N+r}) + k\pi\}/cv_{N+r} + O(v_{N+r}^{-2}),$$

i.e. that

$$\tfrac{1}{2}cv_{N+r} - \{\tan^{-1}(v_{N+r}/u_{N+r}) + k\pi\} = O(v_{N+r}^{-1}),$$

i.e. that

$$\sin\{\tfrac{1}{2}cv_{N+r} - \tan^{-1}(v_{N+r}/u_{N+r})\} = O(v_{N+r}^{-1}),$$

which contradicts (3.13).

The theorem is thus proved.

REFERENCES

1. P. Hartman, A characterization of the spectrum of one-dimensional wave equations, *Amer. J. Math.* **71** (1949), 915–920.
2. ——, The number of L^2-solutions of $x'' + q(t)x = 0$, *Amer. J. Math.* **73** (1951), 635–645.
3. ——, Some examples in the theory of singular boundary value problems, *Amer. J. Math.* **74** (1952), 107–126.
4. J. E. Littlewood, On linear differential equations of the second order with a strongly oscillating coefficient of y, *J. London Math. Soc.* **41** (1966), 627–638.
5. J. B. McLeod, On the spectrum of wildly oscillating functions, *J. London Math. Soc.* **39** (1964), 623–634.
6, 7. ——, On the spectrum of wildly oscillating functions (II) and (III), *J. London Math. Soc.* **40** (1965), 655–661 and 662–666.
8. ——, Some examples of wildly oscillating potentials, *J. London Math. Soc.* **43** (1968), 647–654.

(*Received February 1, 1968*)

On Convolutions and Growth of Typically Real Functions

E. P. Merkes

Department of Mathematics
University of Cincinnati
Cincinnati, Ohio 45221
U.S.A

1. Introduction

A function $f(z) = z + a_2 z^2 + \cdots$, analytic in $E = \{z: |z| < 1\}$, is in the class T of typically real functions in E if $\operatorname{Im} z \operatorname{Im} f(z) > 0$ for all nonreal $z \in E$ [7]. If, furthermore, $w = f(z)$ maps each circle $|z| = r < 1$ onto a curve having the property that each line parallel to the imaginary axis cuts the curve in at most two points, then $f(z)$ is said to be in the class T_c of functions convex in the direction of the imaginary axis [2]. A necessary and sufficient condition for this is that $z f'(z) \in T$ [2].

Some years ago, Robertson [6] obtained the radius of typically-realness for the class of convolution functions $h(z) = f * g(z) = z + \sum_{n=2}^{\infty} a_n b_n z^n$ that arises when $f(z) = z + \sum_{n=2}^{\infty} a_n z^n$ and $g(z) = z + \sum_{n=2}^{\infty} b_n z^n$ range over T. He showed also $h \in T_c$ whenever f and g are on T_c. Elementary proofs of these results of Robertson are presented in Section 2 of this paper. Section 3 contains a growth theorem for T_c that is analogous to the recent result for convex functions by Basgöze, Frank, and Keogh [1]. The corresponding theorem for T is obtained in the final section. It is an analogue for T of the theorem for starlike functions obtained by Keogh [3].

2. Convolutions

The following result is equivalent to a theorem of Robertson [6, Theorem 3].

Theorem 1. *If $f \in T$ and $g \in T_c$, then $h = f * g \in T_c$.*

Proof. If $f(z) = z + \sum_{n=2}^{\infty} a_n z^n \in T$, then there is a nondecreasing function $\alpha(t)$ on $[0, \pi]$ such that

$$a_n = \int_0^\pi \frac{\sin nt}{\sin t}\, d\alpha(t) \qquad (n = 2, 3, \ldots),$$

and $\alpha(\pi) - \alpha(0) = 1$ [5]. If $g(z) = \sum_{n=1}^{\infty} b_n z^n$, $b_1 = 1$, is in T_c, then

(1)
$$h(z) = z + \sum_{n=2}^{\infty} a_n b_n z^n = \int_0^\pi \left(\sum_{n=1}^{\infty} \frac{\sin nt}{\sin t} b_n z^n \right) d\alpha(t)$$

$$= \int_0^\pi \left\{ \sum_{n=1}^{\infty} (e^{int} - e^{-int}) b_n z^n \right\} \frac{d\alpha(t)}{2i \sin t}$$

$$= \int_0^\pi k(z, t)\, d\alpha(t),$$

where

(2)
$$k(z, t) = \begin{cases} zg'(z) & \text{for} \quad t = 0, \\ [g(e^{it}z) - g(e^{-it}z)]/2i \sin t & \text{for} \quad 0 < t < \pi, \\ zg'(-z) & \text{for} \quad t = \pi. \end{cases}$$

The functions $k(g, 0)$ and $k(z, \pi)$ are in T since $-g(-z) \in T_c$ whenever $g(z) \in T_c$. For a $t \in (0, \pi)$, the function $k(z, t)$ is clearly real when $z \in E$ is real. Also if $k(z_0, t)$ is real for some $z_0 \in E$, then by (2)

$$\text{Re } g(z_0 e^{it}) = \text{Re}[g(z_0 e^{-it}) + 2i \sin t\, k(z_0, t)] = \text{Re } g(z_0 e^{-it}).$$

Now $g(z)$ is symmetric relative to the real axis and convex in the direction of the imaginary axis. Hence, the equality of the real parts of g at two distinct points on the circle $|z| = |z_0|$ implies the points are complex conjugates. This shows $z_0 e^{it} = \bar{z}_0 e^{it}$, that is, z_0 is real. Therefore, $k(z, t) \in T$ for $0 < t < \pi$.

For $z \in E$, $\text{Im } z \neq 0$, we obtain from (1) that

$$\text{Im } z\ \text{Im } h(z) = \int_0^\pi \text{Im } z\ \text{Im } k(z, t)\, d\alpha(t) > 0$$

since the integrand is positive for $0 \leq t \leq \pi$ and $\alpha(t)$ has at least one point of increase. Hence, $h(z) \in T$ and the proof is complete.

Corollary 1. *If $f(z) \in T$, then $f(z)$ is convex in the direction of the imaginary axis for $|z| < R_0 = 2 - \sqrt{3}$. This result is sharp.*

Proof. Robertson [6] proved that the Koebe function $g(z) = z/(1 - z)^2$ is convex in the direction of the imaginary axis for $|z| < R$ iff $0 < R \leq R_0$. Thus $g(R_0 z)R_0^{-1} \in T_c$ and we conclude from Theorem 1 that $h(z) = z + \sum_{n=2}^{\infty} na_n R_0^{n-1} z^n$ is in T when $f(z) = z + \sum_{n=2}^{\infty} a_n z^n$ is in T. This proves $zf'(z) = h(z/R_0)R_0$ is typically real for $|z| < R_0$.

Corollary 2 [6]. *If f and g are in T, then $h = f * g$ is typically real for $|z| < R_0 = 2 - \sqrt{3}$. This result is sharp.*

Corollary 3 [6]. *If f and g are in T_c, then $h = f * g$ is in T_c.*

Proof. Apply Theorem 1 to $g(z) \in T_c$ and $zf'(z) \in T$ to show $zh'(z)$ is in T.

    The de la Vallée Poussin mean of order n for a function $f(z)$, analytic in E, is

(3)
$$V_n(f, z) = \frac{n! \, n!}{(2n)!} \frac{1}{2\pi} \int_0^{2\pi} [2 \cos(\theta - \phi)/2]^{2n} f(re^{i\phi}) \, d\phi,$$

where $z = re^{i\theta} \in E$. The de la Vallée Poussin means of the convex function $z/(1 - z)$ in E are themselves convex in E [4]. Combining this with Theorem 1, we obtain the following result.

Corollary 4. *$f(z)$ is in T (in T_c) iff $(n + 1)V_n(f, z)/n$ is in T (in T_c) for $n = 1, 2, \ldots$.*

3. Growth in T_c

    If $f(z)$ is in T_c, then $w = f(z)$ maps the interval $(-1, 1)$ onto an interval of the real w-axis that contains $(-\frac{1}{2}, \frac{1}{2})$. One way to prove this is to show $x/2 < f(x)$ for all $x \in (0, 1)$ and $f \in T_c$. The function $x/2$ is the largest linear function that satisfies such an inequality. Our next theorem is a generalization of this type of inequality.

Theorem 2. *If $f(z) \in T_c$, then $V_n(f, x) < f(x)$, where $0 < x < 1$ and V_n is the mean (3). Moreover, there exist real numbers λ and μ such that, for all $f(z) = z + a_2 z^2 + \cdots$ in T_c, the function $\lambda z + \mu a_2 z^2$ is convex in the direction of the imaginary axis and*

(4)
$$x/2 \leq \lambda x + \mu a_2 x^2 < f(x) \quad (0 < x < 1)$$

iff $\lambda = \mu + \frac{1}{2}, 0 \leq \mu \leq \frac{1}{6}$.

Proof. From the definition of the class T_c, it is clear that for $|z| = r \, (0 < r < 1)$ we have $f(-r) \leqq \operatorname{Re} f(z) \leqq f(r)$ whenever $f \in T_c$. We conclude from (3) that $f(-r) \leqq \operatorname{Re} V_n(f, z) \leqq f(r)$ for $|z| = r$ since $V_n(1, z) = 1$. The first part of the theorem now follows from the above with $z = r = x$, $0 < x < 1$, and from Corollary 4.

The sufficiency of the condition on λ and μ is a consequence of the fact that

$$x/2 \leqq (\mu + \tfrac{1}{2})x + \mu a_2 x^2 \leqq 2x/3 + a_2 x^2/6 = V_2(f, x) < f(x)$$

when $f(z) = z + a_2 z^2 + \cdots$ is in T_c. Suppose now that (4) holds for all $f(z) = z + a_2 z^2 + \cdots$ in T_c. Taking $f(z) = z$ and $f(z) = z/(1 + z)$, respectively, we obtain $\tfrac{1}{2} \leqq \lambda \leqq 1$ and $x/2 \leqq \lambda x - \mu x^2 \leqq \tfrac{1}{2}$ for $0 < x < 1$. Thus $\lambda = \mu + \tfrac{1}{2}$, $\mu \geqq 0$. Since $\lambda z + \mu a_2 z^2$ is convex in the direction of the imaginary axis, we have in addition that $\lambda - 4\mu|a_2| \geqq 0$. This is the case for all $|a_2| \leqq 1$ iff $\mu \leqq \lambda/4$. It follows that $0 \leqq \mu \leqq \tfrac{1}{6}$; $\lambda = \mu + \tfrac{1}{2}$, which completes the proof.

4. Growth in T

We now prove an analogue for T of the latter part of Theorem 2.

Theorem 3. *Let λ and μ be real numbers. Then*

(5) $$x/4 \leqq \lambda x + \mu a_2 z^2 < f(x) \qquad (0 < x < 1)$$

for all $f(z) = z + a_2 z^2 + \cdots$ in T iff $\lambda = 2\mu + \tfrac{1}{4}$, $0 \leqq \mu \leqq \tfrac{1}{16}$.

Proof. Suppose (5) is valid for all $f \in T$. Taking $f(z) = z$ and $f(z) = z/(1 + z)^2$, respectively, we conclude that $\tfrac{1}{4} \leqq \lambda \leqq 1$ and $x/4 \leqq \lambda x - 2\mu x^2 \leqq \tfrac{1}{4}$. Thus, $\lambda = 2\mu + \tfrac{1}{4}$, $\tfrac{1}{4} \leqq \lambda \leqq 1$, $\mu \geqq 0$. For each $s \in (-1, 1)$, the function

$$H(z, s) = \frac{z}{1 + 2sz + z^2} = z - 2sz^2 + \cdots$$

is in T. Let $\mu > 0$ and set $G(x, s) = H(x, s) - \lambda x + 2\mu s x^2$, $0 < x < 1$. For a fixed $x \in (0, 1)$ the function $G(x, s)$ has a minimum value when s satisfies $(1 + 2sx + x^2)^2 = \mu^{-1}$. Since $0 \leqq 1 + 2sx + x^2 < 4$ when $x \in (0, 1)$, $s \in [-1, 1]$, it follows that $\mu > \tfrac{1}{16}$ if $G(x, s)$ is to take on this minimum value when $x \in (0, 1)$, $s \in [-1, 1]$. The minimum value of $G(x, s)$ is

$$\min_s G(x, s) = x[\sqrt{\mu} - \lambda + \mu(1/\sqrt{\mu} - 1 - x^2)] > -(2\sqrt{\mu} - \tfrac{1}{2})^2,$$

where $\lambda = 2\mu + \tfrac{1}{4}$. Therefore, if $\mu > \tfrac{1}{16}$, there exist an $x \in (0, 1)$ and a $s \in [-1, 1]$ such that $G(x, s) < 0$. This contradicts the hypothesis that (5) holds. Hence, $\lambda = 2\mu + \tfrac{1}{4}$, $0 \leqq \mu \leqq \tfrac{1}{16}$.

For the converse, first let $\mu = \frac{1}{16}$, $\lambda = \frac{3}{8}$. From the preceding argument $G(x, s) > 0$ for $x \in (0, 1)$, $s \in [-1, 1]$. Now if $f(z) \in T$, Robertson [5] proved there is an $\alpha(t) \uparrow [-\pi, \pi]$, $\alpha(\pi) - \alpha(-\pi) = 1$, such that

$$f(z) = \int_{-\pi}^{\pi} H(z, -\cos t)\, d\alpha(t).$$

We conclude that for $x \in (0, 1)$

$$f(x) - 3x/8 - a_2 x^2/16 = \int_{-\pi}^{\pi} G(x, -\cos t)\, d\alpha(t) > 0.$$

Finally, for $x \in (0, 1)$, $\mu \in [0, \frac{1}{16}]$, we have $(2\mu + \frac{1}{4})x + \mu a_2 x^2 \leqq 3x/8 + a_2 x^2/16$ and the proof is complete.

REFERENCES

1. T. Basgöze, J. Frank, and F. Keogh, On convex univalent functions, *Canad. J. Math.* **22** (1970), 123–127.
2. L. Fegér, Neue Eigenschaften der Mittlewerte bie den Fourierreihen, *J. London Math. Soc.* **8** (1933), 53–62.
3. F. R. Keogh, A strengthened form of the $\frac{1}{4}$ theorem for starlike univalent functions, this volume, pp. 201–211.
4. G. Pólya and I. J. Schoenberg, Remarks on de la Valleé Poussin means and convex conformal maps of the circle, *Pacific J. Math.* **8** (1958), 295–334.
5. M. S. Robertson, On the coefficients of a typically-real function, *Bull. Amer. Math. Soc.* **41** (1935), 565–572.
6. ———, Applications of a lemma of Fejer to typically-real functions, *Proc. Amer. Soc.* **1** (1950), 555–561.
7. W. Rogosinski, Über positive harmonische Entwicklungen und typischreelle Potenzreihen, *Math. Z.* **35** (1932), 93–121.

(*Received February 20, 1969*)

The Derivative of a Function of Bounded Characteristic[*]

George Piranian
Department of Mathematics
The University of Michigan
Ann Arbor, Michigan 48104
U.S.A.

It has long been known that if a function f is meromorphic in the unit disk D and has bounded characteristic $T(r,f)$, the characteristic $T(r,f')$ of its derivative may nevertheless be unbounded (see [3, Section 138] for Nevanlinna's notation, and [4, p. 557] for a list of relevant papers). Recently, P. B. Kennedy [2] investigated the restrictions on the growth of $T(r,f')$ that are implicit in the assumption that $T(r,f)$ is bounded, and he proved the following two theorems.

Theorem 1. *If f is meromorphic in D and $T(r,f)$ is bounded, then*

$$(1) \qquad \int_0^1 (1-r)\exp 2T(r,f')\, dr < \infty.$$

Theorem 2. *Let the function $\mu(r)$ be positive and increasing in $(0,1)$, and let it satisfy the three conditions*

$$(2) \qquad \int_0^1 (1-r)\exp 2\mu(r)\, dr < \infty,$$

$$(3) \qquad \mu(r) - \mu(\rho) \to \infty \quad as \quad (1-r)/(1-\rho) \to 0,$$

$$(4) \qquad (1-r)\exp \mu(r) \quad \downarrow \quad 0 \quad as \quad r \to 1.$$

[*] This paper was written with support from the National Science Foundation.

*Then there exists a function f, holomorphic and of bounded characteristic in
D, such that for some R in* (0, 1)

(5) $T(r, f') > \mu(r)$ *whenever* $R < r < 1$.

J. Clunie [1, p. 68] has suggested that conditions (3) and (4) in Theorem 2
might be superfluous, and he has shown that if μ satisfies (2), then some
bounded function f satisfies the inequality in (5) on a set that is reasonably
thick near $r = 1$. The following theorem shows that Clunie had sensed the
state of affairs correctly.

Theorem 3. *If an increasing function μ on* (0, 1) *satisfies condition* (2), *then
there exists a holomorphic function f of bounded characteristic such that* (5)
holds for some R.

Our proof strongly resembles Kennedy's proof of Theorem 2; but to
simplify the notation, we introduce the function $\lambda(r) = \exp \mu(r)$. Kennedy's
three conditions then take the form

(2′) $$\int_0^1 (1 - r)\lambda^2 \, dr < \infty,$$

(3′) $\lambda(r)/\lambda(\rho) \to \infty$ as $(1 - r)/(1 - \rho) \to 0$,

(4′) $(1 - r)\lambda(r)$ $\downarrow$ 0 as $r \to 1$.

We shall actually need condition (3′); but we shall now show that each
increasing function λ_1 satisfying (2′) has an increasing majorant satisfying
both (2′) and (3′). Suppose (without loss of generality) that $\lambda_1(r) > 0$; write

$$(1 - r)[\lambda_1(r)]^2 = p_1(r);$$

denote by p_2 the least nondecreasing majorant of p_1 ; and define $\lambda_2(r)$ as the
positive solution of the equation

$$(1 - r)[\lambda(r)]^2 = p_2(r).$$

Since p_2 is a nondecreasing function and

$$\frac{\lambda_2(r)}{\lambda_2(\rho)} = \sqrt{\frac{p_2(r)}{p_2(\rho)}} \sqrt{\frac{1 - \rho}{1 - r}},$$

λ_2 is strictly increasing and has property (3′).

To see that λ_2 satisfies (2′), note that if S is any rectilinear segment joining
the point (1, 0) to some point $P = (r_1, p_1(r_1))$ on the graph of the function p_1,
then to the right of the vertical line $r = r_1$ the graph of p_1 lies above S.
Suppose now that $(r_1, p_1(r_1))$ and $(r_2, p_1(r_2))$ are the endpoints of a horizontal

line segment on the graph of p_2. Then it is geometrically obvious (see Figure 1) that

$$\int_{r_1}^{r_2} p_1(r)\, dr \geq \frac{1}{2} \int_{r_1}^{r_2} p_2(r)\, dr.$$

This implies that $\int_0^1 p_2\, dr \leq 2 \int_0^1 p_1\, dr$, so that λ_2 inherits property (2′) from λ_1. (It can also be shown that if λ_1 has property (4′), then the same is true of λ_2; but we do not need this.)

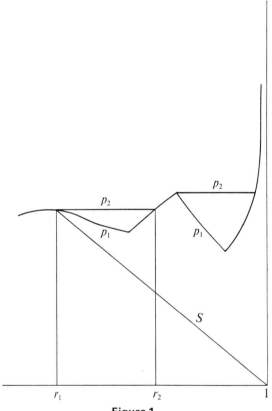

Figure 1.

Conditions (3′) and (4′) both require that the growth of λ be reasonably regular; but while (3′) is harmless, (4′) eliminates some functions λ that are admissible without it. Indeed, if $\lambda_1(r_0) = m_0$, and if a majorant λ_2 of λ_1 satisfies condition (4′), then

$$\int_0^{r_0} (1 - r)[\lambda_2(r)]^2\, dr \geq m_0^2 \int_0^{r_0} \frac{(1 - r_0)^2}{1 - r}\, dr = m_0^2 (1 - r_0)^2\, |\log(1 - r_0)|.$$

Since

$$\int_{r_0}^{1} (1 - r)m_0^2 \, dr = \tfrac{1}{2}m_0^2(1 - r_0)^2,$$

it is easy to construct a step function λ_1 that satisfies (2') but has no majorant satisfying both (2') and (4').

The need for some regularity condition is inherent in Kennedy's construction of a lacunary power series. The important use of hypothesis (4) in [2] occurs in the last sentence on page 339, and there a weaker hypothesis will suffice if we replace the somewhat wasteful inequality in the fourth line above the end page 339 with the inequality

$$(1 - x)^n = \exp(n \log(1 - x)) < \exp(-nx).$$

We shall find our way to the appropriate regularization by observing that the characteristic functions are convex functions of $\log r$ [3, Section 147] and that $\log Cr$ is a linear function of $\log r$. In case $\lambda(r)$ grows too abruptly to the left of certain points r_k ($k = 2, 3, \ldots$), we shall replace λ by functions of the form $\max \{\lambda(r), Cr^\beta\}$, in the corresponding intervals $(0, r_k]$. The details follow.

Condition (2') implies the existence of a function $\alpha(r)$ such that $\alpha(r) > 2$ for $0 \leq r < 1$, $\alpha(r) \uparrow \infty$ as $r \to 1$, and

(6)
$$\int_0^1 (1 - r)\{\alpha(r)\lambda(r)\}^2 \, dr < \infty.$$

We set $n_1 = 1$, and for $k = 1, 2, \ldots$ we define

(7)
$$r_k = 1 - 1/n_k, \qquad n_{k+1} = [\alpha(r_k)n_k],$$

where the brackets indicate the greatest-integer function.

For each r in $[0, r_k]$, we define the quantity

$$\lambda^*(r) = \sup\{\lambda(r), \lambda(r_k)(r/r_k)^{A_1 n_k}, \lambda(r_{k+1})(r/r_{k+1})^{A_1 n_{k+1}}, \ldots\},$$

where A_1 denotes a positive constant (to be chosen later) that is independent of r, k, and λ. Since each of the functions in the braces is strictly increasing, the function λ^* is finite if (6) remains valid after the replacement of λ with λ^*. Elementary considerations give us the inequalities

$$\int_0^{r_k} (1 - r)\{\alpha(r)\lambda(r_k)(r/r_k)^{A_1 n_k}\}^2 \, dr$$

$$< \{\alpha(r_k)\lambda(r_k)\}^2 \left\{ \frac{r_k}{2A_1 n_k + 1} - \frac{r_k^2}{2A_1 n_k + 2} \right\}$$

$$< A_2\{\alpha(r_k)\lambda(r_k)/n_k\}^2$$

$$< A_3 \int_{r_k}^{r_{k+1}} (1 - r)\{\alpha(r_k)\lambda(r_k)\}^2 \, dr.$$

The last estimate implies that the replacement of λ with λ^* increases the integral in (6) by a factor less than $A_3 + 1$.

Since $(r/\rho)^{A_1 n_k} = \exp\{A_1 n_k(\log r - \log \rho)\}$, it is obvious that the function λ^* also satisfies condition (3').

We now choose the coefficients

$$c_k = (3/n_k)\lambda^*(r_{k+1}) \qquad (k = 1, 2, \ldots)$$

and define the function f by the formula

$$f(z) = \sum_1^\infty c_k z^{n_k}.$$

To prove that f is of bounded characteristic, we use the second relation in (7) and deduce that

$$c_k^2 \leqq (3/n_{k+1})^2\{\alpha(r_k)\lambda^*(r_{k+1})\}^2$$

$$< A_4 \int_{r_{k+1}}^{r_{k+2}} (1-r)\{\alpha(r)\lambda^*(r)\}^2 \, dr.$$

This implies that $\sum c_k^2 < \infty$, and therefore $T(r, f)$ is bounded.

To deal with $T(r, f')$, we shall establish a lower bound for $|f'(z)|$ on the circle $|z| = r_k$. Clearly,

$$(8) \qquad |f'(r_k e^{i\theta})| \geqq c_k n_k r_k^{n_k - 1} - \sum_{j=1}^{k-1} c_j n_j - \sum_{j=k+1}^{\infty} c_j n_j r_k^{n_j}.$$

The first term in the right-hand member of (8) is greater than $(3/e)\lambda^*(r_{k+1})$, by the definition of c_k. The sum $\sum_{j=1}^{k-1} c_j n_j$ has the value

$$3 \sum_{j=1}^{k-1} \lambda^*(r_{j+1}),$$

and condition (3') implies that this is $o(\lambda^*(r_{k+1}))$. Concerning the second sum in the right-hand member of (8), we note that, by the definition of λ^*,

$$c_j n_j r_k^{n_j} = 3\lambda^*(r_{j+1}) r_k^{n_j} \leqq 3\lambda^*(r_{k+1})(r_{j+1}/r_{k+1})^{A_1 n_j + 1} r_k^{n_j}$$

$$= 3\lambda^*(r_{k+1}) \exp\left\{A_1 n_{j+1}\left[\log\left(1 - \frac{1}{n_{j+1}}\right) - \log\left(1 - \frac{1}{n_{k+1}}\right)\right]\right.$$

$$\left. + n_j \log\left(1 - \frac{1}{n_k}\right)\right\}$$

$$< 3e^{-A_1}\lambda^*(r_{k+1}) \exp\left\{\frac{4}{3} A_1 \frac{n_{j+1}}{n_{k+1}} - \frac{n_j}{n_k}\right\}.$$

Without loss of generality, we may assume that $\alpha(r_{k+1})/\alpha(r_k) = 1 + o(1)$.

Suppose now that $A_1 = \frac{1}{2}$. Then the content of the last braces is less than $-n_{j+1}/4n_{k+1}$, for all sufficiently large k and for $j > k$. Because $\alpha(r_j) \to \infty$, it follows from (7) that the second sum in (8) is also $o(\lambda^*(r_{k+1}))$, and therefore

$$|f'(z)| \geq \lambda^*(r_{k+1}) \quad \text{on} \quad |z| = r_k$$

for all sufficiently large indices k. Since $T(r, f')$ is an increasing function, f satisfies (5), and Theorem 3 is proved.

REFERENCES

1. J. Clunie, On the derivative of a bounded function, *Proc. London Math. Soc.* (3) **14A** (1965), 58–68.
2. P. B. Kennedy, On the derivative of a function of bounded characteristic, *Quart. J. Math. Oxford Ser.* 2 **15** (1964), 337–341.
3. R. Nevanlinna, *Eindeutige analytische Funktionen*, Springer-Verlag, Berlin, 1936.
4. D. F. Shea, Functions analytic in a finite disk and having asymptotically prescribed characteristic, *Pacific J. Math.* **17** (1966), 549–560.

(*Received October 30, 1967*).

NOTE: The author has discovered that the second sentence on this page is not true. He expresses his profound regrets and apologies, and he hopes to distribute through informal channels a correct proof of Theorem 3. (May 19, 1970).

Quasi-Subordinate Functions

M. S. Robertson

Department of Mathematics
University of Delaware
Newark, Delaware 19711
U.S.A.

1. Introduction

Let $f(z)$ and $F(z)$ be two functions which are single-valued and analytic in the disk $D\{z: |z| < R\}$. The function $f(z)$ is said to be subordinate to $F(z)$ in D if there exists an analytic function $\omega(z)$, bounded and regular in D, $|\omega(z)| \leq |z| < R$, for which $f(z) = F\{\omega(z)\}$ in D. In this case we write $f(z) \prec F(z)$ in D.

If $g(z)$ and $G(z)$ are two functions analytic in D for which $|g(z)| \leq |G(z)|$ in D then $g(z)$ is said to be majorized by $G(z)$ in D.

Several theorems exist in the literature that relate $f(z)$ and $F(z)$ when $f(z) \prec F(z)$ which have their counterparts that relate $g(z)$ and $G(z)$ when $|g(z)| \leq |G(z)|$. For example, if $f(z) \prec F(z)$ and

$$f(z) = \sum_1^\infty a_n z^n, \quad F(z) = \sum_1^\infty A_n z^n, \quad A_1 > 0,$$

in the unit disk $E\{z : |z| < 1\}$ and if $F(z)$ is typically real in E then, as Rogosinski has shown [16], the coefficients a_n satisfy the inequalities

(1.1) $$|a_n| \leq nA_1, \quad n = 1, 2, \ldots .$$

Recently MacGregor [9] has shown that (1.1) also holds if subordination of $f(z)$ by $F(z)$ is replaced by majorization of $f(z)$ by $F(z)$ in E. For another

---

* This research was supported by NSF Grant GP-7439.

example compare Theorem B of the author's paper [14] with Theorem B′ of a recent paper by Lewandowski [6]. For other illustrations see also [2, 5].

In order to establish some semblance of unification in parallel results for subordination and majorization, and to extend further known results for either situation, we are introducing here the concept of quasi-subordination.

Definition. *Let*

$$f(z) = \sum_0^\infty a_n z^n \quad and \quad F(z) = \sum_0^\infty A_n z^n$$

be analytic for $|z| < R$. Let $\phi(z)$ be a function analytic and bounded for $|z| < R$, $|\phi(z)| \leq 1$, such that $f(z)/\phi(z)$ is regular and subordinate to $F(z)$ for $|z| < R$. Then $f(z)$ is said to be quasi-subordinate *to $F(z)$ relative to $\phi(z)$ for $|z| < R$.*

If $f(z)$ is quasi-subordinate to $F(z)$ relative to $\phi(z)$ we shall often say simply that $f(z)$ is quasi-subordinate to $F(z)$ and write $f(z) \prec_q F(z)$ for $|z| < R$. We have

$$f(z) = \phi(z)F\{\omega(z)\}, \qquad |\omega(z)| \leq |z| < R.$$

Two special cases of quasi-subordination are of particular interest:
(1) If $\phi(z) \equiv 1$ then $f(z)$ is subordinate to $F(z)$ for $|z| < R$;
(2) If $\omega(z) \equiv z$ then $f(z)$ is majorized by $F(z)$ for $|z| < R$.
In other words, if either $f(z) \prec F(z)$ or $|f(z)| \leq |F(z)|$ for $|z| < R$, then $f(z) \prec_q F(z)$ for $|z| < R$. Unless otherwise specified we shall assume in what follows that $R = 1$ and D is the unit disk E.

If

$$f(z) = \sum_1^\infty a_n z^n \prec F(z) = \sum_1^\infty A_n z^n$$

in E then it is known [7, 8, 16] that

$$|a_1| \leq |A_1| \quad and \quad |a_2| \leq \max\{|A_1|, |A_2|\}.$$

On the other hand we find that if only $f(z) \prec_q F(z)$ in E then

(1.2) $$|a_1| \leq |A_1|, \quad but \quad |a_2| \leq \tfrac{5}{4} \max\{|A_1|, |A_2|\}.$$

The constant $\tfrac{5}{4}$ is best possible.

It is well known [2, 8, 16], that when $f(z) \prec F(z)$ in E then

(1.3) $$\sum_{k=0}^n |a_k|^2 \leq \sum_{k=0}^n |A_k|^2, \qquad n = 0, 1, \ldots .$$

Inequality (1.3) easily extends to quasi-subordination $f(z) \prec_q F(z)$ in E. But we shall show that if $f(z) \prec_q F(z)$ in E relative to the bounded function $\phi(z)$ and for $F(0) \neq 0$, and if in *some* neighborhood of the origin

$$(1.3) \qquad \left[\frac{f(z)}{\phi(z)}\right]^{1/2} = \sum_0^\infty b_k z^k, \qquad [F(z)]^{1/2} = \sum_0^\infty B_k z^k,$$

then

$$(1.4) \qquad \sum_0^n |b_k|^2 \leq \sum_0^n |B_k|^2, \qquad n = 0, 1, \ldots.$$

This result is crucial to the simple proofs of most of the theorems in this paper. It is perhaps more remarkable since neither

$$\sum_0^\infty b_k z^k \qquad \text{nor} \qquad \sum_0^\infty B_k z^k$$

need converge in the whole unit disk E whenever $f(z)/\phi(z)$ and $F(z)$ have one or more zeros of odd order in $0 < |z| < 1$.

When $F(z)$ is also univalent in E much more can be said. In this case if $f(z) \prec F(z)$ in E and if $F(0) = 0$ then it is known [16] that $|a_n| \leq n|A_1|$, $n = 1$ and 2. MacGregor has recently shown [9] that if $|f(z)| \leq |F(z)|$ in E and if $F(z)$ is univalent in E and $F(0) = 0$ then $|a_n| \leq n|A_1|$ for $n = 1, 2,$ and 3. We shall show more generally that if $f(z) \prec_q F(z)$ in E and if $F(z)$ is univalent in E and $F(0) = 0$ then

$$(1.5) \qquad |a_n| \leq n|A_1|, \qquad n = 1, 2, \text{ and } 3,$$

and either $|a_n| \leq n|A_1|$ for *all* positive integers n, or

$$(1.6) \qquad \overline{\lim_{n \to \infty}} \left|\frac{a_n}{n}\right| < |A_1|.$$

In this connection there are three long-standing and still-unproved conjectures that should be kept in mind.

Conjecture I (Bieberbach, 1916). *If*

$$F(z) = \sum_1^\infty A_n z^n$$

is regular and univalent in E then $|A_n| \leq n|A_1|$, $n = 2, 3, \ldots.$

Conjecture II (Rogosinski, 1943). *If*

$$f(z) = \sum_1^\infty a_n z^n \prec F(z) = \sum_1^\infty A_n z^n, \qquad \text{in } E$$

and if $F(z)$ is regular and univalent in E, then $|a_n| \leq n|A_1|$, $n = 1, 2, \ldots.$

Conjecture III (Robertson, 1936). *If*

$$F(z) = \sum_1^\infty A_n z^n$$

is regular and univalent in E, and if the univalent function $[F(z^2)]^{1/2}$ *has the power series* $\sum_1^\infty D_{2k-1} z^{2k-1}$, *then* $\sum_{k=1}^n |D_{2k-1}|^2 \leqq n|A_1|$, $n = 2, 3, \ldots$.

Conjecture I is known to be true for $n = 2, 3, 4$, and for a given function $F(z)$ for all sufficiently large values of n [3]. Conjecture II [8, 16] is known to be true for $n = 1, 2$, and for special subclasses of univalent functions $F(z)$ for all positive integers n [15, 16]. Conjecture III is known [13] to be correct for $n = 2$ and 3. As the author pointed out in [13], if Conjecture III is true for all positive integers n then so is Conjecture I. We shall show here that if Conjecture III is true, then Conjecture II is true too, even if the concept of subordination in Conjecture II is replaced by the more general concept of quasi-subordination.

For certain kinds of univalent functions $F(z)$, for example when $F(z)$ is spiral-like in E, or in particular starlike in E, then the corresponding coefficients D_{2k-1} of Conjecture III satisfy $|D_{2k-1}| \leqq |D_1| = |A_1|^{1/2}$, $k = 2, 3, \ldots$ so then

$$\sum_{k=1}^n |D_{2k-1}|^2 \leqq n|A_1|, \qquad n = 2, 3, \ldots.$$

So when $f(z) \prec_q F(z)$ and $F(z)$ is spiral-like (or starlike) in E then the following inequalities hold:

(1.7) $|a_n| \leqq n|A_1|, \qquad n = 1, 2, \ldots.$

The author [15] showed that inequalities (1.7) held when $f(z)$ is subordinate to $F(z)$ and $F(z)$ is close-to-convex in E. Although a starlike function is close-to-convex in general a spiral-like function is not. It is an open question whether inequalities (1.7) hold when $f(z)$ is quasi-subordinate to $F(z)$ and $F(z)$ is close-to-convex in E. Inequalities (1.7) would hold, clearly, if the corresponding coefficients D_{2k-1} satisfy $|D_{2k-1}| \leqq |D_1|$ when $F(z)$ is close-to-convex. Although it can be shown that the coefficients c_{2k-1} of an odd, close-to-convex function

$$G(z) = \sum_1^\infty c_{2k-1} z^{2k-1} \qquad (z \in E)$$

are bounded, $|c_{2k-1}| \leqq |c_1|$, $k = 2, 3, \ldots$, it is not true in general that the odd function $[F(z^2)]^{1/2}$ is close-to-convex in E when $F(z)$ is close-to-convex.

From these considerations and other evidence we conjecture that inequalities (1.7) also hold whenever $F(z)$ is univalent in E, $F(0) = 0$, and when $f(z) \prec_q F(z)$ in E.

2. The general case

To begin the presentation we give fundamental considerations for the most part without restrictions on the superordinate function when $f(z) \prec_q F(z)$ in E. We let

$$f(z) = \sum_0^\infty a_n z^n, \qquad F(z) = \sum_0^\infty A_n z^n,$$

$$\phi(z) = \sum_0^\infty \beta_n z^n, \qquad \omega(z) = \sum_1^\infty \alpha_n z^n,$$

$$|\phi(z)| \leq 1, \quad |\omega(z)| \leq |z|, \quad f(z) = \phi(z)F\{\omega(z)\}, \qquad z \in E.$$

Theorem 1. *Let $f(z) \prec_q F(z)$ in E. Let ρ be an arbitrary positive number less than one. Then $f(z) \prec_q F(z)$ for $|z| < \rho$.*

Proof. $f(\rho z) = \phi(\rho z)F\{\omega(\rho z)\}$
$\qquad\qquad = \phi(\rho z)F\{\rho\omega_1(z)\},$

where

$$\omega_1(z) = \frac{1}{\rho}\,\omega(\rho z), \quad \left|\frac{\omega_1(z)}{z}\right| = \left|\frac{\omega(\rho z)}{\rho z}\right| \leq 1 \qquad \text{in } E.$$

Since also $|\phi(\rho z)| \leq 1$ in E we have $f(\rho z) \prec_q F(\rho z)$ in E.
This implies that $f(z) \prec_q F(z)$ for $|z| < \rho$.

Theorem 2. *If in E $f(z) \prec_q F_1(z)$ and $F_1(z) \prec_q F_2(z)$ then $f(z) \prec_q F_2(z)$ in E.*

Proof. $f(z) = \phi_1(z)F_1\{\omega_1(z)\},$
$\qquad\quad F_1(z) = \phi_2(z)F_2\{\omega_2(z)\},$

where $\phi_1, \phi_2, \omega_1,$ and ω_2 are appropriate bounded analytic functions in E. It follows that

$$f(z) = \phi_3(z)F_2\{\omega_3(z)\},$$

where

$$\phi_3(z) = \phi_1(z)\phi_2\{\omega_1(z)\}, \quad |\phi_3(z)| \leq 1,$$

and

$$\omega_3(z) = \omega_2\{\omega_1(z)\}, \quad |\omega_3(z)| = |\omega_2\{\omega_1(z)\}| \leq |\omega_1(z)| \leq |z| < 1,$$

in E. Hence $f(z) \prec_q F_2(z)$ in E.

Theorem 3. *If* $f(z) \prec_q F(z)$ *in* E *and* $F(0) = 0$, *then*

$$f(z) \prec_q \frac{f(z)}{z} \prec_q \frac{F(z)}{z} \qquad in \ E.$$

Proof. Since $f(z) = \phi(z)F\{\omega(z)\}$, $F(0) = 0$, $\omega(0) = 0$, we have $f(0) = 0$. Thus $f(z)/z$ and $F(z)/z$ are regular in E.

$$\frac{f(z)}{z} = \left(\frac{\phi(z)\omega(z)}{z}\right)\frac{F\{\omega(z)\}}{\omega(z)}.$$

Since $|\phi(z)\omega(z)/z| \leq 1$ in E, we have $f(z)/z \prec_q F(z)/z$. Since $|f(z)| \leq |f(z)/z|$ we have $f(z) \prec_q f(z)/z$. By Theorem 2 it follows that $f(z) \prec_q F(z)/z$ in E.

The converse of Theorem 3 is false. This conclusion follows from Theorem 4 and its corollary 1 which we now prove.

Theorem 4. *If* $|f'(0)| = |F'(0)| \neq 0$ *and* $F(0) = 0$ *then* $f(z) \prec_q F(z)$ *if, and only if,* $f(z) \equiv e^{i\mu}F(e^{i\nu}z)$ *for some real* μ *and* ν.

Proof. If $f(z) = \phi(z)F\{\omega(z)\}$, then

$$\sum_1^\infty a_n z^n = \left(\sum_0^\infty \beta_k z^k\right)[A_1(\alpha_1 z + \alpha_2 z^2 + \cdots) + A_2(\alpha_1 z + \alpha_2 z^2 + \cdots)^2 + \cdots];$$

$$a_1 = \beta_0 \alpha_1 A_1, \quad |a_1| = |f'(0)| = |F'(0)| = |A_1| \neq 0.$$

Hence $|\beta_0 \alpha_1| = 1$. Since $|\beta_0| \leq 1$, $|\alpha_1| \leq 1$ we have $\alpha_1 = e^{i\nu}$, $\beta_0 = e^{i\mu}$ for some real μ, ν. In this case $\omega(z) = e^{i\nu}z$, $\phi(z) = e^{i\mu}$ by the maximum modulus theorem. We conclude that $f(z) = e^{i\mu}F(e^{i\nu}z)$. Conversely, if $f(z) = e^{i\mu}F(e^{i\nu}z)$ then $f(z) \prec_q F(z)$ in E.

Corollary 1. *If* $F(0) = 0$, $F'(0) \neq 0$ *and* $f(z)/z \prec F(z)/z$ *in* E *then* $f(z) \prec_q F(z)$ *in* E *if, and only if, for some real* ν, $f(z) \equiv e^{-i\nu}F(e^{i\nu}z)$.

Proof. Since $f(z)/z$ is subordinate to $F(z)/z$ we have

$$f'(0) = \lim_{z \to 0} \frac{f(z)}{z} = \lim_{z \to 0} \frac{F\{\omega(z)\}}{\omega(z)} = F'(0) \neq 0.$$

If $f(z) \prec_q F(z)$ in E and $f(z) = \sum_1^\infty a_n z^n$, $F(z) = \sum_1^\infty A_n z^n$, then $a_1 = \beta_0 \alpha_1 A_1$, where $a_1 = f'(0) = F'(0) = A_1 \neq 0$. Hence $\beta_0 \alpha_1 = 1$ and $\alpha_1 = e^{i\nu}$, $\beta_0 = e^{-i\nu}$ for some real ν. Hence $f(z) = e^{-i\nu}F(e^{i\nu}z)$.

Theorem 5. *Let*

$$f(z) = \sum_1^\infty a_n z^n \prec_q F(z) = \sum_1^\infty A_n z^n, \quad f(z) = \phi(z)F\{\omega(z)\} \qquad in\ E.$$

Let $\phi(0) = \beta_0$, $\omega'(0) = \alpha_1$. *Let* $\mu(\beta_0, \alpha_1) = |\beta_0| + (1 - |\beta_0|^2)|\alpha_1|$.

Then

(1) $|a_1| \leqq |\beta_0 \alpha_1| |A_1| \leqq |A_1|$,

(2) $|a_2| \leqq \mu(\beta_0, \alpha_1)\max\{|A_1|, |A_2|\} \leqq \frac{5}{4}\max\{|A_1|, |A_2|\}$.

(3) *If either* $|\beta_0| = 1$ *or* $|\alpha_1| = (1 + |\beta_0|)^{-1}$ *then* $\mu(\beta_0, \alpha_1) \leqq 1$; *if* $\beta_0 = 1$
$f(z) \prec F(z)$.

(4) *If* $|\alpha_1| \leqq \frac{1}{2}$ *then* $\mu(\beta_0, \alpha_1) \leqq 1$; *if* $|\alpha_1| > \frac{1}{2}$ *then* $\mu(\beta_0, \alpha_1) \leqq |\alpha_1| + 1/4|\alpha_1|$
where $1 < |\alpha_1| + 1/4|\alpha_1| \leqq \frac{5}{4}$.

The inequalities appearing in (1), (2), (3), (4) are all sharp.

Proof. Using the notation of Theorem 4 we have

$$a_1 = \beta_0 \alpha_1 A_1 \quad and \quad a_2 = (\beta_0 \alpha_2 + \beta_1 \alpha_1)A_1 + \beta_0 \alpha_1^2 A_2.$$

Since $|\beta_0| \leqq 1$, $|\alpha_1| \leqq 1$, we have $|a_1| \leqq |A_1|$. Since $|\alpha_2| \leqq 1 - |\alpha_1|^2$, $|\beta_1| \leqq 1 - |\beta_0|^2$ for bounded functions $\omega(z)$ and $\phi(z)$ we have

$$|\beta_0 \alpha_2 + \beta_1 \alpha_1| + |\beta_0 \alpha_1^2| \leqq |\beta_0|(1 - |\alpha_1|^2) + |\alpha_1|(1 - |\beta_0|^2) + |\beta_0 \alpha_1^2|$$
$$= |\beta_0| + (1 - |\beta_0|^2)|\alpha_1| = \mu(\beta_0, \alpha_1).$$

Hence

(2.1) $$|a_2| \leqq \mu(\beta_0, \alpha_1)\max\{|A_1|, |A_2|\}.$$

Inequality (2.1) is sharp. For if $0 \leqq \beta_0 \leqq 1$, $0 \leqq \alpha_1 \leqq 1$ and

$$\phi(z) = \frac{\beta_0 + z}{1 + \beta_0 z} = \beta_0 + (1 - \beta_0^2)z + \cdots,$$

$$\omega(z) = \frac{\alpha_1 z + z^2}{1 + \alpha_1 z} = \alpha_1 z + (1 - \alpha_1^2)z^2 + \cdots,$$

then

$$\beta_0 \alpha_2 + \beta_1 \alpha_1 + \beta_0 \alpha_1^2 = \beta_0 + (1 - \beta_0^2)\alpha_1 = \mu(\beta_0, \alpha_1),$$

$$a_2 = [\beta_0(1 - \alpha_1^2) + (1 - \beta_0^2)\alpha_1]A_1 + \beta_0 \alpha_1^2 A_2.$$

If $A_1 = A_2$ then $a_2 = \mu(\beta_0, \alpha_1)A_1 = \mu(\beta_0, \alpha_1)\max\{|A_1|, |A_2|\}$, and so (2.1) is sharp. If either $|\beta_0| = 1$ or $|\alpha_1| = (1 + |\beta_0|)^{-1}$ then $\mu(\beta_0, \alpha_1) \leqq 1$ and in this case inequality (2.1) becomes

(2.2) $$|a_2| \leqq \max\{|A_1|, |A_2|\}.$$

If $\beta_0 = 1$ then $\phi(z) \equiv 1$ and $f(z) \prec F(z)$ in E. In this situation inequality (2.2) is well known [8, 16].

If $|\alpha_1| \leq \frac{1}{2}$ the inequality $|\alpha_1| \leq (1 + |\beta_0|)^{-1}$ is satisfied and then (2.2) follows. If $|\alpha_1| > \frac{1}{2}$ then $\max_{|\beta_0|} \mu(\beta_0, \alpha_1)$ occurs for $|\beta_0| = 1/2|\alpha_1| < 1$ and has the value $(|\alpha_1| + 1/4|\alpha_1|)$ which lies in the interval $(1, \frac{5}{4})$.

When

$$f(z) = F(z)(1 + 2z)/(2 + z) \text{ in } E$$

and $A_1 = A_2$ then $|a_2| = \frac{5}{4}\max\{|A_1|, |A_2|\}$. Here we have $|f(z)| \leq |F(z)|$ so that $f(z) \prec_q F(z)$ in E.

Theorem 6. *Let $f(z) \prec_q F(z)$ in E. Then for any $p > 0$ and $0 < r < 1$*

$$\int_0^{2\pi} |f(re^{i\theta})|^p \, d\theta \leq \int_0^{2\pi} |F(re^{i\theta})|^p \, d\theta.$$

Proof. Since $f(z)/\phi(z) \prec F(z)$ in E we have [2, 8, 16]

$$\int_0^{2\pi} \left| \frac{f(re^{i\theta})}{\phi(re^{i\theta})} \right|^p \, d\theta \leq \int_0^{2\pi} |F(re^{i\theta})|^p \, d\theta$$

and, since $|\phi(z)| \leq 1$ in E,

$$\int_0^{2\pi} |f(re^{i\theta})|^p \, d\theta \leq \int_0^{2\pi} \left| \frac{f(re^{i\theta})}{\phi(re^{i\theta})} \right|^p \, d\theta.$$

The proof of Theorem 7, which follows, is similar to the familiar corresponding theorem when f is subordinate to F [2, 8, 16]. However, we include it, with minor modifications, to take care of quasi-subordination, for the sake of completeness and for comparison with Theorem 8.

Theorem 7. *Let*

$$f(z) = \sum_0^\infty a_n z^n \prec_q F(z) = \sum_0^\infty A_n z^n \qquad \text{in } E.$$

Then

$$\sum_{k=0}^n |a_k|^2 \leq \sum_{k=0}^n |A_k|^2, \qquad n = 0, 1, \ldots.$$

Proof. Let

$$s_n(z) = \sum_{k=0}^n a_k z^k, \quad S_n(z) = \sum_{k=0}^n A_k z^k, \quad R_n(z) = \sum_{k=n+1}^\infty A_k z^k.$$

Then

$$s_n(z) + \sum_{n+1}^{\infty} a_k z^k = f(z) = \phi(z)F\{\omega(z)\}$$

$$= \phi(z)S_n\{\omega(z)\} + \phi(z)R_n\{\omega(z)\}$$

$$= \phi(z)S_n\{\omega(z)\} + \sum_{n+1}^{\infty} d_k z^k,$$

where $\sum_{n+1}^{\infty} a_k z^k$ and $\sum_{n+1}^{\infty} d_k z^k$ converge for $|z| < 1$. By definition

$$\phi(z)S_n\{\omega(z)\} \prec_q S_n(z).$$

Hence by Theorem 6 with $p = 2$, $z = re^{i\theta}$, $0 < r < 1$,

$$\int_0^{2\pi} |\phi(z)S_n\{\omega(z)\}|^2 \, d\theta \leq \int_0^{2\pi} |S_n(z)|^2 \, d\theta,$$

$$\int_0^{2\pi} \left| s_n(z) + \sum_{n+1}^{\infty} (a_k - d_k)z^k \right|^2 \, d\theta \leq \int_0^{2\pi} |S_n(z)|^2 \, d\theta,$$

$$\sum_{k=0}^{n} |a_k|^2 r^{2k} + \sum_{k=n+1}^{\infty} |a_k - d_k|^2 r^{2k} \leq \sum_{k=0}^{n} |A_k|^2 r^{2k}, \qquad 0 \leq r \leq 1,$$

$$\sum_{k=0}^{n} |a_k|^2 \leq \sum_{k=0}^{n} |A_k|^2, \qquad n = 0, 1, \dots .$$

Theorem 8. *Let*

$$f(z) = \sum_0^{\infty} a_k z^k \prec_q F(z) = \sum_0^{\infty} A_k z_k \qquad in \ E.$$

Let $F(0) \neq 0$ and $f(z) = \phi(z)F\{\omega(z)\}$ in E. In some neighborhood of the origin let

$$\left[\frac{f(z)}{\phi(z)} \right]^{1/2} = \sum_0^{\infty} b_k z^k, \qquad b_0 \neq 0,$$

$$[F(z)]^{1/2} = \sum_0^{\infty} B_k z^k, \qquad B_0 \neq 0.$$

Then

$$\sum_{k=0}^{n} |b_k|^2 \leq \sum_{k=0}^{n} |B_k|^2, \qquad n = 0, 1, \dots .$$

Proof. Let

$$\sigma_n(z) = \sum_{k=0}^{n} b_k z^k, \quad \Sigma_n(z) = \sum_{k=0}^{n} B_k z^k, \quad P_n(z) = \sum_{k=n+1}^{\infty} B_k z^k.$$

Then

$$\sigma_n(z) + \sum_{n+1}^{\infty} b_k z^k = \left[\frac{f(z)}{\phi(z)}\right]^{1/2} = [F\{\omega(z)\}]^{1/2}$$

$$= \sum_n \{\omega(z)\} + P_n\{\omega(z)\}$$

$$= \sum_n \{\omega(z)\} + \sum_{k=n+1}^{\infty} e_k z^k$$

for some neighborhood of the origin. If $f(z)/\phi(z)$ and $F(z)$ have one or more zeros in E, $[f(z)/\phi(z)]^{1/2}$ and $[F(z)]^{1/2}$ cease to be regular throughout E. In this case $\sum_{n+1}^{\infty} b_k z^k$ and $\sum_{k=n+1}^{\infty} e_k z^k$, though convergent in some neighborhood of the origin, may not separately converge throughout E. But since $\omega(z)$, $\sigma_n(z)$, $\sum_n(z)$ are all regular in E the function

$$\sum_n \{\omega(z)\} - \sigma_n(z) = \sum_{k=n+1}^{\infty} (b_k - e_k) z^k$$

is regular for $|z| < 1$. Hence $\sum_{n+1}^{\infty} (b_k - e_k) z^k$ converges in the full unit disk $|z| < 1$. Thus

$$\sigma_n(z) + \sum_{k=n+1}^{\infty} (b_k - e_k) z^k = \sum_n \{\omega(z)\}$$

and $\sum_n \{\omega(z)\}$ is regular and subordinate to $\sum_n(z)$ in E. Therefore

$$\sum_{k=0}^{n} |b_k|^2 \le \sum_{k=0}^{n} |B_k|^2, \qquad n = 0, 1, \dots.$$

Theorem 9. *Let*

$$f(z) = \sum_{k=q}^{\infty} a_k z^k, \quad a_q \ne 0, \quad q \ge 0,$$

be quasi-subordinate to

$$F(z) = \sum_{0}^{\infty} A_k z^k$$

in E and $F(0) \ne 0$. In some neighborhood of the origin let $[F(z)]^{1/2} = \sum_{0}^{\infty} B_k z^k$, $B_0 \ne 0$. *Then*

$$|a_n| \le \sum_{k=0}^{n-q} |B_k|^2, \qquad n = q, q + 1, \dots.$$

Proof. We use the same notation as in Theorem 8. Since $F(0) \ne 0$ and $f(z)/\phi(z) = F\{\omega(z)\}$, it follows that $f(z)/\phi(z)$ cannot vanish at $z = 0$. Let $r_0 = \min\{|a|, 1\}$, where a is a zero of $f(z)/\phi(z)$ closest to the origin, $|a| < 1$.

Let $r_0 = 1$ if $f(z)/\phi(z)$ does not vanish in E. Then for $|z| < r_0$

$$\left[\frac{f(z)}{\phi(z)}\right]^{1/2} = \sum_0^\infty b_k z^k, \qquad b_0 \neq 0,$$

and

$$f(z) = \phi(z)\left(\sum_{k=0}^\infty b_k z^k\right)^2.$$

$\phi(z)$ has a zero at the origin if $f(z)$ has one and of the same order $q \geq 0$. For $n \geq q$ and $r < r_0$ we have

$$a_n = \frac{1}{2\pi i} \oint_{|z|=r} \frac{f(z)}{z^{n+1}}\, dz$$

$$= \frac{1}{2\pi i} \oint_{|z|=r} \phi(z)\left(\sum_{k=0}^\infty b_k z^k\right)^2 \frac{dz}{z^{n+1}}$$

$$= \frac{1}{2\pi i} \oint_{|z|=r} \phi(z)\left(\sum_{k=0}^{n-q} b_k z^k\right)^2 \frac{dz}{z^{n+1}}.$$

Since $\phi(z)(\sum_{k=0}^{n-q} b_k z^k)^2$ is regular in $|z| < 1$ the last integral is independent of r for $0 < r < 1$. For $z = re^{i\theta}$, $0 < r < 1$,

$$|a_n| \leq \frac{1}{2\pi r^n} \int_0^{2\pi} |\phi(z)| \left|\sum_{k=0}^{n-q} b_k z^k\right|^2 d\theta$$

$$\leq \frac{1}{2\pi r^{n-q}} \int_0^{2\pi} \left|\sum_{k=0}^{n-q} b_k z^k\right|^2 d\theta$$

$$= \frac{1}{r^{n-q}} \sum_{k=0}^{n-q} |b_k|^2 r^{2k} \leq \frac{1}{r^{n-q}} \sum_{k=0}^{n-q} |b_k|^2.$$

Letting $r \to 1$ and using Theorem 8 we have

$$|a_n| \leq \sum_{k=0}^{n-q} |b_k|^2 \leq \sum_{k=0}^{n-q} |B_k|^2, \qquad n = q, q+1, \ldots .$$

It will be noticed that there is no restriction in assuming $F(0) \neq 0$ as we have done. For if $F(z)$ has a zero at the origin of order $s \geq 1$ then $f(z)$ also has a zero at the origin of not less than s. By Theorem 3 we then conclude that

$$\frac{f(z)}{z^s} \prec_q \frac{F(z)}{z^s} \qquad \text{in } E.$$

Moreover, $f(z) \prec_q f(z)/z^s$ since $|f(z)| \leq |f(z)/z^s|$ in E. By Theorem 2 it follows that $f(z) \prec_q F(z)/z^s$ in E. In this case Theorem 9 applies with $F(z)$ replaced by $F(z)/z^s$.

Corollary 2. *Let*

$$f(z) = \sum_{k=q}^{\infty} a_k z^k, \qquad a_q \neq 0, \ q \geq 0,$$

be quasi-subordinate to

$$F(z) = \sum_{0}^{\infty} A_k z^k$$

in E. Let $F(0) \neq 0$ and $F(a) = 0$ for some a, $|a| < 1$. In some neighborhood of the origin let

$$\left[\left(\frac{1 - \bar{a}z}{a - z} \right) F(z) \right]^{1/2} = \sum_{0}^{\infty} B_k(a) z^k, \qquad B_0(a) \neq 0.$$

Then

$$|a_n| \leq \sum_{k=0}^{n-q} |B_k(a)|^2, \qquad n = q, q + 1, \ldots .$$

Proof. Since

$$|F(z)| \leq \left| \left(\frac{1 - \bar{a}z}{a - z} \right) F(z) \right| \qquad \text{in } E$$

we have

$$f(z) \prec_q F(z) \prec_q \left(\frac{1 - \bar{a}z}{a - z} \right) F(z) \qquad \text{in } E,$$

and by Theorem 2

$$f(z) \prec_q \left(\frac{1 - \bar{a}z}{a - z} \right) F(z).$$

The Corollary then follows from Theorem 9 on replacing $F(z)$ by

$$\left(\frac{1 - \bar{a}z}{a - z} \right) F(z).$$

It is interesting to notice that either one of the values $\sum_{0}^{n} |B_k|^2$ and $\sum_{0}^{n} |B_k(a)|^2$ can be the smaller when for instance $n = 1$ and

$$F(z) = \frac{a - z}{1 - \bar{a}z}, \qquad 0 < a < 1,$$

depending upon the size of a. We have $|B_0|^2 + |B_1|^2 = (1 + a^2)^2/4a$ and $|B_0(a)|^2 + |B_1(a)|^2 = 1$. $|B_0|^2 + |B_1|^2 < |B_0(a)|^2 + |B_1(a)|^2$ if, and only if, $(1 + a^2)^2 - 4a < 0$, or $a^3 + a^2 + 3a - 1 > 0$. The equation $a^3 + a^2 + 3a - 1 = 0$ has one root in the interval $(0, 1)$.

Theorem 10. *Let*

$$f(z) = \sum_{k=q}^{\infty} a_k z^k, \qquad a_q \neq 0, q \geq 0,$$

be quasi-subordinate to

$$F(z) = \sum_{k=0}^{\infty} A_k z^k$$

in E. Let F(z) have the representation

$$F(z) = \int_0^{2\pi} \mu(z, \phi) \, d\alpha(\phi)$$

where $\mu(z, \phi)$ is regular in E and continuous in ϕ for $0 \leq \phi \leq 2\pi$ with $\mu(0, \phi) \neq 0$ for all ϕ and where $\alpha(\phi)$ is a function of bounded variation in $[0, 2\pi]$ normalized so that $\int_0^{2\pi} d\alpha(\phi) = 1$. In a neighborhood of the origin let

$$h(z, \phi) = [\mu(z, \phi)]^{1/2} = \sum_0^{\infty} c_k(\phi) z^k, \qquad c_0(\phi) \neq 0.$$

Then

$$|a_n| \leq \sum_{k=0}^{n-q} \int_0^{2\pi} |c_k(\phi)|^2 \, |d\alpha(\phi)|, \qquad n = q, q+1, \ldots.$$

Proof. Since $f(z) = \phi(z) F\{\omega(z)\}$ in E and

$$F(z) = \int_0^{2\pi} \mu(z, \phi) \, d\alpha(\phi)$$

it follows that $f(z)$ also has the representation

$$f(z) = \int_0^{2\pi} g(z, \phi) \, d\alpha(\phi) \qquad (z \in E),$$

where $g(z, \phi) = \phi(z)\mu\{\omega(z), \phi\}$. Then for each value of ϕ in $[0, 2\pi]$ the function $g(z, \phi)/\phi(z)$ is subordinate to $\mu(z, \phi)$, or $g(z, \phi) \prec_q \mu(z, \phi)$ relative to the same bounded function $\phi(z)$ for all values of the parameter ϕ. Let

$$g(z, \phi) = \sum_{k=q}^{\infty} \gamma_n(\phi) z^n, \qquad z \in E.$$

Then by Theorem 9

$$|\gamma_n(\phi)| \leq \sum_{k=0}^{n-q} |c_k(\phi)|^2, \qquad n \geq q.$$

Since

$$a_n = \int_0^{2\pi} \gamma_n(\phi) \, d\alpha(\phi)$$

we have

$$|a_n| \leq \sum_{k=0}^{n-q} \int_0^{2\pi} |c_k(\phi)|^2 \, |d\alpha(\phi)|, \qquad n \geq q.$$

3. The Case F Univalent

In the preceding section it was possible for the superordinate function $F(z)$ to have one or more zeros in the domain $0 < |z| < 1$. In this section we shall restrict $F(z)$ to be regular and univalent in E with therefore at most one zero in E.

Theorem 11. *Let*

$$f(z) = \sum_{k=1}^{\infty} a_k z^k \prec_q F(z) = \sum_{k=1}^{\infty} A_k z^k \qquad \text{in } E$$

and let $F(z)$ be univalent in E. Then

$$|a_n| \leq n|A_1|, \qquad n = 1, 2, \text{ and } 3.$$

Proof. Since $A_1 \neq 0$ we may assume $A_1 = 1$ for the sake of simplicity. It is known [11] that

$$[F(z^2)]^{1/2} = z + \sum_{k=2}^{\infty} D_{2k-1} z^{2k-1}$$

is univalent in E and the author [13] has shown that

(3.1) $$\sum_{k=1}^{n} |D_{2k-1}|^2 \leq n, \qquad n = 1, 2, \text{ and } 3,$$

and conjectured at that time (over three decades ago) that inequalities (3.1) hold for *all* positive integers n. Since

$$\left[\frac{F(z)}{z} \right]^{1/2} = \sum_{k=0}^{\infty} D_{2k+1} z^k, \qquad D_1 = 1,$$

and, since by Theorems 3 and 2, $f(z) \prec_q F(z)/z$, it follows from Theorem 9 that

$$|a_n| \leq \sum_{k=0}^{n-1} |D_{2k+1}|^2 = \sum_{k=1}^{n} |D_{2k-1}|^2 \leq n, \qquad n = 1, 2, 3.$$

Corollary 3. *Conjecture I of the introduction is true provided Conjecture III is true.*

Theorem 12. *Let*

$$f(z) = \sum_{k=q}^{\infty} a_k z^k, \qquad a_q \neq 0, q \geq 1,$$

be quasi-subordinate to

$$F(z) = \sum_{k=1}^{\infty} A_k z^k$$

in E. Let F(z) be univalent and spiral-like in E so that for some real α, |α| < π/2,
$\mathrm{Re}[e^{i\alpha} zF'(z)/F(z)] > 0$ *in E (F(z) starlike when α = 0); then*

$$|a_n| \leq (n - q + 1)|A_1| \leq n|A_1|, \quad n = q, q + 1, \dots.$$

Proof. If

$$G(z) = [F(z^2)]^{1/2} = \sum_{k=1}^{\infty} D_{2k-1} z^{2k-1},$$

then in E

$$\mathrm{Re}\left[e^{i\alpha} \frac{zG'(z)}{G(z)} \right] = \mathrm{Re}\left[e^{i\alpha} \frac{z^2 F'(z^2)}{F(z^2)} \right] > 0.$$

Hence $G(z)$ is an odd function, spiral-like in E. If $P(z)$ denotes an even function, regular and having a positive real part in E with $P(0) = 1$, then, letting

$$P(z) = \sum_{k=0}^{\infty} p_{2k} z^{2k}, \quad e^{i\alpha} zG'(z) = [P(z)\cos \alpha + i \sin \alpha]G(z).$$

we have

$$e^{i\alpha}(2n - 1)D_{2n-1} = e^{i\alpha}D_{2n-1} + \cos \alpha \sum_{k=1}^{n-1} p_{2(n-k)} D_{2k-1}$$

$$(n - 1)|D_{2n-1}| \leq \sum_{k=1}^{n-1} |D_{2k-1}|, \quad n = 2, 3, \dots.$$

It follows by induction that $|D_{2n-1}| \leq |D_1|, n \geq 2$. From Theorem 9 we have for $n \geq q$

$$|a_n| \leq \sum_{k=0}^{n-q} |D_{2k+1}|^2 \leq (n - q + 1)|D_1|^2 = (n - q + 1)|A_1| \leq n|A_1|.$$

Theorem 13. *Let*

$$f(z) = \sum_{k=1}^{\infty} a_k z^k \prec_q F(z) = \sum_{k=1}^{\infty} A_k z^k \qquad in \ E.$$

Let $F(z)$ be univalent in E. If $F(z)$ is not the Koebe function $A_1 z(1 - \varepsilon z)^{-2}$, $|\varepsilon| = 1$, then

$$\varlimsup_{n \to \infty} \left| \frac{a_n}{n} \right| < |A_1|.$$

Proof. Since

$$[F(z^2)]^{1/2} = \sum_{k=1}^{\infty} D_{2k-1} z^{2k-1}, \qquad D_1{}^2 = A_1,$$

is univalent in E it follows from a theorem due to Hayman [3] that $\lim_{n \to \infty} |D_{2n-1}| = \alpha$ exists and $\alpha < |D_1|$ except for the Koebe function, $F(z) = A_1 z(1 - \varepsilon z)^{-2}$, $|\varepsilon| = 1$. In this case $|D_{2n-1}| < |D_1|$ for $n > n_0(F)$ and so

$$\varlimsup_{n \to \infty} \left| \frac{a_n}{n} \right| \leqq \lim_{n \to \infty} \frac{1}{n} \sum_{k=1}^{n} |D_{2k-1}|^2 = \alpha^2 < |D_1|^2 = |A_1|.$$

When $F(z)$ is the Koebe function, which is starlike in E, then $|a_n| \leqq n|A_1|$ for all positive integers by Theorem 12.

Theorem 14. *Let* $f(z) \prec_q F(z)$ *in* E. *Let* $F(z)$ *be regular and univalent in* E *with* $F(0) = 0$, $F'(0) = 1$. *Then for* $z \in E$

$$|f'(z)| \leqq \frac{1 + |z|}{(1 - |z|)^3}, \quad |f(z)| \leqq \frac{|z|}{(1 - |z|)^2}.$$

Theorem 14 is known to be true [2] when $f(z) \prec F(z)$ in E.

Proof. $f(z) = \phi(z) F\{\omega(z)\}$

and [11]

$$|\phi'(z)| \leqq \frac{1 - |\phi(z)|^2}{1 - |z|^2}, \quad |\omega'(z)| \leqq \frac{1 - |\omega(z)|^2}{1 - |z|^2}.$$

Since $F(z)$ is univalent in E we have [11]

$$|F\{\omega(z)\}| \leqq \frac{|\omega(z)|}{(1 - |\omega(z)|)^2},$$

$$|F'\{\omega(z)\}| \leqq \frac{1 + |\omega(z)|}{(1 - |\omega(z)|)^3},$$

$$f'(z) = \phi'(z) F\{\omega(z)\} + \phi(z) \omega'(z) F'\{\omega(z)\},$$

$$|f'(z)| \leqq \frac{1 - |\phi(z)|^2}{1 - |z|^2} \cdot \frac{|\omega(z)|}{(1 - |\omega(z)|)^2} + |\phi(z)| \frac{(1 - |\omega(z)|^2)}{1 - |z|^2} \cdot \frac{1 + |\omega(z)|}{(1 - |\omega(z)|)^3}$$

$$= \frac{1 - |\phi(z)|^2}{1 - |z|^2} \cdot \frac{|\omega(z)|}{(1 - |\omega(z)|)^2} + \frac{|\phi(z)|}{1 - |z|^2} \left(\frac{1 + |\omega(z)|}{1 - |\omega(z)|}\right)^2$$

$$\leqq \frac{1 - |\phi(z)|^2}{1 - |z|^2} \cdot \frac{|z|}{(1 - |z|)^2} + \frac{|\phi(z)|}{1 - |z|^2} \left(\frac{1 + |z|}{1 - |z|}\right)^2$$

since $|\omega(z)| \leqq |z| < 1$.

$$|f'(z)| \leqq \frac{|z| + (1 + |z|)^2 |\phi(z)| - |z| \, |\phi(z)|^2}{(1 + |z|)(1 - |z|)^3},$$

$$\max_{|\phi(z)| \leqq 1} [|z| + (1 + |z|)^2 |\phi(z)| - |z| \, |\phi(z)|^2]$$

occurs for $|\phi(z)| = 1$ and has the value $(1 + |z|)^2$. Hence

$$|f'(z)| \leqq \frac{1 + |z|}{(1 - |z|)^3} \qquad (z \in E).$$

An integration gives

$$|f(z)| \leqq \frac{|z|}{(1 - |z|)^2} \qquad (z \in E).$$

4. Some special cases of *F*

In this section we consider superordinate functions $F(z)$ which are not necessarily univalent but have some other restriction imposed on $F(z)$ other than regularity in E.

In Theorem 15 $F(z)$ is typically real in E. In other words $F(z)$ is real for z real, and Im $F(z) > 0$ for Im $z > 0$ when $z \in E$. The author has shown [12] that when $F'(0) = 1$, $F(z)$ has the representation

$$F(z) = \int_0^\pi \frac{z \, d\alpha(\phi)}{1 - 2z \cos \phi + z^2}, \quad \int_0^\pi d\alpha(\phi) = 1 \qquad (z \in E),$$

where $\alpha(\phi)$ is a nondecreasing function in $[0, \pi]$.

Theorem 15. *Let*

$$f(z) = \sum_1^\infty a_k z^k \prec_q F(z) = z + \sum_2^\infty A_k z^k \qquad in \ E.$$

Let F(z) be typically real in E. Then

$$|a_n| \leqq n, \qquad n = 1, 2, \ldots$$

and $|a_n| = n$ for all n when $f(z) = F(z) = z(1 - z)^{-2}$.

Proof. The function

$$\mu_0(z, \phi) = \frac{z}{1 - 2z \cos \phi + z^2}$$

is starlike in *E* since when $z = re^{i\theta}, r < 1$,

$$\operatorname{Re}\left[\frac{z\mu_0'(z, \phi)}{\mu_0(z, \phi)}\right] = \operatorname{Re}\left[\frac{1 - z^2}{1 - 2z \cos \phi + z^2}\right]$$

$$= \frac{(1 - r^2)(1 - 2r \cos \theta \cos \phi + r^2)}{|1 - 2z \cos \phi + z^2|^2} > 0.$$

Also $[\mu_0(z^2, \phi)]^{1/2}$ is an odd function, univalent and starlike in *E*. If

$$\left[\frac{\mu_0(z, \phi)}{z}\right]^{1/2} = \sum_{k=0}^{\infty} c_k^{(0)}(\phi)z^k, \qquad c_0(\phi) = 1,$$

then (as in Theorem 12 with $\alpha = 0$) $|c_k^{(0)}(\phi)| \leqq 1, k = 0, 1, \ldots$. Using Theorem 10 (the interval $[0, \pi]$ now replaces the interval $[0, 2\pi]$) we conclude that

$$|a_n| \leqq \sum_{k=0}^{n-1} \int_0^{\pi} |c_k^{(0)}(\phi)|^2 \, d\alpha(\phi) \leqq \sum_{k=0}^{n-1} \int_0^{\pi} d\alpha(\phi) = n, \qquad n \geqq 1.$$

If

$$F(z) = z + \sum_{2}^{\infty} A_k z^k$$

is regular, univalent and convex in *E* it is known that $\operatorname{Re}[F(z)/z] > \frac{1}{2}$ in *E* [10, 17]. The converse is false as the simple example $F(z) = z + z^2/2$, which maps the circle $|z| = 1$ into a cardioid, clearly shows. For functions $F(z)$, $\operatorname{Re}[F(z)/z] > \frac{1}{2}$ in *E*, we have the following Theorem 16.

Theorem 16. *Let*

$$f(z) = \sum_{1}^{\infty} a_k z^k \prec_q F(z) = z + \sum_{2}^{\infty} A_k z^k \qquad in\ E.$$

Let $\text{Re}[F(z)/z] > \frac{1}{2}$ *in E. Then for* $n = 1, 2, \ldots$

$$|a_n| \le 1 + \sum_{k=1}^{n-1} \left[\frac{1 \cdot 3 \cdot 5 \cdot \,\cdots\, \cdot (2k-1)}{2 \cdot 4 \cdot 6 \cdot \,\cdots\, \cdot (2k)} \right]^2 \sim \frac{1}{\pi} \log_e n.$$

For each n there exists a function $f_n(z) \prec_q z/(1-z)$ *for which the equality signs hold.*

Proof. Since $\text{Re}[F(z)/z] > \frac{1}{2}$ in E we write

$$\frac{F(z)}{z} = \frac{1 + P(z)}{2}, \quad P(z) = \int_0^{2\pi} \frac{1 + ze^{i\phi}}{1 - ze^{i\phi}} \, d\alpha(\phi), \quad P(0) = 1,$$

where $\alpha(\phi)$ is a nondecreasing function in the interval $[0, 2\pi]$ and $\int_0^{2\pi} d\alpha(\phi) = 1$. Hence

$$F(z) = \int_0^{2\pi} \mu_1(z, \phi) \, d\alpha(\phi), \quad \mu_1(z, \phi) = z(1 - ze^{i\phi})^{-1},$$

$$\left[\frac{\mu_1(z, \phi)}{z} \right]^{1/2} = (1 - ze^{i\phi})^{-1/2} = 1 + \sum_{k=1}^{\infty} \left[\frac{1 \cdot 3 \cdot 5 \cdot \,\cdots\, \cdot (2k-3)}{2 \cdot 4 \cdot 6 \cdot \,\cdots\, \cdot (2k-2)} \right] e^{ni\phi} z^n$$

$$= \sum_0^{\infty} c_k^{(1)}(\phi) z^k, \quad z \in E.$$

Again, using Theorem 10, we have for $n \ge 1$

$$|a_n| \le \sum_{k=0}^{n-1} \int_0^{2\pi} |c_k^{(1)}(\phi)|^2 \, d\alpha(\phi)$$

(4.1)
$$= 1 + \sum_{k=1}^{n-1} \left[\frac{1 \cdot 3 \cdot 5 \cdot \,\cdots\, \cdot (2k-1)}{2 \cdot 4 \cdot 6 \cdot \,\cdots\, \cdot (2k)} \right]^2 = G_{n-1}.$$

There exists (see Landau [4]; also [1, p. 442]) a bounded function $\phi_n(z)$, which depends upon n, whose power-series expansion about the origin has a partial sum of order n that takes on the value G_n at $z = 1$. Thus the function

$$f_n(z) = \phi_{n-1}(z) \cdot \frac{z}{1-z} \prec_q \frac{z}{1-z}$$

and the equality signs hold for each n in (4.1). The sharpness occurs when $f(z)$ is majorized by the convex function $z(1-z)^{-1}$. This was observed by MacGregor [9].

Theorems 11, 12 with $q = 1$, 15, and 16 were established in the particular case that $f(z)$ is majorized by $F(z)$ in a recent paper by MacGregor [9] by other methods of proof. It was in a large measure his paper [9] that stimulated the author to introduce the quasi-subordinate concept as has been done here.

REFERENCES

1. P. Dienes, *The Taylor Series*, Clarendon Press, Oxford, 1931; and Dover, New York, 1957.
2. G. M. Goluzin, *Geometrische Funktionentheorie*, Deutsche Verlag der Wissenschaften, Berlin, 1957.
3. W. K. Hayman, *Multivalent Functions*, Cambridge University Press, Cambridge, England, 1958.
4. E. Landau, *Darstellung und Begrundung einiger neurer Ergebnisse der Funktionentheorie*, Springer, Berlin, 1916; Chelsea, New York.
5. Z. Lewandowski, Starlike majorants and subordination, *Ann. Univ. Mariae Curie-Skłodowska, Sect. A* **15** (1961), 79–84.
6. ———, Some remarks on a paper of M. S. Robertson, *Ann. Univ. Mariae Curie-Skłodowska, Sect. A* **17** (1963), 43–46.
7. J. E. Littlewood, On inequalities in the theory of functions, *Proc. London Math. Soc.* (2) **23** (1925), 481–519.
8. ———, *Lectures on the Theory of Functions*, Oxford University Press, London, 1944.
9. T. H. MacGregor, Majorization by univalent functions, *Duke Math. J.* **34** (1967), 95–102.
10. A. Marx, Untersuchungen über schlichte Abbildungen, *Math. Ann.* **107** (1932), 40–67.
11. Z. Nehari, *Conformal Mapping*, McGraw-Hill, New York, 1952.
12. M. S. Robertson, On the coefficients of a typically-real function, *Bull. Amer. Math. Soc.* **41** (1935), 565–572.
13. ———, A remark on the odd schlicht functions, *Bull. Amer. Math. Soc. Ser. 2* **42** (1936), 366–370.
14. ———, Applications of the subordinate principle to univalent functions, *Pacific J. Math.* **11** (1961), 315–324.
15. ———, The generalized Bieberbach conjecture for subordinate functions, *Michigan Math. J.* **12** (1965), 421–429.
16. W. Rogosinski, On the coefficients of subordinate functions, *Proc. London Math. Soc.* (2) **48** (1943), 48–82.
17. E. Strohhäcker, Beiträge zur Theorie der schlichten Funktionen, *Math. Z.* **37** (1933), 356–380.

(Received August 18, 1967)

Convex Meromorphic Functions

W. C. Royster

Department of Mathematics
University of Kentucky
Lexington, Kentucky 40506
U.S.A.

1. Introduction

Let Σ denote the class of univalent meromorphic functions in the unit disk E. If the function $f(z)$ is a member of the class Σ then $f(z)$ has at most a simple pole in E. Denote by Σ_0 the subclass of Σ consisting of functions $f(z)$ having a Laurent expansion $f(z) = z^{-1} + \sum_{n=0}^{\infty} b_n z^n$. The class Σ_0 has been studied extensively (see [1, 3, 11]). Denote by Σ_p the subclass of Σ consisting of functions $f(z)$ which have a pole at $z = p$, $0 < |p| < 1$ and are normalized by the conditions $f(0) = 0$, $f'(0) = 1$. In the neighborhood of the origin the functions have a series expansion $f(z) = z + a_2 z^2 + a_3 z^3 + \cdots$. It is known that $|a_n| \leq p(p^{-n} - p^n)(1 - p^2)^{-1}$, $n = 2, 3, 4, 6$; in fact, Jenkins [6] proved if the Bieberbach conjecture is valid for $n \leq N$, $N \geq 2$, then $|a_n| \leq p(p^{-n} - p^n)(1 - p^2)^{-1}$, $n = 2, 3, \ldots, N$, $0 < p < 1$. For earlier references to the study of Σ_p see [8].

In this paper we study a class of functions K_p which consists of functions belonging to Σ_p and which map E onto a domain such that its complement with respect to the extended plane is a convex domain, possibly degenerating into a straight-line segment. We shall define the class K_p in the following way. The function $f(z)$ belongs to K_p if (a) it is regular in E except for a simple pole at $z = p$, $0 < |p| < 1$; (b) there exists a positive real number ρ, $|p| < \rho < 1$ such that

$$(1.1) \qquad \operatorname{Re}\left\{1 + z\,\frac{f''(z)}{f'(z)}\right\} \leq 0, \qquad z = re^{i\theta},$$

for all r such that $\rho < r < 1$; and (c) $f(z)$ is normalized so that $f(0) = 0$, $f'(0) = 1$ and p is real and positive. Condition (c) does not restrict generality. Indeed, if $p = Re^{i\alpha}$, $0 < R < 1$, α real and $f(z)$ satisfies conditions (a) and (b) then $F(z) = e^{-i\alpha}[f(e^{i\alpha}z) - f(0)]/f'(0)$ belongs to Σ_p and satisfies conditions (a) and (b).

If $f \in K_p$ then the function

(1.2) $$P(z) = -\left\{1 + z\frac{f''(z)}{f'(z)} + \frac{z + p}{z - p} - \frac{1 + pz}{1 - pz}\right\}$$

is regular in E and has positive real part in E; also $p(0) = 1$, and hence $P(z)$ belongs to the class $\mathscr{P}$ of normalized analytic functions with positive real part in E.

We shall use equation (1.2) to obtain an integral representation for members of K_p, to study the coefficient problem and to obtain bounds on the length of the images of $|z| = r$ under the mapping by a member of K_p. Finally, we relate the class K_p to the class of meromorphic starlike functions.

2. Sufficient Condition

We have seen that if $f \in K_p$ then equation (1.2) holds. We shall show if $f(z)$ satisfies (1.2) and $0 < p < 2 - \sqrt{3}$ then $f(z) \in K_p$ and this is best possible in the sense that there is a function $f_0(z)$ which satisfies (1.2); but for $p > 2 - \sqrt{3}$, $\text{Re}\{1 + zf_0''(z)/f_0'(z)\} \geq 0$ for $|z| > p$.

Let

(2.1) $$Q(z) = \frac{1 + pz}{1 - pz} - \frac{z + p}{z - p}.$$

Then

$$\text{Re } Q(re^{i\theta}) = 2p(1 - r^2)\left[\frac{p(1 + r^2) - r(1 + p^2)\cos\theta}{(1 - 2pr\cos\theta + r^2p^2)(r^2 - 2r\cos\theta + p^2)}\right],$$

$z = re^{i\theta}$ and

(2.2) $$-\text{Re}\left\{1 + z\frac{f''(z)}{f'(z)}\right\} = \text{Re}\{P(z) - Q(z)\}.$$

A simple calculation shows if $f(z)$ has as its only singularity a simple pole at $z = p$ and satisfies (1.2) then $P(p) = (1 + p^2)(1 - p^2)^{-1}$.

Lemma 1 [9]. *Let* $P(z) \in \mathscr{P}$, $P(p) = (1 + p^2)(1 - p^2)^{-1}$, $0 < p < 1$. *Then*

(2.3) $$\min_{|z| = r} \text{Re}\{P(z)\} = \frac{(1 + p^2)(1 - r^2)}{(1 + rp)^2 + (r + p)^2}.$$

Proof. Let $\omega(z) = (P(z) - 1)(P(z) - 1)^{-1}$. Then $\omega(z)$ is regular and bounded, $|\omega(z)| < 1$ in E, and $\omega(0) = 0$, $\omega(p) = p^2$. In order to use the information we have about $\omega(z)$ at $z = p$ define new functions

(2.4)
$$g(z) = \frac{\omega(z) - p^2}{1 - p^2\omega(z)} \cdot \frac{1 - pz}{z - p},$$

$$h(z) = \frac{g(z) - p}{1 - pg(z)}.$$

Then $h(z)$ is regular and $|h(z)| < 1$ in E, $h(0) = 0$, and hence $|h(z)| < |z|$ in E. Letting $z = (\zeta + p)(1 + p\zeta)$, we get

$$P(z) = \frac{1 + p^2}{1 - p^2} \cdot \frac{1 + \zeta g(z)}{1 - \zeta g(z)}$$

so that

(2.5)
$$\min_{|z| = r} \mathrm{Re}\{P(z)\} = \min_{|z| = r} \frac{1 - |g\zeta|^2}{|1 - g\zeta|^2} \cdot \frac{1 + p^2}{1 - p^2}.$$

A simple geometric argument shows the minimum occurs when $g = a + R$ and $\zeta = -(a + R)$ where $a = p(1 - r^2)(1 - r^2p^2)^{-1}$, $R = r(1 - p^2)(1 - r^2p^2)^{-1}$. That is

$$\min_{|z| = r} \mathrm{Re}\{P(z)\} = \frac{1 - (a + R)^2}{1 + (a + R)^2} \cdot \frac{1 + p^2}{1 - p^2} = \frac{(1 + p)^2(1 - r^2)}{(1 + pr)^2 + (r + p)^2}.$$

Equality is attained at $z = -r$ with $h(z) = -z$.

Lemma 2. *There exists a positive real number* $R(p)$, $R(p) < 1$, *such that if* $0 < p < 2 - \sqrt{3}$ *and* $R(p) < r < 1$, *then*

(2.6)
$$\max_{0 \le \theta \le 2\pi} \mathrm{Re}\{Q(re^{i\theta})\} = \mathrm{Re}\{Q(r, \pi)\} = \frac{2p(1 - r^2)}{(r + p)(1 + rp)}.$$

Proof. On the circle $|z| = r$, $r \ne p$,

$$\mathrm{Re}\{Q(r, \theta)\} = 2p(1 - r^2)q(r, \theta),$$

where

(2.7)
$$q(r, \theta) = \frac{p(1 + r^2) - r(1 + p^2)\cos\theta}{(r^2 + p^2 - 2rp\cos\theta)(1 + r^2p^2 - 2pr\cos\theta)}.$$

The derivative of $q(r, \theta)$ with respect to θ is given by

$$-A\alpha\beta\gamma\sin\theta\left(\cos^2\theta - \frac{2}{\alpha}\cos\theta - \frac{\alpha - \beta - \gamma}{\alpha\beta\gamma}\right),$$

where $A > 0$ and $\alpha = r(1 + p^2)p^{-1}(1 + r^2)^{-1}$, $\beta = 2rp(1 + p^2r^2)^{-1}$, $\gamma = 2rp(r^2 + p^2)^{-1}$. Hence its zeros are at $\theta = 0$, π, and $\cos\theta_1 = \delta_1$, $\cos\theta_2 = \delta_2$, where

$$\delta_k = \frac{2p^2(1 + r^2) + (-1)^{k-1}(1 - p^2)\sqrt{(r^2 - p^2)(1 - r^2p^2)}}{2pr(1 + p^2)}, \qquad k = 1, 2.$$

It follows that $q(r, \pi)$ will be a maximum value of $q(r, \theta)$ if and only if $(1 + \delta_1)(1 + \delta_2) < 0$. This condition holds if the expression

$$r^2 - \frac{p^4 - 6p^2 + 1}{p(p^2 + 1)} r + 1$$

is negative. That is, if r lies between r_1 and r_2, where

$$r_k = \frac{1 - 6p^2 + p^4 + (-1)^k(1 - p)^2\sqrt{p^4 - 14p^2 + 1}}{2p(p^2 + 1)}, \quad k = 1, 2$$

and since r_k must be real, we see that p satisfies $0 < p < 2 - \sqrt{3}$. It is easy to see that $r_2 > 1$ and $r_1 < 1$ for $0 < p < 2 - \sqrt{3}$. The number $R(p)$ in the lemma is r_1 and the lemma is proved. These bounds are attained by the function

$$f_0(z) = \frac{p}{1 + p^2} \frac{(p^2 + 1)z - 2p^2z}{(p - z)(1 - pz)}.$$

It is easy to show if $p > 2 - \sqrt{3}$ then $\mathrm{Re}\{1 + zf_0''(z)/f_0'(z)\} \geq 0$, for some $|z| > p$.

For $2 - \sqrt{3} < p < 1$, $r > p$, $q(r, \theta)$ takes its maximum value at $\cos\theta_2 = \delta_2$, the maximum value is

$$q(r, \theta_2) = \frac{(1 + p^2)^2}{2p(1 - p^2)} \frac{1}{\sqrt{1 - p^2r^2} + \sqrt{r^2 - p^2}}.$$

Theorem 1. *Let $f(z)$ be univalent meromorphic in E, have a pole at $z = p$, and satisfy (1.2) for $|z| > p$, $0 < p < 2 - \sqrt{3}$. Then $f \in K_p$.*

Proof. If f satisfies the hypothesis of the theorem then by Lemma 1 and Lemma 2 we have

$$\mathrm{Re}\{P(re^{i\theta}) - Q(re^{i\theta})\} \geq (1 - r^2)\left[\frac{1 + p^2}{(1 + pr)^2 + (r + p)^2} - \frac{2p}{(r + p)(1 + rp)}\right]$$

$$\geq -A\left(r^2 - \frac{p^4 - 6p^2 + 1}{p(1 + p^2)} r + 1\right),$$

where $A > 0$. This last expression is positive for $0 < p < 2 - \sqrt{3}$ and $R(p) < r < 1$.

The function $f_0(z)$ shows that the bound on p is best possible. This function maps E onto the exterior of a vertical slit centered at $w = -p(1 + p^2)^{-1}$ of length $4p(1 - p^2)^{-2}$.

3. Coefficient Problem

We mentioned in Section 1 that if $f \in \Sigma_p$ and has a Taylor's expansion $f(z) = z + a_2 z^2 + \cdots$ about the origin then the upper bounds for $|a_2|$, $|a_3|$, $|a_4|$, and $|a_6|$ are known. The function $f_1(z) = pz(p - z)^{-1}(1 - pz)^{-1}$ which yields the upper bounds satisfies the hypotheses of Theorem 1 and hence belongs to K_p for $0 < p < 2 - \sqrt{3}$. Hence

$$|a_n| \leq \frac{p}{1 - p^2} \left(\frac{1}{p^n} - p^n \right), \qquad n = 2, 3, 4, 6.$$

We shall prove the following.

Theorem 2. *Let $f \in K_p$, $0 < p < 2 - \sqrt{3}$. Then*

(3.1)
$$\frac{1 + p^4}{p(1 + p^2)} \leq |a_2| \leq p + \frac{1}{p}.$$

Both bounds are sharp.

Proof. The inequality on the right is known. To prove the inequality on the left note if $f \in K_p$ then $P(z)$ in (1.2) has the expansion

(3.2)
$$P(z) = 1 + 2\left(p + \frac{1}{p} - a_2 \right) z + \cdots.$$

Using the functions $\omega(z)$, $g(z)$, and $-h(z)$ in Lemma 1 we get $g(0) = p$, $g'(0) = (1 - p^2)(1 - (1 + p^2)\omega'(0)/p)$, $h'(0) = -g'(0)/(1 - p^2)$. Hence

$$\omega'(0) = \frac{p}{1 + p^2} (1 + h'(0))$$

and since $h(z)$ satisfies the Schwarz lemma we get

$$|\omega'(0)| \leq \frac{2p}{1 + p^2}, \qquad |p'(0)| \leq \frac{4p}{1 + p^2}$$

with equality for

$$p(z) = \frac{1 - z^2}{1 - 4pz/(1 + p^2) + z^2}.$$

Hence

$$\left| \frac{1}{p} + p - a_2 \right| \leqq \frac{2p}{1 + p^2},$$

and the left-hand inequality in (3.1) is proved. The bound is sharp for $f_0(z)$.

4. The Arc Length Problem

For $f \in K_p$, let $L_r(f) = \int_{|z|=r} |f'(z)||dz|$ denote the arc length of the image of the circle $|z| = r$ under the mapping $w = f(z)$. We find an upper bound for $L_r(f)$ and a sharp upper bound for $L_r(f)$ in the case $p = 0$.

Consider $P(z)$ given by (1.2) which satisfies $\text{Re}\{P(z)\} \geqq 0$ in E, $P(0) = 1$, $P(p) = (1 + p^2)(1 - p^2)^{-1}$. This function can be represented by a Herglotz integral,

$$P(z) = \int_{-\pi}^{\pi} (e^{it} + z)(e^{it} - z)^{-1} \, d\mu(t),$$

where $\mu(t)$ is a nondecreasing real function in $-\pi \leqq t \leqq \pi$ subject to the conditions

(4.1)   (a) $\displaystyle\int_{-\pi}^{\pi} d\mu(t) = 1,$ (b) $\displaystyle\int_{-\pi}^{\pi} (e^{it} - p)^{-1} \, d\mu(t) = \frac{p}{1 - p^2}.$

Integrating the differential equation (1.2), where $P(z)$ is replaced by its integral representation, yields

(4.2) $$f(z) = \int_0^z \frac{p^2}{(z - p)^2 (1 - pz)^2} \left[\exp \int_{-\pi}^{\pi} 2 \log(1 - ze^{-it}) \, d\mu(t) \right] dz,$$

$z \neq p$, where $\mu(t)$ is nondecreasing and satisfies (4.1). Condition (b) of (4.1) implies the only singularity of $f(z)$ at $z = p$ is a simple pole. It follows that if $f \in K_p$ then f can be represented by the integral formula (4.2). Each function f representable by (4.2) need not belong to K_p even when $\mu(t)$ satisfies (4.1). However, if $0 < p < 2 - \sqrt{3}$ then $f(z)$ represented by (4.2) belongs to K_p.

Now

(4.3) $$L_r(f) = r \int_{-\pi}^{\pi} \frac{p^2 \left| \exp \displaystyle\int_{-\pi}^{\pi} 2 \log(1 - re^{i(\theta-t)}) \, d\mu(t) \right|}{|re^{i\theta} - p|^2 \, |1 - rpe^{i\theta}|^2} \, d\theta.$$

An application of the continuous form of the arithmetic-geometric mean inequality to the above integral gives

(4.4) $$L_r(f) \leqq rp^2 \int_{-\pi}^{\pi} \frac{|1 - re^{i(\theta-t)}|^2 \, d\theta}{|re^{i\theta} - p|^2 \, |1 - rpe^{i\theta}|^2}$$

$$\leqq r \max_{-\pi \leqq t \leqq \pi} I(t),$$

where $I(t)$ is the integral on the right. It is easy to show $I(t) \leq I(\pi)$, $-\pi \leq t \leq \pi$. Hence

$$L_r(f) \leq rp^2 A \int_{-\pi}^{\pi} \frac{1 + \gamma \cos \theta}{(1 - \alpha \cos \theta)(1 - \beta \cos \theta)} \, d\theta,$$

where $\alpha = 2rp(r^2 + p^2)^{-1}$, $\beta = 2rp(1 + r^2 p^2)^{-1}$, $\gamma = 2r(1 + r^2)^{-1}$, $A = (1 + r^2)(r^2 + p^2)^{-1}(1 + r^2 p^2)^{-1}$. A long but straightforward computation yields

(4.5) $$L_r(f) \leq 2\pi \left(\frac{1 + p}{1 - p} \right) \frac{p^2 r(1 + r^2)}{(r^2 - p^2)(1 - r^2 p^2)}, \qquad r > p.$$

The above bound is sharp for the class of functions represented by (4.2). It is not sharp for the class K_p. Indeed, the function whose derivative is given by $(1 + z)^2(z - p)^{-2}(1 - pz)^{-2}$ yields equality but does not belong to K_p.

The sharp upper bound for $L_r(f)$ is attained probably for a mass function $\mu(t)$ with two jump points in $-\pi \leq t \leq \pi$. It is easy to see using (4.1b) that the extreme value cannot occur when $\mu(t)$ has only one jump point in $-\pi \leq t \leq \pi$. It can be shown [9] using a generalization of the Golusin variational formula that $\mu(t)$ has at most three jump points.

For the case $p = 0$ we can find exact bounds on $L_r(f)$.

Theorem 3. *Let $f(z)$ be univalent convex meromorphic in E with a pole at $z = 0$, residue equal to one. Then for each r, $0 < r < 1$, $L_r(f) \leq L_r(f^*)$, where $f^*(z) = z + z^{-1} + b$ and b is a constant. Equality can occur only (except for an additive constant) for $f(z) = e^{i\alpha}(f^*(e^{-i\alpha}z))$, α real.*

Proof. If $f \in K_0$ then

(4.6) $$f(z) = \int_a^z \frac{1}{z^2} \exp \left[\int_{-\pi}^{\pi} 2 \log(1 - ze^{-it}) \, d\mu(t) \right] dz,$$

where $a \neq 0$, $|a| < 1$ and $\int_{-\pi}^{\pi} e^{-it} \, d\mu(t) = 0$. Hence

(4.7) $$L_r(f) = \frac{1}{r} \int_{-\pi}^{\pi} \exp \left[\int_{-\pi}^{\pi} 2 \log(1 - re^{i(\theta - t)}) \, d\mu(t) \right] d\theta.$$

It can be shown using a variational formula from [9] that to maximize the integral in (4.7) $\mu(t)$ must be chosen a step function with jump points at t_1 and t_2 in the interval $-\pi \leq t \leq \pi$ [9] which when considered with the constraint in (4.6) gives

$$\lambda_1 e^{-it_1} + \lambda_2 e^{-it_2} = 0, \qquad \lambda_1 + \lambda_2 = 1.$$

Thus $\lambda_1 = \lambda_2 = \frac{1}{2}$, $e^{-it_1} = -e^{-it_2}$ and $t_2 = t_1 + \pi$. It follows that

$$L_r(f) \leq \frac{1}{r} \max_{-\pi \leq t \leq \pi} \int_{-\pi}^{\pi} |1 - re^{i(\theta - t)}| \, |1 + re^{i(\theta - t)}| \, d\theta$$

$$= \frac{1}{r} \int_{-\pi}^{\pi} |1 - r^2 e^{2i\theta}| \, d\theta$$

$$= \frac{4}{r} [2E(r^2) - (1 - r^4)K(r^2)],$$

where $K(r^2)$ and $E(r^2)$ are the complete elliptic integrals of the first and second kind, respectively [5, p. 387].

It is interesting to note that the arithmetic-geometric mean inequality which has been used extensively to study arc length problems [2, 7] does not yield the best result. This inequality is equivalent to the assumption the integral is maximized by a step function $\mu(t)$ with one jump point.

5. Concluding Remarks

The class K_p can be related to the class of meromorphic starlike functions with pole at the origin. If $f \in K_p$ then $g(z) = (p^2 z)^{-1}(z - p)^2(1 - pz)^2 f'(z)$ is meromorphic starlike with respect to the origin, and if g is meromorphic starlike with respect to the origin and $pg'(p)/g(p) = (1 + p^2)(1 - p^2)^{-1}$ then $f \in K_p$. Thus extremal problems pertaining to starlike functions in Σ_0 can be related to extremal problems in K_p.

The class K_p can be studied from the standpoint of the Schwartz-Christoffel transformation. If $f \in K_p$ then the image of E can be approximated by mappings of the form

$$(5.1) \qquad f(z) = A_1 \int_0^z \frac{\prod_{j=1}^{n} (t - a_j)^{\gamma_j}}{(t - p)^2 (1 - pt)^2} \, dt,$$

where $a_j = e^{i\theta_j}$, $0 \leq \theta_j \leq 2\pi$, $j = 1, 2, \ldots, n$, $\theta_j < \theta_{j+1}$, $\gamma_j \geq 0$, $\sum_{j=1}^{n} \gamma_j = 2$, and A_1 is chosen so that f is normalized properly and where γ_j and a_j are subject to the condition $\sum_{j=1}^{n} \gamma_j(a_j - p)^{-1} = 2p(1 - p^2)^{-1}$. The function in (5.1) maps E onto the exterior of a convex polygon. The Schwartz-Christoffel transformation has been used in studying many classes and subclasses of univalent functions (see [12 and 4]).

REFERENCES

1. Z. Birnbaum, Beiträge zur Theorie der schlichten Funktionen, *Studia Math.* **1** (1929), 159–190.
2. P. Duren, An arc length problem for close-to-convex functions, *J. London Math. Soc.* **39** (1964), 757–761.
3. G. Golusin, Einige Koeffizientenabschätzungen für schlichte Funktionen, *Rec. Math.* (*Mat. Sbornik*) *N.S.* **3** (1938), 321–330.
4. A. Goodman, On the Schwarz-Christoffel transformation and p-valent functions, *Trans. Amer. Math. Soc.* **68** (1950), 204–223.
5. I. Gradshteyn and I. Ryzhik, *Tables of integrals, series, and products*, 4th ed. prepared by Yu. V. Geronimus and M. Yu. Tseytlin, transl. edited by A. Jeffrey, Academic Press, New York, 1965 (translation).
6. J. Jenkins, On a conjecture of Goodman's concerning meromorphic univalent functions, *Michigan Math. J.* **9** (1962), 25–27.
7. F. Keogh, Some inequalities for convex and star-shaped domains, *J. London Math. Soc.* **29** (1954), 121–123.
8. Y. Komatu, Note on the theory of conformal representation by meromorphic functions I and II, *Proc. Japan Acad.* **21** (1945), 269–284.
9. J. Pfaltzgraff and B. Pinchuk, A variational method for classes of meromorphic functions, to appear, *J. Analyse Math.* (1970).
10. J. Pfaltzgraff and W. Royster, Unpublished notes.
11. C. Pommerenke, Über einige Klassen meromorphic schlichter Funktionen, *Math. Z.* **78** (1962), 263–284.
12. M. Robertson, On the theory of univalent functions, *Ann. of Math.* **37** (1936), 374–408.

(*Received July 4, 1968 and, in revised form, September 24, 1968*)

Nonmaximal Prime Ideals in the Ring of Holomorphic Functions*

L. A. Rubel

Department of Mathematics
University of Illinois
Urbana, Illinois 61801
U.S.A.
and
Institute for Advanced Study
Princeton, New Jersey 08540
U.S.A.

In [1], Henriksen gave a proof by Kaplansky that in the ring of all entire functions, there exists a prime ideal that is not maximal. The result and the proof easily extend to the ring A of all functions analytic on a region G in the complex plane. We give here a different proof, which is somewhat more explicit.

Let $\{w_n\}$, $n = 1, 2, 3, \ldots$ be a sequence of distinct points of G that has no limit point in G, and for each $f \in A$ let $\phi_f(n)$ be defined as the multiplicity of the zero of f at the point w_n. In particular, $\phi_f(n) = 0$ means that $f(w_n) \neq 0$, while $\phi_f(n) = \infty$ implies that f is the zero function. Now let U be an ultra-filter on the set of positive integers that contains every set whose complement is finite. Let Φ be the class of functions $\phi : \{1, 2, 3, \ldots\} \to \{\infty, 0, 1, 2, \ldots\}$ such that for each integer k, $E_k(\phi) \in U$, where

$$E_k(\phi) = \{n : \phi(n) \geq k\}.$$

* This research was partially supported by the United States Air Force Office of Scientific Research Grants, numbers AF OSR 460-63 and 460-67.

Let I be the set of functions $f \in A$ such that $\phi_f \in \Phi$. We shall show that I is a nonmaximal prime ideal.

First, Φ contains any function ϕ such that $\phi(n) \to \infty$ as $n \to \infty$, since then $E_k(\phi)$ is co-finite. Hence I contains a nonzero function, by the Weierstrass factorization theorem. Now I is an ideal since if $f, g \in I$ then $f - g \in I$ because

$$\phi_{f-g} \geqq \min(\phi_f, \phi_g)$$

and thus

$$E_k(\phi_{f-g}) \supseteq E_k(\phi_f) \cap E_k(\phi_g).$$

Also, if $f \in I, g \in A$ then $fg \in I$ since

$$\phi_{fg} \geqq \phi_f, \quad E_k(\phi_{fg}) \supseteq E_k(\phi_f).$$

Now we show that I is prime. Suppose $f, g \in I$. Then

$$E_{2k}(\phi_{fg}) = E_{2k}(\phi_f + \phi_g) = \bigcup_{j=0}^{2k} \{E_j(\phi_f) \cap E_{2k-j}(\phi_g)\}.$$

Hence, there is some j with $0 \leqq j \leqq 2k$ with

$$E_j(\phi_f) \cap E_{2k-j}(\phi_g) \in U.$$

Since either $j \geqq k$ or $2k - j \geqq k$. we see that either $E_j(\phi_f) \in U$ for some $j \geqq k$ or $E_j(\phi_g) \in U$ for some $j \geqq k$. Since $j \geqq k$ implies that $E_j \subseteq E_k$, we see that for each k, either $E_k(\phi_f) \in U$ or $E_k(\phi_g) \in U$. It follows that one of these holds for infinitely many k, hence for all k, and thus that either f or g belongs to I, and that I is consequently prime.

Finally, I is not maximal. For take $F \in A$ such that F has a simple zero at each point w_n. Then $F \notin I$ since $E_2(\phi_F) = \varnothing \notin U$. Let $J = (I, F)$ be the ideal generated by I and F. The ideal J properly contains I, yet $J \neq A$, since if $1 = f + gF$, $f \in I$, $g \in A$, then $\lambda \in I$, where $\lambda = 1 - gF$. This is impossible, since $\phi_\lambda(n) = 0$ for all n. This completes the proof.

REFERENCES

1. M. Henriksen, On the prime ideals of the ring of entire functions, *Pacific J. Math.* **3** (1953), 711–720.

(Received September 18, 1967)

Meromorphic Functions with One Valiron Deficient Value

S. M. Shah*

Department of Mathematics
University of Kentucky
Lexington, Kentucky 40506
U.S.A.

H. Shankar

Department of Mathematics
Ohio University
Athens, Ohio 45701
U.S.A.

1. Introduction

In this paper we shall be concerned with a class of functions $f(z)$ meromorphic in $|z| < \infty$, and satisfying certain growth conditions (see sections 2,3). Let $T(r,f)$, $n(r, a)$ and $N(r, a)$ have their usual meanings [8, pp. 6–12]. We write

$$\delta(a,f) = 1 - \limsup_{r \to \infty} N(r, a)/T(r,f),$$

$$\Delta(a,f) = 1 - \liminf_{r \to \infty} N(r, a)/T(r,f).$$

If $\delta(a,f) > 0$ then a is said to be a *Nevanlinna deficient value* of $f(z)$, and if $\Delta(a,f) > 0$ then a is said to be a *Valiron deficient value*. There can be at most countably many Nevanlinna deficient values for a function $f(z)$ [5,

* The research of this author was supported by the National Science Foundation under Grant GP-7544, Presented to Amer. Math. Soc. meeting, 23 January 1970.

pp. 42–44], but there can be uncountably many Valiron deficient values (cf. [6, 16, 17]). If $f(z)$ is of zero order then it can have at most one Nevanlinna deficient value [14; 5, pp. 114–115], but it may have more than one Valiron deficient value ([9, 15]). In sections 2 and 3, we consider functions of zero order and satisfying some growth condition, and show that for these functions $T(r, f) \sim N(r, a)$ except for one a.

If $f(z)$ is an entire function we show that

$$(1.1) \qquad \log M(r, f) \sim T(r, f) \sim N(r, a)$$

for every finite a, where $M(r, f) = \max_{|z|=r} |f(z)|$. Thus for these functions we have only one Valiron deficient value.

In Section 4, we utilize the properties of a proximate order of $f(z)$ to obtain a relation satisfied by the counting function $n(r, a)$ when $f(z)$ is not necessarily of zero order.

We shall assume throughout this paper that $f(z)$ is a transcendental meromorphic function. It is shown [5, pp. 6–7] that for such functions $T(r, f)/\log r \to \infty$ as $r \to \infty$. We shall denote by b, c, $c_1 \ldots$ finite positive constants and by a, α, β, $\ldots$ finite or infinite complex constants. These constants are not necessarily the same at each occurrence.

2. Functions of Zero Order

Theorem 1. *Let $f(z)$ be a meromorphic function of zero order.*

Part (a). Let $\Psi(r)$ and $\theta(r)$ be two functions tending to infinity with r, and let $\Phi(r)$ be any function also tending to infinity (however slowly) with r. Let $c > 1$. Suppose that for all large r, $\Psi(r\theta(r)) \leqq c\Psi(r)$, $\theta(r)$ is nondecreasing, and either (i) $\Psi(r)$ is nondecreasing or (ii) $\Psi(r)/r^b$ is nonincreasing for some b in (0, 1). If ultimately

$$(2.1) \qquad \Psi(r)\Phi(r)/\log \theta(r) \leqq T(r, f) \leqq \Psi(r),$$

and if

$$(2.2) \qquad N(r, \alpha) = o(T(r, f))$$

for some constant α, $|\alpha| \leqq \infty$, then

$$(2.3) \qquad T(r, f) \sim N(r, a)$$

for every $a \neq \alpha$ and

$$(2.4) \qquad \limsup_{r \to \infty} n(r, a)\Phi(r)/T(r, f) \leqq c$$

for every a.

Part (b). Let $L(r)$ be a slowly oscillating function [7], and write

$$\xi(r) = (L(r))^{-1} \int_1^r L(t)/t \, dt.$$

Also let $\Phi(r)$ *be any function tending to infinity with r such that* $\lim_{r \to \infty} \Phi(r)/r = 0$, *and suppose that for all large r, $\Phi(r)/r$ and $1/\xi(r)$ are nonincreasing. If ultimately*

(2.5) $L(r)(\xi(r))^{-1}\Phi(\xi(r)) \leqq n(r, \beta) \leqq L(r),$

and if

(2.6) $N(r, \alpha) = o(N(r, \beta))$

for some distinct constants α and β, then the conclusion (2.3) *holds. Furthermore,*

(2.7) $\limsup_{r \to \infty} n(r, \alpha)\Phi(\xi(r))/T(r, f) = 0,$

(2.8) $\limsup_{r \to \infty} n(r, \beta)\Phi(\xi(r))/T(r, f) \leqq 1,$

and for $a \neq \alpha, \beta$,

(2.9) $\limsup_{r \to \infty} n(r, a)\Phi(\xi(r))/T(r, f) \leqq e.$

Part (c). If for some constants α and β $N(r, \beta)$ is a slowly oscillating function of r and if $N(r, \alpha) = o(N(r, \beta))$, then the conclusion (2.3) holds. If b be any positive number less than one, then for every a

(2.10) $\limsup_{r \to \infty} n(r, a)\kappa(r)/T(r, f) \leqq c_1(f) < \infty,$

where $\kappa(r) = \{\sup_{t \geqq r} (n(t, \beta)/N(t, \beta))\}^{-b}$, *and* $\kappa(r)$ *tends monotonically to infinity as* $r \to \infty$.

Proof (a). We may suppose without loss of generality that $\alpha = 0$, for if $\alpha \neq \infty$ we consider the function $f^*(z) = f(z) - \alpha$ and if $\alpha = \infty$ we consider $f^*(z) = 1/f(z)$. Next, we may assume that $f(0) = 1$. Thus $f(z) = f_1(z)/f_2(z)$, where

(2.11)
$$f_1(z) = \prod_1^\infty (1 - z/a_i)$$
$$f_2(z) = \prod_1^\infty (1 - z/b_j)$$

and a_i and b_j are the zeros and poles respectively* of $f(z)$. Since $\psi(r\theta(r)) \leqq$

---

* We have assumed that $f_1(z)$ has an infinity of zeros. If $f_1(z)$ is a polynomial or a constant, then the conclusions (2.14) and (2.15) are trivially true and the rest of the proof remains the same.

$c\psi(r)$, we have

(2.12) $$n(r, 1/f) = o(\psi(r)/\log \theta(r))$$

and

(2.13) $$\log M(r, f_1) \leqq N(r, 1/f_1) + o\left(r \int_r^\infty \psi(t)\,dt/t^2 \log \theta(t)\right).$$

Let r_0 be so large that $c/\theta(r_0) \leqq \frac{1}{2}$. If $\psi(r)$ be nondecreasing for $r > r_1 \geqq r_0$, then the integral in the inequality (2.13) is less than $2c\psi(r)/\log \theta(r)$ (see [13, p. 183]). If $\psi(r)/r^b$ is nonincreasing for $r > r_1$, then the integral

$$I = \int_r^\infty \psi(t) t^{-2} (\log \theta(t))^{-1}\,dt$$

$$= \sum_{p=0}^\infty \int_{r(\theta(r))^p}^{r(\theta(r))^{p+1}} (\psi(t)/t^b) t^{b-2} (\log \theta(t))^{-1}\,dt,$$

$$< \sum_{p=0}^\infty (\psi(r)/\log \theta(r))(r(1-b))^{-1}(c/\theta(r))^p$$

$$< 2(r(1-b))^{-1}\psi(r)/\log \theta(r).$$

Hence

(2.14) $$T(r, f_1) \leqq \log M(r, f_1) = o(T(r, f))$$

and

(2.15) $$T(r, f_2) \sim T(r, f).$$

Consequently we have for all large r

(2.16) $$\psi(r)\Phi(r)/2 \log \theta(r) \leqq T(r, f_2) \leqq \log M(r, f_2) \leqq 3T(2r, f_2)$$
$$\leqq \max\{4c, 2^{b+2}\}\psi(r).$$

Since $f_2(z)$ is an entire function we deduce (see [13, Theorem 1(a)]) that

(2.17) $$\log M(r, f_2) \sim T(r, f_2) \sim N(r, 1/f_2).$$

Furthermore

$$\limsup_{r \to \infty} \{n(r, f)\Phi(r)/T(r, f)\}$$

$$= \limsup_{r \to \infty} \{n(r, 1/f_2)\Phi(r)/T(r, f)\}$$

(2.18) $$\leqq \limsup_{r \to \infty} \{(c\psi r)/\log \theta(r))(\Phi(r)/T(r, f))\} \leqq c$$

and

(2.19) $$\limsup_{r \to \infty} \{n(r, 1/f)\Phi(r)/T(r, f)\} = 0.$$

Consider now $f(z) - a$, where $0 < |a| < \infty$. Then

$$f(z) - a = (f_1(z) - af_2(z))/f_2(z) \equiv F(z)/f_2(z)$$

and

(2.20) $$T(r, F) \sim T(r, f_2) \sim T(r, f).$$

Hence the entire function $F(z)$ satisfies the inequalities of the type (2.16) and we deduce that $N(r, 1/F) \sim T(r, F)$. This implies that for $a \neq 0$, ∞, $N(r, 1/(f - a)) \sim N(r, f) \sim T(r, f)$. Since for $a \neq 0$, ∞,

(2.21) $$n(r, 1/(f - a)) < \{(1 + o(1))c\psi(r)\}/\log \theta(r),$$

the inequality (2.4) follows from the inequalities (2.18), (2.19), and (2.21). This completes the proof of Part (a).

We omit the proofs of Part (b), and Part (c) (cf. [13, pp. 183–184]. There is a misprint in line 12, p. 184; replace $\Phi(r)$ by $\sqrt{\Phi(r)}$).

The following are the two corollaries to Theorem 1.

Corollary 1. *If* $T(r, f) = O((\log r)^2)$ *and* $N(r, \alpha) = o(T(r, f))$ *then the conclusions* (2.3) *and* (2.4) *hold.*

Proof. Let us denote

$$\limsup_{r \to \infty} T(r, f)/(\log r)^2 = A_1,$$

$$\eta(r) = \inf_{t \geq r} \{T(t, f)/\log t\}.$$

Then $\eta(r)$ increases to infinity and also $\eta(r) \leq T(r, f)/\log r$. Let $A < A_1$. In Theorem 1, Part (a) we take $\psi(r) = A(\log r)^2$, $\theta(r) = r$, $c = 4$, $\Phi(r) = \eta(r)/A$. For another proof of the conclusion (2.3) see [14].

Corollary 2. *Let* $f(z)$ *be a function of zero order. Suppose that* (i) $n(r, \beta) = O(\log r)$ *and that either* (ii$_a$) $N(r, \alpha) = o(N(r, \beta))$ *or* (ii$_b$) $N(r, \alpha) = O((\log r)^2)$ *and* $\delta(\alpha) = 1$. *Then the conclusions* (2.3) *and* (2.7)–(2.9) *follow.*

Proof. We may suppose that $\alpha = 0$, $\beta = \infty$ and $f(z) = f_1(z)/f_2(z)$ where $f_1(z)$ and $f_2(z)$ are given by the expressions (2.11). Then the hypothesis (ii$_b$) implies that

$$\log M(r, f_1) \sim T(r, f_1) \sim N(r, 1/f_1) = o(T(r, f))$$

and so $T(r, f) \sim T(r, f_2)$. Since $n(r, \infty) = O(\log r)$,

$$\log M(r, f_2) \sim T(r, f_2) \sim N(r, 1/f_2)$$

and hence $N(r, 1/f) = o(T(r, f)) = o(N(r, f))$. Therefore we need to consider the case when $n(r, \infty) = O(\log r)$, $N(r, 0) = o(N(r, \infty))$. Let us now denote

$$\limsup_{r \to \infty} n(r, \infty)/\log r = A_1,$$

and choose in Theorem 1, Part (b) $L(r) = A \log r$, $(A_1 < A)$. Hence $\xi(r) = (\log r)/2$. We can now construct a function $\Phi(r)$ such that $\Phi(r) \leqq n(r, \infty)/2A$ and also such that ultimately $\Phi(r)/r$ decreases to zero and $\Phi(r)$ increases to infinity as r tends to infinity.

Then

$$\{L(r)\Phi(\xi(r))\}/\xi(r) \leqq 2A\Phi(r) \leqq n(r, \infty) \leqq L(r),$$

and the conclusions (2.3) and (2.7)–(2.9) follow.

Remarks. (i) If $f(z)$ is an entire function then in each Part (a)–(c) of Theorem 1, $\alpha = \infty$ and (1.1) holds for every (finite) a (cf. [13]). (ii) A part of this theorem is stated without proof by one of us in [14].

3. Proximate Orders

If $f(z)$ is an entire function of finite positive order ρ, it is possible [2, pp. 54–58; 10] to determine a positive continuous function $\rho(r)$ with the following properties:
(i) $\rho(r)$ is differentiable for $r \geqq r_0$ except possibly at isolated points, at which $\rho'(r - 0)$ and $\rho'(r + 0)$ exist;
(ii) $\lim_{r \to \infty} \rho(r) = \rho$;
(iii) $\lim_{r \to \infty} \rho'(r)r \log r = 0$;
where $\rho'(r)$ is either $\rho'(r - 0)$ or $\rho'(r + 0)$ where these are unequal;
(iv) $\limsup_{r \to \infty} \log M(r)/r^{\rho(r)} = 1$.
Such a function is called a proximate order for $f(z)$. Earl [3] has given a construction for proximate orders for entire functions $f(z)$ satisfying the condition $\log \log \log M(r, f) = O(\log \log r)$. His method can be applied to meromorphic functions and yields the following.

Theorem A. *If $f(z)$ is a transcendental meromorphic function for which*

(3.1) $\limsup_{r \to \infty} \log \log T(r, f)/\log \log r = \gamma, \qquad 0 \leqq \gamma < \infty,$

then there is a continuous monotonic function $\rho(r)$ for which
(i) *$\rho'(r - 0)$ and $\rho'(r + 0)$ exist everywhere and are equal in adjacent open intervals:*

(ii) $\{\rho'(r)r \log r\}/\rho(r) \to \gamma - 1$, as $r \to \infty$, where $\rho'(r)$ is either $\rho'(r - 0)$ or $\rho'(r + 0)$ when these are unequal;
(iii) $\lim \sup_{r\to\infty} T(r,f)/r^{\rho(r)} = 1$.

For a transcendental meromorphic function $f(z)$, $\rho(r)\log r$ tends to infinity with r. If $f(z)$ is also of zero order then $\rho(r)$ decreases to zero as $r \to \infty$. We use this proximate order in next two theorems.

Theorem 2. *Let $f(z)$ be a transcendental meromorphic function of zero order. If*

$$(3.2) \qquad \liminf_{r \to \infty} T(r,f)/\rho(r)r^{\rho(r)}\Phi(1/\rho(r)) > 0$$

for some function $\Phi(r)$ which increases to infinity with r, and if

$$(3.3) \qquad N(r, \alpha) = o(T(r,f))$$

for some α, then for every $a \neq \alpha$,

$$(3.4) \qquad T(r,f) \sim N(r, a);$$

and for every a,

$$(3.5) \qquad \limsup_{r \to \infty} n(r, a)\Phi(r)/T(r,f) \leq c_2(f)$$

where $c_2(f)$ is a finite constant depending on $f(z)$.

Proof of Theorem 2

We take functions $\psi(r) = 2r^{\rho(r)}$, $r \geq r_0$, and $\log \theta(r) = 1/\rho(r)$. It is easily verified that $\psi(r)/r^b$, $(0 < b < 1)$, is a decreasing function of r for $r > r_1 \geq r_0$. Furthermore as $r \to \infty$, $\theta(r)$ increases to infinity, $\psi(r)$ tends to infinity and $\psi(r\theta)/\psi(r) < c$, ($\theta$ denotes $\theta(r)$), and $c > 1$. Since $T(r,f) > c_1\rho(r)r^{\rho(r)}\Phi(1/\rho(r))$ we have for all large r

$$(3.6) \qquad \psi(r)\Phi^*(r)/\log \theta(r) < T(r,f) < 2r^{\rho(r)} = \psi(r),$$

where $\Phi^*(r) = \dfrac{c_1}{2}\phi(1/\rho(r))$. Hence the hypotheses of Theorem 1, Part (a) are satisfied and conclusions (3.4) and (3.5) follow.

Corollary. *Let $f(z)$ be a transcendental entire function of zero order, and suppose that (3.2) holds. Then the relation (1.1) holds, and the inequality (3.5) holds for every (finite) a.*

The proof of the corollary immediately follows from the theorem, for $\log M(r,f) < 3T(2r,f) < c_3 r^{\rho(r)} < c_4 r^{\rho(r)}$.

4. The Counting Function $n(r, a)$.

If $f(z)$ is an entire function of positive finite order ρ and finite type T, then it is known ([1, p. 16; 2, p. 71]) that

$$(4.1) \qquad \limsup_{r \to \infty} \frac{n(r, a)}{r^\rho} \leqq e\rho T.$$

This result is sharp in the sense that there are entire functions $f(z)$ (see [12, Theorem 2(i); 11, Theorem 4]) for which there is an equality sign in (4.1). We now prove a similar relation satisfied by a wider class of functions.

Theorem 3. *Let $f(z)$ be a transcendental meromorphic function such that*

$$\limsup_{r \to \infty} \frac{\log \log T(r, f)}{\log \log r} = \gamma, \qquad 0 \leqq \gamma < \infty.$$

Then for every a,

$$(4.2) \qquad \limsup_{r \to \infty} n(r, a)/\rho(r)r^{\rho(r)} \leqq e^\gamma.$$

Proof. Let $b = b(r) > 1$. Since $T(r, f) < (1 + \varepsilon)r^{\rho(r)}$, $r > r_1$, we have $n(r, a) < (1 + \varepsilon)(br)^{\rho(br)}/\log b$.

Let $b = \exp(1/\rho(r))$, $r > r_2 \geqq r_1$. Then

$$n(r, a)/\rho(r)r^{\rho(r)} < (1 + 2\varepsilon)\exp\{\rho(br)(\log b + \log r) - \rho(r)\log r\}$$
$$< (1 + 2\varepsilon)\exp\{\gamma + o(1)\},$$

and the theorem is proved.

5. Example

We now estimate the comparison functions $\psi(r)$, $\theta(r)$, ... for the function $f(z) = \prod_{n=1}^{\infty} (1 + z/2^n)/\prod_{n=1}^{\infty} (1 - z/\exp(n^{1/k}))$ where $k \geqq 2$. This function has all its zeros located on the negative real axis, and all its poles are located on the positive real axis. We have $\alpha = 0$, $\beta = \infty$,

$$n(r, \infty) = [(\log r)^k];$$
$$n(r, 0) \sim \log r/\log 2;$$
$$T(r, f) \sim (\log r)^{k+1}/k + 1;$$
$$\kappa(r) \sim \{(\log r)/(k + 1)\}^b$$

We take, for $r > r_0$,

$$\psi(r) = c_1(\log r)^{k+1}/k + 1, \; c_1 > 1;$$
$$\theta(r) = r;$$
$$\Phi(r) = \log r/\log \log r;$$
$$L(r) = (\log r)^k;$$
$$\xi(r) = \log r/(k + 1).$$

We observe that the conditions on the comparison functions ψ, Φ, θ, ξ are all satisfied. Furthermore $N(r, \infty)$ is a slowly oscillating function and the inequalities (2.1), (2.2), (2.5), (2.6), and (3.2) hold with the above choice of comparison functions.

REFERENCES

1. R. P. Boas, *Entire functions*, Academic Press, New York, 1954.
2. M. L. Cartwright, *Integral functions*, Cambridge University Press, Cambridge, England, 1956.
3. J. P. Earl, Note on the construction of proximate orders, *J. London Math. Soc.* **43** (1968), 695–698.
4. A. Edrei and W. H. J. Fuchs, The deficiencies of meromorphic functions of order less than one, *Duke Math. J.* **27** (1960), 233–249.
5. W. K. Hayman, *Meromorphic functions*, Clarendon Press, Oxford, England, 1964.
6. A. Hyllengren, Valiron deficient values of meromorphic functions in the plane, to appear.
7. J. Korevaar, T. van Aardenne- Ehrenfest, and N. G. de Bruijn, A note on slowly oscillating functions, *Nieuw Arch. Wisk.* **23** (1949), 77–86.
8. R. Nevanlinna, *Le theoreme de Picard-Borel et la theorie des functions meromorphes*, Gauthier-Villars, Paris, 1929.
9. S. M. Shah, Note on a theorem of Valiron and Collingwood, *Proc. Nat. Acad. Sci. India* **12** (1942), 9–12.
10. ———, On proximate orders of integral functions, *Bull. Amer. Math. Soc.* **52** (1946), 326–328.
11. ———, A note on uniqueness sets for entire functions, *Proc. Indian Acad. Sci.* **28** (1948), 1–8.
12. ———, The maximum term of an entire series (III), *Quart. J. Math. Oxford* **19** (1948), 220–223.
13. ———, Entire functions with no finite deficient value, *Arch. Rat. Mech. Anal.* **26** (1967), 179–187.
14. ———, Meromorphic functions with one deficient value, *J. Austral, Math. Soc.* **10** (1969), 355–358.
15. D. F. Shea, On the Valiron deficiencies of meromorphic functions of finite order, *Trans. Amer. Math. Soc.* **124** (1966), 201–227.
16. G. Valiron, Sur les zéros des fonctions entières d'ordre fini, *Rend. Circ. Mat. Palermo* **43** (1919), 255–268.
17. ———, Sur les valuers asymptotiques de quelques fonctions méromorphes, *Rend. Circ. Mat. Palermo* **49** (1925), 415–421.

(Received June 15, 1969)

The Basis of Vitali's Theorem

J. M. Whittaker

Department of Pure Mathematics
University of Sheffield
Sheffield
England

1.

Amongst the many topics of complex variable theory to which Macintyre contributed is the well-known theorem of Vitali (stated for simplicity in the case of a circle):

Theorem A. *If the functions $f_n(z)$ are regular and uniformly bounded in $|z| < 1$ and $\lim_{n \to \infty} f_n(a_k)$ exists at a set of points $a_1, a_2, \ldots$ having a limit point in $|z| < 1$, then there is a function $f(z)$, regular in $|z| < 1$, to which $f_n(z)$ converges uniformly in $|z| \leqq \rho < 1$.*

In a valuable paper Bowen and Macintyre [2] gave a simple proof of this result based on an inequality due to the writer [3] and extended it in various ways. They showed, for example, that $\{a_k\}$ need not have a limit point in $|z| < 1$, provided that $\prod |a_k|$ diverges, and they generalized theorems on functions regular in a strip which can be deduced from Vitali's theorem.

2.

My object here is to pursue this line of argument in other directions. It might be supposed, for example, that the vital element in Vitali's theorem is basically dependent on the hypothesis of uniform boundedness of the functtions. This is not the case. The essential principle of Theorem A is contained in the following result, in which no hypothesis of the nature of uniformity is made:

353

Theorem 1. *Let $f_n(z)$ be regular in $|z| < 1$ and let $\lim_{n \to \infty} f_n(a_k)$ exist at a set of points $\{a_k\}$ such that $\prod |a_k|$ diverges. Let $\rho < 1$. Then*

(1)
$$\frac{\max_{|z| \le \rho} |f_m(z) - f_n(z)|}{1 + \max_{|z| < 1} |f_m(z) - f_n(z)|} \to 0 \qquad as \quad m, n \to \infty.$$

The writer's inequality, in the slightly modified form used by Bowen and Macintyre, is

(2)
$$|F(z)| \le M \prod_{k=0}^{N} \left| \frac{z - a_k}{1 - \bar{a}_k z} \right| + L \frac{6\Delta(2\rho)^N}{1 - \rho^4},$$

where $F(z)$ is regular and $|F(z)| \le M$ in $|z| < 1$, $L = \max\{|F(a_0)|, |F(a_1)|, \ldots, |F(a_N)|\}$, and

$$\Delta = \sum_{k=0}^{N} {\prod_{i}}' |a_i - a_k|^{-1},$$

where the prime means that $i = k$ is to be omitted.

Bowen and Macintyre showed that the divergence of $\prod |a_k|$ implies that

(3)
$$\prod_{k=0}^{N} \left| \frac{z - a_k}{1 - \bar{a}_k z} \right|$$

diverges to zero as $N \to \infty$, uniformly in $|z| \le \rho$; so that, given ε, we can find N such that the expression in (3) does not exceed ε for $|z| \le \rho$. This fixes $a_0, \ldots, a_N$ and so Δ, and we can then choose p so that

$$\left| f_m(a_k) - f_n(a_k) \right| \frac{6\Delta(2\rho)^N}{1 - \rho^4} < \varepsilon, \qquad (m, n > p)$$

for $k = 0, 1, \ldots, N$. Hence applying (2) to $F(z) = f_m(z) - f_n(z)$, we obtain

$$\max_{|z| \le \rho} |f_m(z) - f_n(z)| \le \varepsilon \max_{|z| < 1} |f_m(z) - f_n(z)| + \varepsilon$$

for $m, n > p$ and this is the result stated. Vitali's theorem, as extended by Bowen and Macintyre, follows immediately; since, if $|f_n(z)| \le M$, the denominator of (1) does not exceed $1 + 2M$.

Theorem 1 can be generalized in the usual way, a domain D taking the place of $|z| < 1$ and an interior closed domain $\bar{D}_1$ of $|z| \le \rho$.

3.

Analyzing the proof of Theorem 1 we see that the two terms on the right of the fundamental inequality were treated separately. First N was chosen so as to make the first term small and the existence of the limit at $a_0, \ldots, a_N$

then sufficed to make the second term small. More delicate results can be obtained by making the two terms vary simultaneously. As an example of this type of reasoning I prove a result related to one previously given [4] in connection with the Riemann zeta-function. Given $f(s)$, where $s = \sigma + it$, defined in a strip $(a < \sigma < b, t > t_0)$, denote as usual by $\mu(\sigma)$ the lower bound of numbers ξ such that

$$|f(\sigma + it)| = O(t^\xi).$$

Define also the function $\lambda(\alpha, \beta)$ to be the lower bound of numbers ξ such that

$$\min_{\alpha \leq \sigma \leq \beta} |f(\sigma + it)| = O(t^\xi) \qquad (a < \alpha < \beta < b).$$

Clearly

$$\lambda(\alpha, \beta) \leq \mu(\sigma) \qquad (\alpha \leq \sigma \leq \beta).$$

Theorem 2. *Let*
    (i) $f(s)$ be *regular in* $(a < \sigma < b, t > t_0)$,
    (ii) $|f(s)| < Kt^{\gamma+\varepsilon} (a < \sigma < b, t > t_\varepsilon)$,
    (iii) $\mu(c) = \gamma$ *for some* c $(a < c < b)$.
Then, if $\delta > 0$,

$$\lambda(a + \delta, b - \delta) = \gamma.$$

A quality of uniformity, of the same general nature as that in Vitali's theorem, will be recognized. We are saying that if $|f(s)|$ is sometimes as large as t^γ on a certain vertical line, then it is as large as t^γ at all points on a sequence of horizontal lines nearly crossing the strip.

Consider first the case $a = -1, b = 1, c = 0$. If the result is false, then for some $h, \omega > 0$,

$$\lambda(-h, h) = \gamma - 2\omega,$$

and so

(4) $$\min_{-h \leq \sigma \leq h} |f(\sigma + it)| < t^{\gamma-\omega} \qquad (t \geq t_h).$$

Choose ρ so that $h < \rho < 1$ and consider the circles C_1, C_2 of center $(0, t)$ and radii $\rho, 1$. The circle C_1 has in common with the strip $-h \leq \sigma \leq h$ a rectangle of sides $2h, \sqrt{(\rho^2 - h^2)} = 2\rho\theta$ (say). Divide this rectangle into N (which will be chosen later) equal strips by drawing lines parallel to the σ-axis. In view of (4) there are points $a_0, a_1, \ldots, a_N$, one on each of the lines so drawn, such that

$$|f(a_k)| < (2t)^{\gamma-\omega}.$$

Now it is clear that, if d is the width of a strip,

$$|a_i - a_k| \geq (k - i)d \qquad (i = 0, 1, \ldots, k - 1),$$
$$\geq (i - k)d \qquad (i = k + 1, \ldots, N),$$

so that

$$\Delta \leq \frac{1}{d^N} \sum_{k=0}^{N} \frac{1}{k(k-1) \cdot \cdots \cdot 2 \cdot 1 \cdot 1 \cdot 2 \cdot \cdots \cdot (N-k)}$$

$$= \frac{2^N}{d^N N!} < \left(\frac{2e}{dN}\right)^N = \left(\frac{2e}{2\rho\theta}\right)^N.$$

Now define a positive number δ by the equation

(5)
$$\frac{2\delta}{1-\delta} = \log \frac{1}{\rho} \bigg/ \log \frac{2e}{\theta},$$

and make N depend on t in such a way that

(6)
$$\rho^{-N} \sim t^{2\delta\omega}.$$

We can now apply (2) to $f(s)$ in C_2, and obtain for the value at the center the inequality

$$|f(it)| \leq K t^{\gamma + \delta\omega} \rho^{N+1} + (2t)^{\gamma - \omega} \frac{6}{1 - \rho^4} \left(\frac{2e}{\theta}\right)^N.$$

Moreover, using (5), (6),

$$t^{\gamma + \delta\omega} \rho^N \sim t^{\gamma + \delta\omega - 2\delta\omega} = t^{\gamma - \delta\omega},$$

$$t^{\gamma - \omega} \left(\frac{2e}{\theta}\right)^N = t^{\gamma - \omega} \left(\frac{1}{\rho}\right)^{N(1-\delta)/2\delta}$$

$$\sim t^{\gamma - \omega} t^{2\omega(1-\delta)/2\delta} = t^{\gamma - \delta\omega}.$$

Thus

$$|f(it)| \leq K' t^{\gamma - \delta\omega} \qquad (t \geq t_1),$$

and this implies that $\mu(0) \leq \gamma - \delta\omega < \gamma$, which contradicts (iii). By making a linear transformation, the proof extends to the case when c is the midpoint of (a, b). In the general case, if c is not the midpoint of (a, b) let it be in (say) the left half, so that $a < c < \frac{1}{2}(a + b)$. Apply the result already established to the interval $(a, 2c - a)$ of which c is the midpoint. We deduce that

$$\lambda(a + \delta, 2c - a - \delta) = \gamma,$$

and so $\mu(\sigma) \geq \gamma$ for points σ in $(a + \delta, 2c - a - \delta)$. If $\frac{1}{2}(a + b)$ is not such a point, apply the result to the interval $(c, 3c - 2a - 2\delta)$ and continue in this way till $\frac{1}{2}(a + b)$ is crossed. Then

$$\mu\{\tfrac{1}{2}(a + b)\} = \gamma,$$

and the special case can be applied to the whole interval.

4.

Results of generalized Vitali type, but involving mean values, can be obtained by integrating the fundamental inequality. Theorems of this kind have been given by Whittaker [4, 5] and Bowen [1].

5.

The inequality can thus be used to establish a family of theorems, akin in their general nature but differing widely in their detailed content and method of proof. Comparing Theorems 1 and 2 it is seen that in the latter not only do the second and third terms of the inequality vary together, but they tend to infinity instead of zero. Again the hypotheses of Theorem 1 affect the second and third terms of the inequality, those of Theorem 2 the first and second.

REFERENCES

1. N. A. Bowen, On mean value theorems for bounded regular functions, *Proc. Edinburgh Math. Soc.* **14** (1964), 103–109.
2. N. A. Bowen and A. J. Macintyre, Interpolatory methods for theorems of Vitali and Montel, *Proc. Roy. Soc. Edinburgh* (*A*) **64** (1954), 71–79.
3. J. M. Whittaker, *Interpolatory function theory*, Cambridge University Press, Cambridge, England, 1935, p. 57.
4. ———, An inequality for the Riemann zeta-function, *Proc. London Math. Soc.* **41** (1936), 544–552.
5. ———, A mean value theorem for analytic functions, *Proc. London Math. Soc.* **42** (1937), 186–195.

(*Received February 15, 1968*)

Coefficient Density and the Distribution of Virtually Isolated Singularities on the Circle of Convergence

R. Wilson

Greenroyd
Rannoch Road
Crowborough
Sussex, England

1. Introduction

In 1940 A. J. Macintyre and the author published a paper [2] dealing with the effects on the coefficient density of the Taylor series of a function due to the distribution of singular points on its circle of convergence. More precisely, let $f(z)$ be regular about the origin with the Taylor series

(1) $$f(z) = c_0 + c_1 z + c_2 z^2 + \cdots + c_n z^n + \cdots$$

and assume, for convenience, that the circle of convergence is of unit radius. Then, if γ be any positive number, we considered the densities ([4, pp. 556–561)] of the sequences of n for which $|c_n| > e^{-\gamma n}$ with various distributions of singularities.

The case of two singularities was first discussed and these were either both isolated noncritical points or both almost isolated singularities of finite exponential order or both easily approachable singularities.* It was found that

* See [3, p. 221; 5, pp. 734–737 and 777; 6, section 2] for definitions of these singularities.

values for the coefficient densities between 0 and 1 could arise only if both singularities were identical in form and placed at two of the vertices of a regular polygon inscribed in the circle of convergence.

Although the general approach was similar, it was found that the critical part of the argument used for two singularities could not be extended to the case of three. Consequently, in the case of three singularities this essential part had to be dealt with in an entirely different way. Even so the case of three easily approachable singularities is difficult and requires a more elaborate treatment.*

However, none of these methods will enable us to deal with virtually isolated singularities on the circle of convergence. The reason is that in all the other cases an explicit form of the coefficient function† is available which is exclusively associated with the type of singularity concerned. For the virtually isolated singularity no such form of the coefficient function is known [6, Section 5]. In fact, A. J. Macintyre, in developing the key theorems in the coefficient theory of the virtually isolated singularity, was obliged to use the coefficient function of an easily approachable singularity in dealing with the virtually isolated singularity (see [1, pp. 8–10]), and we shall here use the coefficient function of an almost isolated singularity for this purpose.

In dealing with the problem of coefficient density with virtually isolated singularities on the circle of convergence I shall draw extensively on the techniques of the earlier paper except for the critical part of the argument for which entirely new methods are necessary.

2. The Virtually Isolated Singularity

We shall be concerned with the case when $f(z)$ of (1) has one or more virtually isolated singularities on its circle of convergence. Such a singularity consists of an isolated singular point. If $f(z)$ is uniform in the neighborhood of this point then the singularity is either a pole or an isolated essential point. Otherwise the singularity is a critical point which may be of a very general character. In this case the series (1) represents a branch of $f(z)$ for which the singularity is an isolated point on a sheet of the Riemann surface associated with this branch, but on other sheets it may not be so (see [6, Section 5]). The sole criterion is that $f(z)$ shall be regular in the neighborhood of the singularity for an angle exceeding 2π radians.

Suppose that $f(z)$ has on its circle of convergence only one singularity and that a virtually isolated singularity at the point $z = e^{i\alpha}$. The key theorem on the coefficient properties of $f(z)$ of (1), due to A. J. Macintyre [1, Theorems A and B], is stated below.

* R. Wilson, unpublished notes.
† That is the interpolating function of the coefficients.

Theorem A. *If $f(z)$ has on its circle of convergence only one singularity and that a virtually isolated singularity at $z = e^{i\alpha}$, then to every assigned positive γ,*

$$(2) \qquad\qquad |c_n| > e^{-\gamma n}$$

for a sequence of n of density 1, and for every given positive ε

$$(3) \qquad\qquad \left| \frac{c_{n+1}}{c_n} - e^{-i\alpha} \right| < \varepsilon$$

for a sequence of n of density 1.

It follows that there exists a common sequence of n of density 1 for which (2) and (3) hold jointly.

For subsequent use, it will be convenient to separate off the singularity by an appropriate process of dissection of $f(z)$. By treating the singularity as an almost isolated singularity, of which the virtually isolated singularity is a special case, we can use a method of dissection due to Pólya [5, pp. 738–741], which effectively preserves the properties (2) and (3) for the modified function.

3. Separation of Singularities

Now suppose that $f(z)$ has on its circle of convergence m virtually isolated singularities at the points $e^{i\alpha_v}$, $(v = 1, 2, \ldots, m)$, and no other singularities thereon. Then there exists a number $k > 1$ such that there are no singularities inside or on the circumference of $|z| = k$ other than the m above mentioned. We take for the contour of the Cauchy integral of $f(z)$ the circle $|z| = k$ and m loops each containing one of the intervals $(e^{i\alpha_v}, ke^{i\alpha_v})$, $(v = 1, 2, \ldots, m)$, but excluding the origin. Thus

$$(4) \quad f(z) = \frac{1}{2\pi i} \int_C \frac{f(w)}{w - z}\, dw = \sum_{v=1}^{m} \frac{1}{2\pi i} \int_{\Gamma_v} \frac{f(w)}{w - z}\, dw + \frac{1}{2\pi i} \int_{C'} \frac{f(w)}{w - z}\, dw,$$

where the Γ_v are the m loops and C' is the circle $|z| = k$ (see [5, pp. 768–771]). Hence, from (4),

$$(5) \qquad\qquad f(z) = \sum_{v=1}^{m} f_v(z) + f^*(z).$$

Here each $f_v(z)$ has the same singularity as $f(z)$ at $z = e^{i\alpha_v}$ and possibly a virtually isolated singularity at $z = ke^{i\alpha_v}$, while $f^*(z)$ has the negative of such of these singularities as exist together with all other singularities of $f(z)$ except those on $|z| = 1$. If the singularity of $f_v(z)$ at $z = e^{i\alpha_v}$ is noncritical then there is no singularity at $z = ke^{i\alpha_v}$ but otherwise there is one. In any case we can treat the combined singularities of $f_\nu(z)$ at $z = e^{i\alpha_\nu}$ and at $z = ke^{i\alpha_\nu}$ as an almost isolated singularity and write

(6) $$f_\nu(z) = \sum_0^\infty G_\nu(n)e^{-i\alpha_\nu n}z^n, \qquad (\nu = 1, 2, \ldots, m),$$

where $G_\nu(z)$ is regular and at most of order 1, minimum type, in the right half-plane [2, pp. 63–64; 5, Satz III, p. 744]. Hence, from (5) and (6),

(7) $$c_n = \sum_{\nu=1}^m G_\nu(n)e^{-i\alpha_\nu n} + c_n{}^*,$$

where $\overline{\lim}_{n \to \infty} |c_n{}^*|^{1/n} < 1$.

Since each $f_\nu(z)$ has on its circle of convergence $|z| = 1$ only one singularity and that virtually isolated, the results of Theorem A apply and so, for every positive γ, however small,

(8) $$|G_\nu(n)| > e^{-\gamma n}$$

for a sequence of n of density 1, and for every positive ε,

(9) $$\left| \frac{G_\nu(n+1)}{G_\nu(n)} - 1 \right| < \varepsilon,$$

for a sequence of n of density 1, since the c_n of Theorem A is replaced by $G_\nu(n)e^{-i\alpha_\nu n}$ from (6).

It follows that there exists a common sequence of density 1 for which (8) and (9) hold jointly. Further, if p be a positive integer,

(10)
$$\lim_{n' \to \infty} \frac{G_\nu(n+p)}{G_\nu(n)} = \lim_{n' \to \infty} \prod_0^{p-1} \frac{G_\nu(n+\mu+1)}{G_\nu(n+\mu)} = \prod_0^{p-1} \lim_{n' \to \infty} \frac{G_\nu(n+\mu+1)}{G_\nu(n+\mu)} = 1,$$

where (n') is the common sequence of n of density 1 for which (9) holds in the p consecutive cases.

4. Two Virtually Isolated Singularities

We first suppose that $f(z)$ has on its circle of convergence two virtually isolated singularities at the points $e^{i\alpha_1}$, $e^{i\alpha_2}$ and is otherwise regular on $|z| = 1$. Then, as in (7),

(11) $$c_n = G_1(n)e^{-i\alpha_1 n} + G_2(n)e^{-i\alpha_2 n} + c_n{}^*,$$

where the $G_\nu(z)$ are each regular and at most of order 1, minimum type in the right half-plane and satisfy therein conditions (8) and (9), and

$$\overline{\lim}_{n \to \infty} |c_n{}^*|^{1/n} < 1,$$

so that $|c_n{}^*| < e^{-\gamma n}$ for some $\gamma > 0$ and $n > n_0$.

Now suppose that $|c_n| < e^{-\gamma n}$ for a sequence of integers $\{n_\nu\}$ of positive upper density. Then for any integer h, sufficiently large,

$$\varlimsup_{\nu \to \infty} \frac{n_\nu}{\nu} > \frac{1}{h}.$$

We thus have, from (11), with $\gamma > \gamma' > 0$

(12) $$|G_1(n) + e^{i(\alpha_1 - \alpha_2)n} G_2(n)| < e^{-\gamma' n}$$

for a set of integers of upper density greater than $1/h$.

We consider with each integer n the expressions

$$|G_1(n + p) + e^{i(\alpha_1 - \alpha_2)(n+p)} G_2(n + p)|, \qquad (p = 1, 2, \ldots, h).$$

In the sequence of intervals

$$0 \le n < h, \qquad h \le n < 2h, \qquad 2h \le n < 3h, \ldots$$

it must happen that an increasing sequence $\{n_\mu\}$ of positive upper density exists such that the inequality (12) holds for two values of n at least in every

$$n_\mu h \le n < (n_\mu + 1)h$$

since otherwise the upper density of integers for which (12) is satisfied would be at most $1/h$.

Since the two selected values can be chosen in only $\frac{1}{2}h(h - 1)$ different ways in each interval, it follows that there must be a subsequence of $\{n_\mu\}$ of positive upper density such that both

(13) $$|G_1(n) + e^{i(\alpha_1 - \alpha_2)n} G_2(n)| < e^{-\gamma' n}$$

and

(14) $$|G_1(n + p) + e^{i(\alpha_1 - \alpha_2)(n+p)} G_2(n + p)| < e^{-\gamma'(n+p)} < e^{-\gamma' n}$$

hold for one of the values $p = 1, 2, \ldots, h - 1$ at least.

It follows from (13) and (14) that the inequality on the modulus of the determinant

(15)
$$\left| \begin{array}{cc} G_1(n) & G_2(n) \\ G_1(n + p) & e^{i(\alpha_1 - \alpha_2)p} G_2(n + p) \end{array} \right| < e^{-\gamma' n} \{|G_2(n)| + |G_2(n + p)|\} < e^{-\gamma'' n},$$

with $\gamma' > \gamma'' > 0$, holds for some p in a set of n of positive upper density, for $|G_\nu(n)| < e^{(\gamma' - \gamma'')n}$ with n sufficiently large since $\varlimsup_{n \to \infty} |G_\nu(n)|^{1/n} = 1$.

If we write $G(n)$ for the determinant in (15) then $\sum_0^\infty G(n)z^n$ has a virtually isolated singularity at $z = 1$ and no other singularity on its circle of convergence. For it is the algebraic sum of Hadamard products of $\sum_0^\infty G_\nu(n)z^n$

and $z^{-p} \sum_p^\infty G_v(n)z^n$, $(v = 1, 2)$, each of which has a virtually isolated singularity at $z = 1$ and no other singularity on $|z| = 1$ [6, Theorem 2].

But from (2) of Theorem A, $|G(n)| > e^{-\gamma''n}$ for a sequence of n of density 1 and hence it is impossible for (15) to hold unless $G(z) \equiv 0$. Thus, from (15),

$$(16) \qquad G(n) \equiv \begin{vmatrix} G_1(n) & G_2(n) \\ G_1(n + p) & e^{i(\alpha_1 - \alpha_2)p}G_2(n + p) \end{vmatrix} \equiv 0$$

and since, from (10), there is a common sequence (n') of n of density 1 for which both

$$(17) \qquad \lim_{n' \to \infty} \frac{G_1(n + p)}{G_1(n)} = 1, \quad \lim_{n' \to \infty} \frac{G_2(n + p)}{G_2(n)} = 1,$$

hold, it follows from (16) and (17) that

$$(18) \qquad e^{i(\alpha_1 - \alpha_2)p} = 1 \quad \text{or} \quad \alpha_1 - \alpha_2 = 2m\pi/p.$$

Returning now to (12) it is clear that the element $e^{i(\alpha_1 - \alpha_2)n}$ takes, for successive values of n, the repeated set of the pth roots of unity, from (18). However, since the inequality (12) holds for a sequence of n of positive upper density, at least one of the roots of unity representing $e^{i(\alpha_1 - \alpha_2)n}$ in (12) must arise for a sequence of n of positive upper density. Thus, setting for convenience $e^{i(\alpha_1 - \alpha_2)n}$ in (12) as $-\omega^{-1}$, formula (12) becomes

$$(19) \qquad |G_1(n) - \omega^{-1}G_2(n)| < e^{-\gamma'n}$$

for a set of integers of positive upper density. However, if we now write $G(z) \equiv G_1(z) - \omega^{-1}G_2(z)$, then it follows that $\sum_0^\infty G(n)z^n$ has a virtually isolated singularity at $z = 1$ and no other singularity on its circle of convergence. Hence by (2) of Theorem A the inequality (19) can hold for a set of positive upper density only if $G(z) \equiv 0$, that is if

$$G_2(z) = \omega G_1(z) \quad \text{and} \quad G_2(n) = \omega G_1(n).$$

In this case $c_n - c_n{}^*$ in (11) will vanish for that set of values of n for which $e^{i(\alpha_1 - \alpha_2)n} = -\omega^{-1}$ and for such values $c_n = c_n{}^*$ will be "small" in the sense that for some positive γ and $n > n_0$, $|c_n{}^*| < e^{-\gamma n}$.

We state these results in the following theorem.

Theorem 1. *If $f(z)$ has on its circle of convergence $|z| = 1$ only two singularities, each of virtually isolated type, then the density of small coefficients in the series (1) is zero unless the singularities are identical in form and are placed at points which subtend an angle $2m\pi/p$ at the center, where p is an integer* *greater than* 1, *and in this case the density exists and is equal to* $1/q$, *where* s/q *is* m/p *in its lowest terms.*

Thus, if the upper density of the vanishing $c_n - c_n^*$ is positive, it follows that the nonvanishing $c_n - c_n^*$ have a density $(q - 1)/q$ and that the function $f(z)$ can be dissected into the sum

$$f(z) \equiv f_1(z) + \omega f_1(z\omega') + f^*(z),$$

where ω, ω' are qth roots of unity, $f_1(z)$ represents the expansion due to a single virtually isolated singularity and $f^*(z)$ is regular inside $|z| = k$, with $k > 1$.

It will be noticed that the singularities are at two of the vertices of a p-sided polygon but that an infinite number of such polygons exists. If that with the smallest number of sides has q then the density of small coefficients is equal to $1/q$. In particular, when the points are at the opposite ends of a diameter $q = 2$.

5. Three Singular Points: The Extreme Case

We now consider the case in which $f(z)$ has three virtually isolated singularities at the points $e^{i\alpha_v}$ ($v = 1, 2, 3$) and no other singularities on its circle of convergence $|z| = 1$. As in (7) we can express the coefficients c_n in the form

(20)
$$c_n = \sum_{v=1}^{3} G_v(n)e^{-i\alpha_v n} + c_n^*,$$

where each $G_v(z)$ is regular and at most of order 1, minimum type, in the right half-plane and satisfies (8) and (9) therein and $\overline{\lim}_{n \to \infty} |c_n^*|^{1/n} < 1$, so that

(21)
$$|c_n^*| < e^{-\gamma n}$$

for some $\gamma > 0$ and $n > n_0$.

Now suppose that the upper density of small coefficients c_n is greater than $\frac{1}{2}$. Then for a set of n of positive upper density one of the sets c_n, c_{n+1}, c_{n+2}; c_n, c_{n+2}, c_{n+3}; c_n, c_{n+1}, c_{n+3} must have its three terms all small. Otherwise from the four terms c_n, c_{n+1}, c_{n+2}, c_{n+3} at most two would be small (except possibly for a sequence of n of zero density), and the upper density of small coefficients in (1) would then be at most $\frac{1}{2}$, contrary to hypothesis.

Suppose then that $|c_n|, |c_{n+1}|, |c_{n+2}| < e^{-\gamma n}$ for a sequence of n of positive upper density. Then from (20) and (21),

$$|G_1(n)e^{-i\alpha_1 n} + G_2(n)e^{-i\alpha_2 n} + G_3(n)e^{-i\alpha_3 n}| < e^{-\gamma' n}$$

for the sequence $\{n_\mu, n_\mu + 1, n_\mu + 2\}$. Since each $G_\nu(z)$ is at most of order 1, minimum type, in the right half-plane it follows that the modulus of the determinant

$$\begin{vmatrix} G_1(n)e^{-i\alpha_1 n} & G_2(n)e^{-i\alpha_2 n} & G_3(n)e^{-i\alpha_3 n} \\ G_1(n+1)e^{-i\alpha_1(n+1)} & G_2(n+1)e^{-i\alpha_2(n+1)} & G_3(n+1)e^{-i\alpha_3(n+1)} \\ G_1(n+2)e^{-i\alpha_1(n+2)} & G_2(n+2)e^{-i\alpha_2(n+2)} & G_3(n+2)e^{-i\alpha_3(n+2)} \end{vmatrix}$$

is less than $e^{-\gamma'' n}$ for the sequence $\{n_\mu\}$, where $\gamma' > \gamma'' > 0$ and n is sufficiently large, as may be proved by adding the second and third columns to the first and expanding in terms of the elements of the new first column.

On dividing the columns by $e^{-i\alpha_1 n}$, $e^{-i\alpha_2 n}$, $e^{-i\alpha_3 n}$, respectively, each of which is of modulus unity, we see that the modulus of the determinant

$$(22) \qquad G(n) \equiv \begin{vmatrix} G_1(n) & G_2(n) & G_3(n) \\ G_1(n+1)e^{-i\alpha_1} & G_2(n+1)e^{-i\alpha_2} & G_3(n+1)e^{-i\alpha_3} \\ G_1(n+2)e^{-2i\alpha_1} & G_2(n+2)e^{-2i\alpha_2} & G_3(n+2)e^{-2i\alpha_3} \end{vmatrix}$$

is less than $e^{-\gamma'' n}$ for the sequence $\{n_\mu\}$ of positive upper density.

However, the function represented by $\sum_0^\infty G(n)z^n$ has on its circle of convergence only one singularity and that a virtually isolated singularity. For the series is the algebraic sum of successive Hadamard products derived from the series $\sum_0^\infty G_\nu(n)z^n$, $z^{-1}\sum_1^\infty G_\nu(n)z^n$, $z^{-2}\sum_2^\infty G_\nu(n)z^n$, $(\nu = 1, 2, 3)$, each of which has on its circle of convergence only one singularity and that a virtually isolated singularity at $z = 1$ (see [6, Theorem 2]). Hence, from (2) of Theorem A, $|G(n)| > e^{-\gamma'' n}$ for a sequence of n of density 1, unless $G(z) \equiv 0$. From the contradiction with (22) it follows that $G(z) \equiv 0$ and $G(n) \equiv 0$. But from (9) and (11) there exists a common sequence (n') of n of density 1 for which

$$\lim_{n' \to \infty} \frac{G_\nu(n+1)}{G_\nu(n)} = 1, \quad \lim_{n' \to \infty} \frac{G_\nu(n+2)}{G_\nu(n)} = 1, \qquad (\nu = 1, 2, 3),$$

and so, from $G(n) \equiv 0$ and (22), we deduce that

$$\begin{vmatrix} 1 & 1 & 1 \\ e^{-i\alpha_1} & e^{-i\alpha_2} & e^{-i\alpha_3} \\ e^{-2i\alpha_1} & e^{-2i\alpha_2} & e^{-2i\alpha_3} \end{vmatrix} = 0.$$

The value of this determinant is $\prod(e^{-i\alpha_2} - e^{-i\alpha_3})$, and so it cannot vanish if the points $e^{i\alpha_1}$, $e^{i\alpha_2}$, $e^{i\alpha_3}$ are distinct.

On the supposition that c_n, c_{n+1}, c_{n+3}, are small for a sequence of n of positive upper density we get similarly

$$(23) \qquad \begin{vmatrix} 1 & 1 & 1 \\ e^{-i\alpha_1} & e^{-i\alpha_2} & e^{-i\alpha_3} \\ e^{-3i\alpha_1} & e^{-3i\alpha_2} & e^{-3i\alpha_3} \end{vmatrix} = (e^{-i\alpha_1} + e^{-i\alpha_2} + e^{-i\alpha_3})\prod(e^{-i\alpha_2} - e^{-i\alpha_3}) = 0.$$

The vanishing of this determinant can only arise from the condition $e^{-i\alpha_1} + e^{-i\alpha_2} + e^{-i\alpha_3} = 0$, which in turn means that the three points $e^{-i\alpha_1}$, $e^{-\alpha_{i_2}}$, $e^{-i\alpha_3}$ are the vertices of an equilateral triangle.

On the assumption that c_n, c_{n+2}, c_{n+3} are small for a sequence of n of positive upper density we get in the same way

$$\begin{vmatrix} 1 & 1 & 1 \\ e^{-2i\alpha_1} & e^{-2i\alpha_2} & e^{-2i\alpha_3} \\ e^{-3i\alpha_1} & e^{-3i\alpha_2} & e^{-3i\alpha_3} \end{vmatrix} \equiv e^{-3i(\alpha_1+\alpha_2+\alpha_3)}(e^{i\alpha_1} + e^{i\alpha_2} + e^{i\alpha_3}) \prod (e^{i\alpha_2} - e^{i\alpha_3}) = 0,$$

so that again the three singularities are at the vertices of an equilateral triangle.

We have thus obtained the result that, if $|c_n| < e^{-\gamma n}$ for a set of n of upper density greater than $\frac{1}{2}$, then

(24) $$c_n = [G_1(n) + \omega^n G_2(n) + \omega^{2n} G_3(n)]e^{-i\alpha_1 n} + c_n^*,$$

so that two of the three inequalities

$$|G_1(n) + G_2(n) + G_3(n)| < e^{-\gamma' n},$$
$$|G_1(n) + \omega G_2(n) + \omega^2 G_3(n)| < e^{-\gamma' n},$$
$$|G_1(n) + \omega^2 G_2(n) + \omega G_3(n)| < e^{-\gamma' n}$$

must be satisfied for a set of n positive upper density. But each of the expressions within the modulus signs is the coefficient of z^n in the Taylor series of a function having on its circle of convergence only one singularity and that a virtually isolated singularity at $z = 1$, and from (2) of Theorem A this is impossible unless two of the three functions out of

$$G_1(z) + G_2(z) + G_3(z), \ G_1(z) + \omega G_2(z) + \omega^2 G_3(z), \ G_1(z) + \omega^2 G_2(z) + \omega G_3(z),$$

are identically zero and hence that $G_1(z)$, $G_2(z)$, $G_3(z)$ are proportional either to 1, ω, ω^2 or to 1, ω^2, ω or to $1, 1, 1$.

Finally we deduce that the lower density of the nonsmall coefficients of $f(z)$ is at least $\frac{1}{2}$ unless $f(z)$ takes one of the forms

$$f_1(z) + f_1(\omega z) + f_1(\omega^2 z) + f^*(z),$$
$$f_1(z) + \omega f_1(\omega z) + \omega^2 f_1(\omega^2 z) + f^*(z),$$
$$f_1(z) + \omega^2 f_1(\omega z) + \omega f_1(\omega^2 z) + f^*(z),$$

where $f_1(z)$ has a virtually isolated singularity at $z = e^{i\alpha_1}$, and $f^*(z)$ is regular inside $|z| = k$ with $k > 1$. In each of these cases the density exists and is equal to $\frac{2}{3}$. Hence we have

Theorem 2. *If $f(z)$ has on its circle of convergence only three singular points, each virtually isolated, then the upper density of small coefficients is at most $\frac{1}{2}$ unless the singularities are virtually identical in form and occur at the vertices of an equilateral triangle. In this case the density exists and is equal to $\frac{2}{3}$.*

As in the earlier paper [2, pp. 77–79], similar methods enable us to extend the results of Theorem 2 to the case when $f(z)$ has m virtually isolated singularities on its circle of convergence and no others, as with (7). These are set out in the theorem below.

Theorem 3. *If $f(z)$ has on its circle of convergence exactly m singular points, each virtually isolated, then the upper density of small coefficients is at most $(m-2)/(m-1)$ unless the singularities occur at the vertices of an m-sided regular polygon inscribed in the circle of convergence and the singularities are identical in form. In this case the density exists and is equal to $(m-1)/m$.*

6. Three Singularities: All Cases

We now consider the case when $f(z)$ has three virtually isolated singularities at $e^{i\alpha_1}$, $e^{i\alpha_2}$, $e^{i\alpha_3}$ and no other singularities on its circle of convergence under the more general assumption that the coefficients c_n are small for a sequence of upper density greater than $1/l$. Thus, out of $2l$ consecutive terms $c_n, c_{n+1}, \ldots, c_{n+2l-1}$ at least three must be small for a sequence of n of positive upper density. In consequence of this assumption there must exist integers p, q, with $p < q < 2l$, such that c_n, c_{n+p}, c_{n+q} are all three small for a sequence of n of positive upper density. An argument similar to that of Section 5 leads, in place of (23) to the result

$$
(25) \qquad
\begin{vmatrix}
1 & 1 & 1 \\
e^{-ip\alpha_1} & e^{-ip\alpha_2} & e^{-ip\alpha_3} \\
e^{-iq\alpha_1} & e^{-iq\alpha_2} & e^{-iq\alpha_3}
\end{vmatrix} = 0.
$$

As in the earlier paper [2, p. 76] it follows that *if s is the highest common factor of p and q, then the $\alpha_\mu - \alpha_\nu$ are all of the form $2\pi(u/p + v/s)$, or of the form $2\pi(u/q + v/s)$, or of the form $2\pi[u/(q-p) + v/s]$, where u and v are integers or zero.*

It is evident from this that every $e^{i(\alpha_\mu - \alpha_\nu)}$ is either a pth root of unity or a qth root of unity or a $(q-p)$th root of unity. It may, in particular, be an sth root of unity. If we write ω_2, ω_3 for $e^{i(\alpha_1 - \alpha_2)}$ and $e^{i(\alpha_1 - \alpha_3)}$ respectively, then the determinant in (25) is of rank 2 unless p, q have a common factor $s > 2$ and each $\alpha_\mu - \alpha_\nu$ is an sth root of unity. For in this case the necessary conditions $\omega_2^p = 1 = \omega_3^p$ and $\omega_2^q = 1 = \omega_3^q$ are fulfilled.

Returning to the fundamental assumption that c_n is small for a sequence of n of positive upper density, we find, as in (24), that

$$c_n = \{G_1(n) + \omega_2^n G_2(n) + \omega_3^n G_3(n)\}e^{-\alpha_1 n} + c_n{}^*,$$

where $|c_n|$ is small for some sequence $\{n_v, n_v + p, n_v + q\}$ of positive upper density. Since, however, $\omega_2{}^n$ and $\omega_3{}^n$ can take only a finite number of different values, it follows that, for one pair of values ω_2', ω_3' at least of $\omega_2{}^n$, $\omega_3{}^n$, the relations

$$|G_1(n) + \omega_2' G_2(n) + \omega_3' G_3(n)| < e^{-\gamma' n},$$

(26)
$$|G_1(n) + \omega_2{}^p \omega_2' G_2(n) + \omega_3{}^p \omega_3' G_3(n)| < e^{-\gamma' n},$$

$$|G_1(n) + \omega_2{}^q \omega_2' G_2(n) + \omega_3{}^q \omega_3' G_3(n)| < e^{-\gamma' n},$$

must be satisfied for a set of n of positive upper density. But each of the expressions within the modulus signs is the coefficient of z^n in the Taylor series of a function having on its circle of convergence only one singularity and that a virtually isolated singularity at $z = 1$. Hence, from (2) of Theorem A, relations (26) are impossible unless two of the three functions out of

$$G_1(z) + \omega_2' G_2(z) + \omega_3' G_3(z),$$

(27)
$$G_1(z) + \omega_2{}^p \omega_2' G_2(z) + \omega_3{}^p \omega_3' G_3(z),$$

$$G_1(z) + \omega_2{}^q \omega_2' G_2(z) + \omega_3{}^q \omega_3' G_3(z),$$

are identically zero, and therefore, when the determinant (25) is of rank 2, exactly two of the relations (27) are independent and so $G_1(z)$, $G_2(z)$, $G_3(z)$ are proportional to 1, Ω_2, Ω_3, say.

It follows that the function $f(z)$ is of the form

$$f_1(z) + \Omega_2 f_1(\omega_2 z) + \Omega_3 f_1(\omega_3 z) + f^*(z),$$

where $f_1(z)$ has a unique virtually isolated singularity at $z = e^{i\alpha_1}$ and $f^*(z)$ is regular inside $|z| = k$ with $k > 1$.

If the determinant (25) is of rank 1 then p and q have a common factor $s > 2$ and each $\alpha_\mu - \alpha_\nu$ is an sth root of unity, so that relations (27) each reduce to

$$G_1(z) + \omega_2' G_2(z) + \omega_3' G_3(z) = 0.$$

Thus

$$c_n - c_n{}^* = \{G_1(n) + \omega_2' G_2(n) + \omega_3' G_3(n)\}e^{-i\alpha_1 n}$$

vanishes for every n for which $\omega_2{}^n = \omega_2'$, $\omega_3{}^n = \omega_3'$.

These results are put together in the theorem below.

Theorem 4. *If $f(z)$ has on its circle of convergence only three singularities, each of virtually isolated type, then the upper density of small coefficients is positive only if either* (a) *the three singularities are virtually identical or* (b) *one of the singularities is virtually the sum of the other two, as the case may be, and if the singularities are placed at three of the vertices of some regular polygon inscribed in the circle of convergence. In such cases the density exists.*

Theorems 1–4 include Theorems 1, 4, 5, and 6 of the earlier paper as special cases since the virtually isolated singularity includes the isolated essential point as a special case.

The possibility of extending Theorem 4 to the case of m singularities as Theorem 3 extends the case of Theorem 2, depends on the ability to evaluate the determinant of order m of the form (25) which arises and to organize logically the different cases occuring in accordance with the rank of the determinant.

REFERENCES

1. A. J. Macintyre, An overconvergence theorem of G. Bourion and its applications to the coefficients of certain power series, *Ann. Acad. Sci. Fenn. Ser. A I* 250/23 (1958), 3–11.
2. A. J. Macintyre and R. Wilson, Coefficient density and the distribution of singular points on the circle of convergence, *Proc. London Math. Soc.* (2) **47** (1940), 60–80.
3. ———, Some converses of Fabry's theorem, *J. London Math. Soc.* **16** (1941), 220–229.
4. G. Pólya, Untersuchungen über Lücken und Singularitäten von Potenzreihen, *Math. Z.* **29** (1929), 549–640.
5. ———, Untersuchungen über Lücken und Singularitäten von Potenzreihen, zweite Mitteilung, *Ann. of Math.* (2), **34** (1933), 731–777.
6. R. Wilson, On Hadamard composition and the virtually isolated singularity, this volume, pp. 371–377.

(Received April 8, 1968)

On Hadamard Composition and the Virtually Isolated Singularity

R. Wilson

*Greenroyd
Rannoch Road
Crowborough
Sussex, England*

1. Introduction

Let $f(z)$, $g(z)$ be two functions, each regular about the origin, with the respective Taylor series

(1) $$f(z) = a_0 + a_1 z + a_2 z^2 + \cdots + a_n z^n + \cdots,$$

(2) $$g(z) = b_0 + b_1 z + b_2 z^2 + \cdots + b_n z^n + \cdots.$$

Then $h(z)$, the Hadamard composition function of $f(z)$, $g(z)$ is regular about the origin and has the Taylor expansion

(3) $$h(z) = a_0 b_0 + a_1 b_1 z + a_2 b_2 z^2 + \cdots + a_n b_n z^n + \cdots.$$

The classical theorem on Hadamard multiplication is as follows [1, p. 346].

Theorem A. *If α and β denote vertices of the star-domain of $f(z)$ and $g(z)$, respectively, then $h(z)$ is regular in the star-domain whose vertices are at the points $\alpha\beta$.*

As Pólya has pointed out [4, p. 731], this does not tell us where the singularities of $h(z)$ must lie but only where they may lie.

In the case of a multiform function, the series (1)–(3) relate to specific branches of $f(z)$, $g(z)$, $h(z)$, as the case may be. This is important because singularities such as poles and isolated essential points will occur only on those sheets of the Riemann surface on which they are defined, while a critical point will occur on several sheets, or maybe all sheets, of the Riemann surface of the function concerned. The same is true if, in place of (3), the integral form of the composition function [1, p 347]

$$(4) \qquad h(z) = \frac{1}{2\pi i} \int_C f(u) g\left(\frac{z}{u}\right) \frac{du}{u} = \frac{1}{2\pi i} \int_C f\left(\frac{z}{u}\right) g(u) \frac{du}{u}$$

is used, since the contour is chosen to exclude the singularities of both functions and these take on C the values of the branches concerned.

2. The Virtually Isolated Singularity

We shall usually be concerned with the case in which $f(z)$ and $g(z)$ have each on their respective circles of convergence only one singularity. For convenience this will be taken to be at the point $z = 1$. In many cases the singularity will be of virtually isolated type.

In a joint paper with the author [3, p. 221], A. J. Macintyre defined the virtually isolated singularity as follows: suppose that $f(z)$ is regular in the neighborhood of $z = 1$ throughout the angle $-\phi < \arg(1 - z) < \phi$. Then, following Pólya [4, pp. 735–736; 3, p. 221], if $\phi > \frac{1}{2}\pi$ the singularity at $z = 1$ is said to be *easily approachable*; if $\phi = \pi$ so that $f(z)$ is regular near $z = 1$ except possibly on the real axis beyond this point, the singularity is said to be *almost isolated*. Finally, if $\phi > \pi$ the singularity is said to be *virtually isolated*. This means that $f(z)$ can be continued analytically in the neighborhood of $z = 1$ from the upper half-plane into a definite angle $0 > \arg(z - 1) > -(\phi - \pi)$ of the lower half-plane, and similarly from the lower half-plane into an equal angle of the upper half-plane.

Clearly poles and isolated essential points belong to this category. For, by continuing to encircle the point $z = 1$, either way, the angle ϕ may be taken to be as large as we please. A similar situation arises with isolated critical points which remain such on all sheets of the Riemann surface associated with this singularity,* except that each extension by an angle of 2π takes us onto a different sheet of the Riemann surface. Both these are special cases of the virtually isolated singularity, which can take highly complicated forms. We give below a simple example in which the angle $\phi(>\pi)$ is bounded; in fact we shall take ϕ to be less than $\frac{3}{2}\pi$.

---

* Such as algebraic-logarithmic points.

On the sheet of the Riemann surface for which the series (1) represents the branch of $f(z)$ under consideration, the point $z = 1$ is an isolated critical point so that $f(z)$ can be continued around the point $z = 1$ from $\arg(z - 1) = 0$ through an angle 2π and onto the next sheet of the Riemann surface. Suppose that on this sheet there is a set of poles along the line $\arg(z - 1) = \phi - \pi$ having a limiting point at $z = 1$. Similarly, continuing negatively from $\arg(z - 1) = 0$ onto the previous sheet of the Riemann surface, suppose that there is a set of poles on the line $\arg(z - 1) = \pi - \phi$ having a limiting point at $z = 1$. Then $f(z)$ is regular in the neighborhood of $z = 1$ for an angle of less than 3π. In fact, the singularity is a virtually isolated singularity only for the branch under consideration. For each of the branches of $f(z)$ associated with the other two sheets of the Riemann surface mentioned above, the singularity at $z = 1$ is easily approachable.

3. Hadamard Multiplication

The history of the development leading to the Hadamard composition of two virtually isolated singularities is given in a classical paper by Pólya [4, p. 732]. First, Faber proved that the Hadamard product of two isolated singularities, when neither is a critical point,* is effectively singular. Secondly, Faber proved that the Hadamard product of two isolated singularities, of which at least one is a noncritical point, is effectively singular. The proof contains a hiatus, later made good by Pólya. In order to deal with the Hadamard product of two isolated singular points, each of which is critical, Pólya proved the more general theorem that the Hadamard product of an almost isolated singularity and an isolable singularity is effectively singular [4, pp. 733–762].

The proof of this theorem is long and difficult but the theorem contains the result that the Hadamard product of two virtually isolated singularities is effectively singular. For the virtually isolated singularity is a special case of the almost isolated singularity and, being isolated on that sheet of the Riemann surface on which it is defined, it is also isolable there.[†]

In this paper we shall be concerned in proving that the Hadamard product of two virtually isolated singularities is a virtually isolated singularity, using an important result of A. J. Macintyre, and with consequences of this result.

---

* Each is therefore either a pole or an isolated essential point.

[†] A function is said to have an isolable singularity on its circle of convergence, say at the point $z = 1$, if it is possible to continue the function analytically either way along a simple closed curve round $z = 1$ lying inside any circle about $z = 1$, however small.

4. The First Theorem

We first prove the following theorem:

Theorem 1. *Let $f(z)$, $g(z)$ each have on their respective circles of convergence only one singularity and that a virtually isolated singularity at the point $z = 1$. Then $h(z)$ is also singular at the point $z = 1$.*

The appropriate instrument for dealing with this situation is to be found in a paper by A. J. Macintyre which was read at the International Mathematical Congress held at Helsinki in 1957. The result required is stated in the theorem below [2, Theorem A].

Theorem B. *Let $f(z)$ have on its circle of convergence $|z| = 1$ only one singularity and that of virtually isolated type. Then, for every assigned positive γ, $|a_n| > e^{-\gamma n}$ for a sequence of n of density 1.*

Applying this result to $f(z)$ and $g(z)$ under the hypotheses of Theorem 1 and using the notation of (1) and (2) we have, with $\gamma > 0$ given,

$$(5) \qquad\qquad |a_n| > e^{-\gamma n}, \qquad |b_n| > e^{-\gamma n},$$

for a common sequence of n of density 1. It follows from (5) that

$$(6) \qquad\qquad |a_n b_n| > e^{-2\gamma n}$$

for a sequence of n of density 1. Since γ can be taken to be as small as we like, it follows from (6) that $\overline{\lim}_{n \to \infty} |a_n b_n|^{1/n} = 1$, and so from (3), that the circle of convergence of $h(z)$ about the origin is of unit radius. Hence $h(z)$ has at least one singularity on $|z| = 1$.

However, since the only vertices of the star-domains of either $f(z)$ or $g(z)$ on $|z| = 1$ are at the point $z = 1$, it follows from Theorem A that the only possible singularity of $h(z)$ on $|z| = 1$ is at the point $z = 1$. Hence $h(z)$ is singular at $z = 1$ and this completes the proof of Theorem 1.

5. The Second Theorem

We next prove the theorem below.

Theorem 2. *Let $f(z)$, $g(z)$ each have on their respective circles of convergence only one singularity and that a virtually isolated singularity at the point $z = 1$. Then $h(z)$ has a virtually isolated singularity at $z = 1$.*

Results like that of Theorem 2 are easy to prove when, as in the case of poles and isolated essential points, interpolating functions of the coefficients a_n, b_n of $f(z)$ and $g(z)$ exist which are of a character exclusively associated with the singularity concerned. In fact, in these cases we can go further and give the order of the pole at $z = 1$ of the composition function $h(z)$ or the order and type of the isolated essential point of $h(z)$ there, as the case may be [1, pp. 348–349; 6].

In the proof of Theorem B given in an earlier paper [3, pp. 224–226], we made use of such an explicit form for the interpolating function of the coefficients arising from a virtually isolated singularity but, unfortunately, the proof contained an error, not easily ascertainable. In the event, Macintyre, in properly establishing Theorem B, had to be content with treating the virtually isolated singularity as an easily approachable singularity for the purpose of interpolating the coefficients. His proof, in fact, depended not on the peculiar properties of the interpolating function but on the use of Bourion's theory of overconvergence applied to the special properties of the virtually isolated singularity [2, Theorem A].

Theorem 1 tells us that $h(z)$ is singular at $z = 1$. To prove that this singularity is virtually isolated we have to establish that it is isolated from the field of singularities of $h(z)$, that is, that there is no other singularity of $h(z)$ in the close neighborhood of $z = 1$.

For this purpose it will be convenient to treat the virtually isolated singularities of $f(z)$ and $g(z)$ as almost isolated singularities. A method of dissection separating off the almost isolated singularity has been given by Pólya [4, pp. 738–741], and we shall use a modified version of this. Suppose that inside and on the circumference of $|z| = k$, with $k > 1$, there are no singularities of either $f(z)$ or $g(z)$ except the virtually isolated singularities at $z = 1$.* Then, in the Cauchy formula representing $f(z)$, we take the contour C to consist of $|z| = k$ and a loop enclosing the interval $(1, ke^{i\psi})$ but excluding the origin so that the contour C has a double point at $z = ke^{i\psi}$.

For convenience we take ψ to be a small positive angle. Thus

(7) $$f(z) = \frac{1}{2\pi i} \int_C \frac{f(w)}{w - z}\, dw = \frac{1}{2\pi i} \int_\Gamma \frac{f(w)}{w - z}\, dw + \frac{1}{2\pi i} \int_{C'} \frac{f(w)}{w - z}\, dw,$$

where Γ is the loop and C' the circle $|z| = k$. Hence, from (7),

(8) $$f(z) = f_1(z) + f^*(z),$$

where $f_1(z)$ has the same singularity as $f(z)$ at $z = 1$ and possibly a virtually isolated singularity at $z = ke^{i\psi}$, while $f^*(z)$ has the negative of this singularity, when it exists, and all the other singularities of $f(z)$ except that at $z = 1$.

---

* Here $f(z)$ and $g(z)$ refer, as in most cases arising in this paragraph, to the branches of $f(z)$ and $g(z)$ represented by the series (1) and (2).

In the special case in which the singularity of $f(z)$ at $z = 1$ is a pole or isolated essential point there will be no singularity at the point of separation $z = ke^{i\psi}$ of the two contours but if the singularity at $z = 1$ is a critical point, as will generally be the case, then there will be a corresponding critical singularity at $z = ke^{i\psi}$ ([4, p. 770; 5, pp. 425–426]).

A similar situation holds with regard to $g(z)$, so that, as in (8),

$$(9) \qquad\qquad g(z) = g_1(z) + g^*(z),$$

where $g_1(z)$ has the same singularity as $g(z)$ at $z = 1$ and possibly a virtually isolated singularity at $z = ke^{i\psi'}$, while $g^*(z)$ has the negative of this singularity, when it exists, and all other singularities of $g(z)$ except that at $z = 1$. For convenience, we have taken ψ' to be different from ψ.

We now apply Theorem A to the functions $f(z)$ and $g(z)$ of (8) and (9) to determine the location of all possible singularities of $h(z)$. In accordance with Theorem A these are situated at the points 1, $ke^{i\psi}$, $ke^{i\psi'}$, $k^2 e^{i(\psi + \psi')}$ and at a set of points lying outside $|z| = k$. Since $k > 1$ it is clear that there can be no other singularity of $h(z)$ in the immediate neighborhood of $z = 1$. Since the point $z = 1$ is singular for $h(z)$, from Theorem 1, and it is isolated from the field of singularities of $h(z)$, it must be a virtually isolated singularity. To establish Theorem 2 in the general sense in which it is stated, we have to be sure that the singularity of $h(z)$ at $z = 1$ is not in all cases a special kind of virtually isolated singularity.* We do this by means of an example.

Let $f(z)$ have at $z = 1$ any virtually isolated singularity whatsoever and let $g(z)$ have at $z = 1$ a simple pole of residue unity. This is, of course, a very special type of virtually isolated singularity. Then, using (4), we have

$$h(z) = \frac{1}{2\pi i} \int_C f(u) \frac{1}{1 - z/u} \frac{du}{u} = \frac{1}{2\pi i} \int_C \frac{f(u)}{u - z} du = f(z).$$

Thus the singularity of $h(z)$ at $z = 1$ is the same as that of $f(z)$ there and is equally general and so the product singularity of $h(z)$ at $z = 1$ is not in any way specialized. This completes the proof of Theorem 2.

6. Conclusion

If the virtually isolated singularities of $f(z)$ and $g(z)$ in Theorem 2 are taken at the points $z = \alpha$, $z = \beta$ on their respective circles of convergence $|z| = |\alpha|$, $|z| = |\beta|$, then simple transformations can be used to reduce the problem to the case already discussed. We thus have

---

* Such, for example, as an isolated essential point or a special type of critical point.

Theorem 3. *Let $f(z)$, $g(z)$ each have on their respective circles of convergence only one singularity and that of virtually isolated type at the respective points $z = \alpha$, $z = \beta$. Then $h(z)$ has at the point $z = \alpha\beta$ a virtually isolated singularity.*

Following Pólya, we can extend the result of Theorem 3 to the more general theorem stated below, using processes similar to those given by Pólya [4, Satz IX, pp. 768–771].

Theorem 4. *Let the boundaries of the star-domains of the series* (1) *and* (2) *each consist of a countable set of points, having no finite limiting point, and let every singular point be a vertex of the star-domain. If at the points $z = \alpha_\mu$, $z = \beta_\nu$ there are virtually isolated singularities of $f(z)$ and $g(z)$, respectively, then $h(z)$ has a virtually isolated singularity at the point $z = \alpha_\mu \beta_\nu$, provided that $\alpha_\mu \beta_\nu$ is different from every other product $\alpha_{\mu'} \beta_{\nu'}$ of the singular points of $f(z), g(z)$.*

REFERENCES

1. P. Dienes, *The Taylor Series*, Clarendon Press, Oxford, 1931; Dover, New York, 1957.
2. A. J. Macintyre, An overconvergence theorem of G. Bourion and its application to the coefficients of certain power series, *Ann. Acad. Sci. Fenn. Ser. A I* 250/23 (1958), 3–11.
3. A. J. Macintyre and R. Wilson, Some converses of Fabry's theorem, *J. London Math. Soc.* **16** (1941), 220–229.
4. G. Pólya, Untersuchungen über Lücken und Singularitäten von Potenzreihen, zweiten Mitteilung, *Ann. of Math.* (2), **34** (1933), 731–777.
5. R. Wilson, Functions with dominant singularities of the generalized algebraic-logarithmic type (II): On the order of the Hadamard product, *Proc. London Math. Soc.* (2), **43** (1937), 417–438.
6. ———, On the Hadamard product of two isolated essential points of finite exponential order, *J. London Math. Soc.* **28** (1953), 490–494.

(*Received April 8, 1968*)